Engineering Materials

This series provides topical information on innovative, structural and functional materials and composites with applications in optical, electrical, mechanical, civil, aeronautical, medical, bio- and nano-engineering. The individual volumes are complete, comprehensive monographs covering the structure, properties, manufacturing process and applications of these materials. This multidisciplinary series is devoted to professionals, students and all those interested in the latest developments in the Materials Science field, that look for a carefully selected collection of high quality review articles on their respective field of expertise.

More information about this series at http://www.springer.com/series/4288

Nabisab Mujawar Mubarak · Mohammad Khalid ·
Rashmi Walvekar · Arshid Numan
Editors

Contemporary Nanomaterials in Material Engineering Applications

Springer

Editors
Nabisab Mujawar Mubarak
Department of Chemical Engineering
Faculty of Engineering and Science
Curtin University
Miri, Sarawak, Malaysia

Rashmi Walvekar
Department of Chemical Engineering
School of Energy and Chemical Engineering
Xiamen University Malaysia
Sepang, Selangor, Malaysia

Mohammad Khalid
Graphene & Advanced 2D Materials
Research Group (GAMRG)
School of Science and Technology
Sunway University
Petaling Jaya, Selangor, Malaysia

Arshid Numan
Graphene & Advanced 2D Materials
Research Group (GAMRG)
School of Science and Technology
Sunway University
Petaling Jaya, Selangor, Malaysia

ISSN 1612-1317 ISSN 1868-1212 (electronic)
Engineering Materials
ISBN 978-3-030-62760-7 ISBN 978-3-030-62761-4 (eBook)
https://doi.org/10.1007/978-3-030-62761-4

This Springer imprint is published by the registered company Springer Nature Switzerland AG
The registered company address is: Gewerbestrasse 11, 6330 Cham, Switzerland

Preface

It was 1964 when Dr. Richard Feynman (Noble Laureate in Physics) delivered his famous lecture, "There is Plenty of Room at the Bottom." In this lecture, he gave the concept of nanomaterials and predicted the possibility to put the entire 24 volumes of the Encyclopedia of Britannica under the head of a pin. Since then, there is rapid development in the field of nanotechnology and today, what was considered impossible is now a reality.

In the past few decades, nanomaterials and nanotechnology emerged from fundamental concepts into a vast range of applications covering every aspect of life. The unique properties of nanomaterials, especially contemporary nanomaterials (CNMs), have played a prominent role in the nanotechnology revolution. The most eye-catching features of these structures are their electronic, mechanical, optical, and chemical characteristics, which open numerous avenues to future applications. However, continuous discovery of new CNMs and rapid development in nanotechnology make it difficult for the industries to stay up-to-date with the fast-changing technology landscape. Further, it is a challenging task to bring CNMs from laboratory to market. Therefore, the latest developments and findings of CNMs need to be highlighted to bridge the gap between the latest research and its technical realization. Currently, the literature on CNMs is scattered into original articles and review papers. This book would categorize this subject and provides an overview and critical commentary that will be an invaluable reference for researchers, scientists, and students, studying and working with nanomaterials for engineering applications.

This book intends to introduce readers to a broad view of nanomaterials, from fundamental principles to fabrication techniques, and from characterization to applications. It is primarily intended as a reference book for graduate students in materials science and other related engineering fields. It will also serve as a reference book for professionals that are interested in this field.

Miri, Sarawak, Malaysia — Nabisab Mujawar Mubarak
Petaling Jaya, Malaysia — Mohammad Khalid
Sepang, Malaysia — Rashmi Walvekar
Petaling Jaya, Malaysia — Arshid Numan

Contents

Importance of Nanomaterials in Engineering Application

Lau Yien Jun, Fahad Saleem Ahmed Khan, Nabisab Mujawar Mubarak, Lau Sie Yon, Chua Han Bing, Mohammad Khalid, and E. C. Abdullah

Abstract Nanomaterials have gained prominence in technological advancement due to their exceptional tunable properties and enhanced performance over their bulk counterparts. These nanomaterials are revolutionising various industrial applications due to their outstanding, and unique characteristics. In this chapter, a summary of the fundamental engineering concept of nanomaterial, particularly carbon nanomaterials are included. Besides, this chapter also highlights the engineering applications of nanomaterials in diverse industrial fields, such as biomedical, food processing, biotechnology, environmental remediation, construction, renewable energy, electronics, and energy storage. Lastly, the potential effects of nanomaterials on human health are briefed. In short, nanomaterials are expected to bring breakthrough development field of the leading cutting edge of nanotechnology.

Keywords Nanotechnology · Nanomaterials · Engineering applications

1 Introduction

Nanomaterials are defined as the materials that possess one or more external dimensions with a dimensional size ranging from 1 to 100 nm. Richard Feynman, who was a Nobel Prize winning American physicist, was broadly credited with the kick-starting

L. Y. Jun · F. S. A. Khan · N. M. Mubarak (✉) · L. S. Yon · C. H. Bing
Department of Chemical Engineering, Faculty of Engineering and Science, Curtin University, 98009 Miri, Sarawak, Malaysia
e-mail: mubarak.mujawar@curtin.edu.my; mubarak.yaseen@gmail.com

M. Khalid
Graphene & Advanced 2D Materials Research Group (GAMRG), School of Science and Technology, Sunway University, No. 5, Jalan Universiti, Bandar Sunway, 47500 Petaling Jaya, Selangor, Malaysia

E. C. Abdullah
Department of Chemical Process Engineering, Malaysia-Japan International Institute of Technology (MJIIT) Universiti Teknologi Malaysia (UTM), Jalan Sultan Yahya Petra (Jalan Semarak), 54100 Kuala Lumpur, Malaysia

N. M. Mubarak et al. (eds.), *Contemporary Nanomaterials in Material Engineering Applications*, Engineering Materials,
https://doi.org/10.1007/978-3-030-62761-4_1

of the modern interest in nanotechnology [21]. In the year 1959, he gave a visionary talk at an annual meeting of an American Physical Society, entitled "There's plenty of room at the bottom". In his speech, he laid the conceptual foundations of manipulating individual atoms and molecules at the atomic level using tiny and precise tools. In the year 1974, a Japanese Professor Norio Taniguchi had termed this field as "nanotechnology", which describes the level of precision in manufacturing materials at the nanometre scale [87].

In the past decades, nanomaterials are growing explosively worldwide as an important product of nanotechnologies owing to their extraordinary physiochemical properties. Nanomaterials offer unique functional applications as they possess a high surface to volume ratio, as well as superior mechanical, chemical, and physical properties. Figure 1 illustrated the comparison of the sizes of nanomaterials.

It is essential to fabricate nanomaterials with high functionality and suitable properties for various applications. Generally, there are two approaches to synthesise nanomaterials, which are the "Top-down" and "Bottom-up" methods. For "Top-down" approach, it involves making structures and devices from bulk materials without specific control at the atomic level. One of the main benefits of "Top-down" method is the large scale production in the industrial setting [29]. Nevertheless, the main shortcomings of this method are high cost and energy consumption, lengthy processing time, and the deformation of surface structures [88]. The examples of "Top-down" method including mechanical milling, cutting, chemical etching, laser ablation, electro-explosion, and lithography.

On the contrary, the "Bottom-up" method involves the assembly of atomic scale materials to generate a controlled nanostructure. The advantages of this method are economical technique, controllable particle size, surface properties, and deposition parameters [8]. Nonetheless, the disadvantages of this method are difficulty in mass production, and chemical purification of nanomaterials is required [38]. The physical

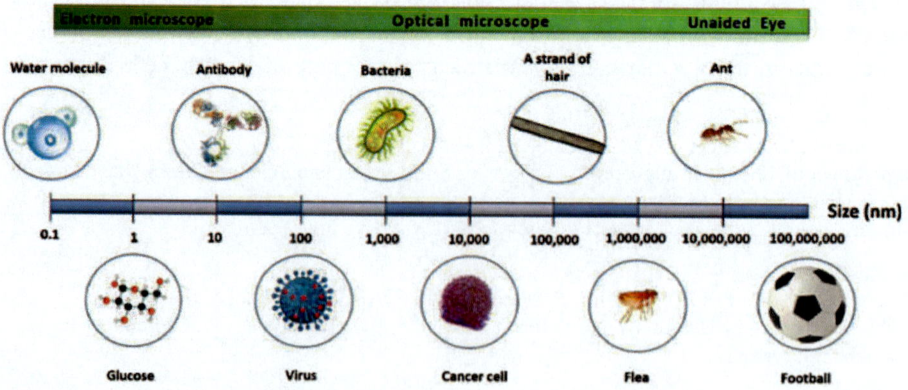

Fig. 1 Comparison of nanomaterial sizes

and chemical processing "Bottom-up" methods are by spinning, atomic layer deposition, vapour-phase deposition, electrolytic deposition, self-assembled monolayer, spray pyrolysis, and sol–gel method.

2 Type of Nanomaterials

They are various type of nanomaterials, which can be classified into organic and inorganics categories, as shown in Fig. 2. For organic groups, it consists of carbon-based nanomaterials, such as fullerene, single-walled carbon nanotube (CNT), graphene and buckyball. On the other hand, the inorganics group can be further categorised into metal, metal oxide and quantum dots categories. The commonly used metal and metal oxide nanomaterials are gold (Au), aluminium (Al), silver (Ag), copper (Cu), zinc (Zn), aluminium oxide (Al_2O_3), silicon oxide (SiO_2), iron oxide (Fe_2O_3), titanium dioxide (TiO_2), and copper oxide (CuO). Moreover, the examples of quantum dots are cadmium selenide (CdSn) and zinc sulphide (ZnS).

Nanomaterials can be classified further into several classes based on their shapes and number of dimensions. For instances, zero-dimensional (0-D), one-dimensional (1-D), two-dimensional (2-D) and three-dimensional (3-D). 0-D nanomaterials exhibit in spherical and clusters forms, such as fullerene, gold and silver nanoparticles. Besides, 1-D nanomaterials show in tube, wire, fibre or rod forms, for example, carbon nanotube (CNT). Additionally, 2-D nanomaterials are in films or sheets forms, such as graphene or graphene oxide (GO). Furthermore, 3-D nanomaterials are in 3-D structure, such as diamond and graphite. These nanomaterials can be exhibit in amorphous or crystalline structures. Figure 3 displayed the types of nanomaterials in various dimensions. In this section, carbon-based nanomaterials are highlighted due to their wide applicability in diverse fields.

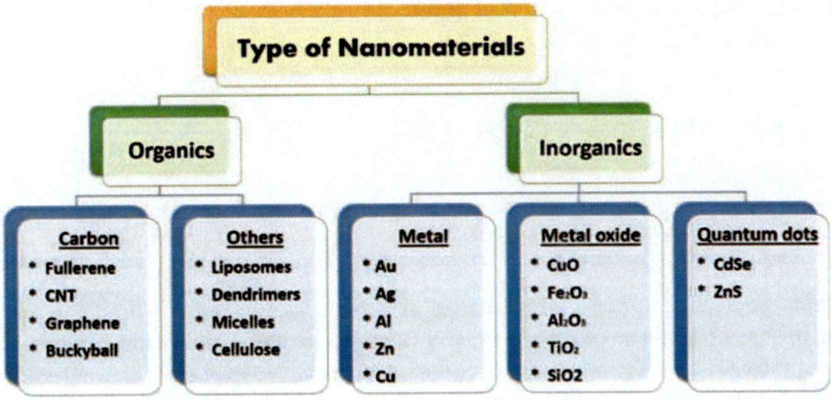

Fig. 2 Types of nanomaterial

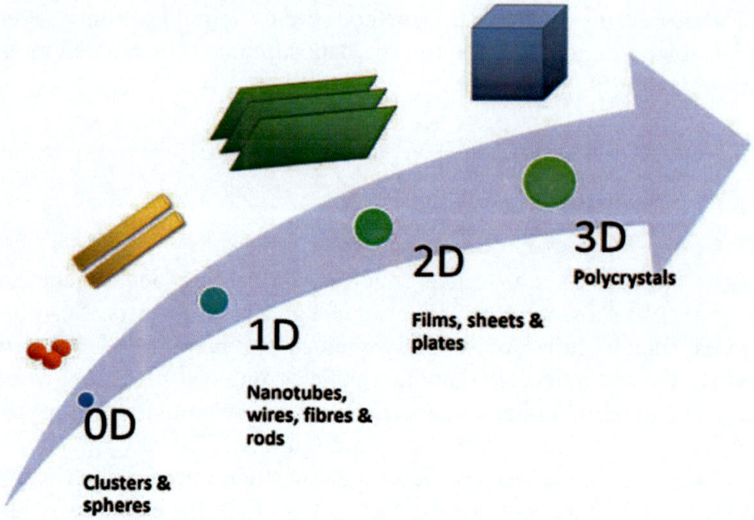

Fig. 3 Type of nanomaterials with different dimensions

2.1 *Fullerene (0-D)*

Fullerenes are molecular composed of carbon, which are also known as Buckyball or Buckminister fullerene (C_{60}). Their first discovery in the year 1985 by Kroto et al. [48], was awarded Nobel prize in chemistry 10 years later. They are made of carbon atoms in the form of tube, ellipsoid or hollow spherical shapes. Fullerene exhibited in several unique properties such as exceptional durability, high electron affinity, extraordinary radical scavengers, and ease of modification. Nevertheless, the insolubility of fullerene in the polar solvent has restricted its applicability. The attention of fullerene has reduced gradually recently with the increasing focus on the research of CNT and graphene-based materials.

2.2 *Carbon Nanotubes (1-D)*

CNTs are cylindrical carbon structures with diameters in the range of a nanometre. CNTs are mainly categorised as single-walled carbon nanotube (SWCNT) and multi-walled carbon nanotube (MWCNT) based on their structure and geometry. SWCNT is formed by rolling up a single sheet of graphene, while MWCNT is formed by rolling up multiple sheets of graphene. CNTs have received considerable attention in diverse fields of science since their discovery. This is because of their exceptional and desirable characteristics, such as extraordinary mechanical, thermal, and electrical properties. Besides, the surface properties of the CNTs can be engineered and tailored easily according to the application required [34, 42]. A wide range of application possibilities has been explored with CNTs, such as their use in pharmaceutical, medicine, biosensor, bioremediation, biofuel cell development, agriculture and food processing industries [43, 55, 61, 96].

2.3 Graphene and Its Derivatives (2-D)

Graphene is a single layer of sp^2 hybrid bonded carbon atoms packed in the form of a two-dimensional and hexagonal honeycomb structure. The discovery of graphene by Konstantin Novoselov and Andre Geim in 2004 has received immense attention to exploring its potential applications [67]. The most commonly used method for synthesis of graphene and its derivatives are chemical vapour deposition (CVD), epitaxial growth, and mechanical exfoliation methods [50]. Graphene and its derivatives have unique chemical, physical, electronic, optical, and thermal properties [93]. Due to these excellent features, graphene and its derivatives have gained attention in a broad spectrum of applications, prominently in the biomedical, electronic, biosensor, energy storage and environmental pollution control [36, 99].

2.4 Graphite (3-D)

Graphite is a layered honeycomb structure, where multiple layers of graphene sheets are bonded through weak Van der Waals forces, with sp^2 hybridised crystal structure. It is also known as one of the world's softest minerals. Pristine graphite can be classified into three major types, which are a crystalline flake, vein or lump, and amorphous graphite [92]. Nevertheless, pristine graphite is not often used due to their brittle nature, shear-planes and inconsistent mechanical properties [25]. The most common application of graphite is in lithium-ion batteries, lubricants, refractories, and lead pencils [39, 44].

3 Nanomaterial for Engineering Application

For the past few decades, the rapid development of nanotechnology has made a massive revolution in various industrial applications [20, 91]. The emergent features of nanomaterials, such as superior mechanical, chemical, optical, thermal and electrical properties, make them received tremendous attention among researchers in multidisciplinary fields [55, 61, 96]. Today, the use of nanomaterials is increasing with widespread commercial applications in consumer products, such as cosmetics, food agriculture, electronics and automobile, and so forth. Also, the technical applications of nanotechnology have extended to techniques in various fields, such as

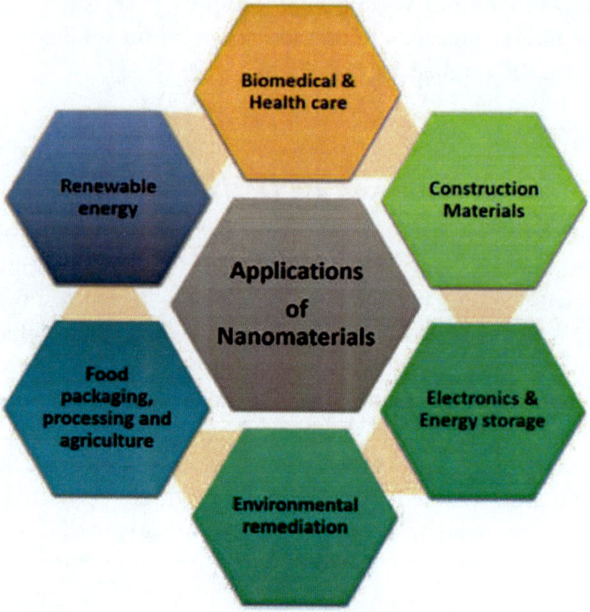

Fig. 4 Application of nanomaterials in various industries fields

biotechnology, construction, renewable energy, and biomedical. Moreover, nanotechnology also provides solutions to environmental challenges, particularly in the areas of wastewater treatment and air pollution treatment. Figure 4 displayed the applications of nanomaterials in various industry sectors.

Nanomaterials have a noticeable impact on various applications in biomedical fields, such as therapy, biosensor, diagnosis, novel approaches for treatment, as well as prevention. In addition, nanomaterials can be used in various forms, such as thin film coating, reinforced or filler materials, or connecting materials. As compared to conventional materials, nanomaterials can produce lighter, stronger, bacteria-resistant, and corrosion-resistant materials. Examples of nanomaterials used in everyday applications are sports equipment, cosmetics, and automobiles. Moreover, nanomaterials play an important role in electrical and electronics applications as it can provide a larger storage capacity with lower power consumption. In recent years, nanotechnology is employed in environmental remediation applications, particularly in the fields of wastewater treatments. Also, in the food industry, nanomaterials are used for antimicrobial and antioxidant agents, to prolong the shelf life and the freshness of food. Furthermore, nanotechnology has also made a significant impact on renewable energy, particularly energy conversion and storage. Few of the nanomaterials applications are briefly described in the next section, whereas the application of nanomaterials based on various fields is listed in Table 4.

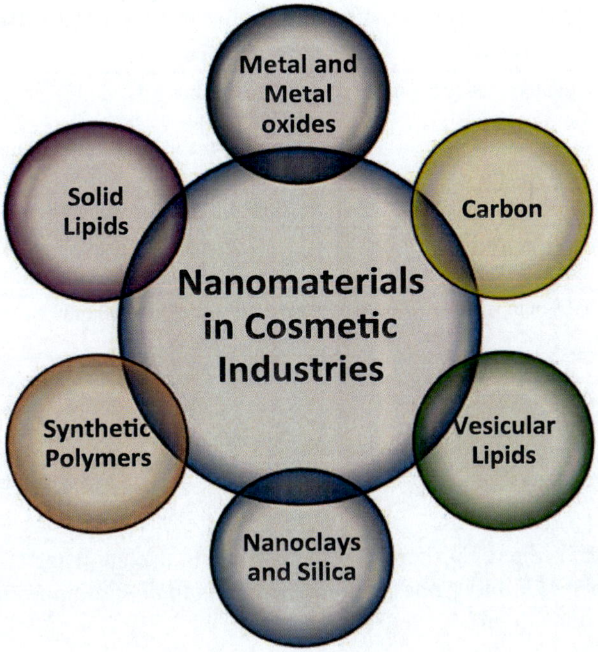

Fig. 5 Nanomaterials employed in cosmetic products

3.1 Cosmetics and Personal Products

The primary use of nanomaterials in cosmetics is to improve the distribution of cosmetic constituents into the skin. The minor size of lipids vesicles can allow these materials to be immersed more rapidly into skin. Scientific researches to validate these statements have given differing outcomes. These can be because of the study surroundings, for instance, size and strength of nanoparticles that make it challenging to attain steady results even within the similar lab [35]. Varying physico-chemical properties of vesicles can also cause changes in the interaction with the skin. Few vesicles that cannot infiltrate the skin can conveniently release their constituents onto the skin's surface in a way that eases uptake and penetration into the skin's layers [65]. Moreover, nanomaterials can be employed to impart stability to formulations that have constituents that can decompose because of oxidation and various other reasons. Nevertheless, carrier nanoparticles might have strength concerns upon application to the skin.

Nanoparticles of zinc oxide (ZnO) and titanium oxide (TiO_2) are commonly employed in sunscreen drug products. However, they can also be used in cosmetic goods (where the subsequent product can be cosmetic or drug) [68]. These nanomaterials effectively block ultra-violet (UV) rays but also result in a translucent formulation that is attractive to the customer (Fig. 5).

Silver nanoparticles have been used as preservative/anti-bacterial agents in customer products. Cosmetic goods in the United States cannot provide anti-bacterial claims as this claim is linked with physiological function; therefore, it uses in drugs

Table 1 Recent studies on cosmetic product using nanomaterials

Nanomaterial	Description	pH	Penetration	References
Quantum dot-655/COOH	Shape: Ellipsoid Diameter: 6–12 nm Hydrodynamic diameter: 18 nm Skin: Pig	9	Negative	[78]
Quantum dot-655/PEG	Skin: Human stratum corneum	8.3	Negative	[18]
TiO_2 nanparticles	Particle size: 200 nm Skin: Pig	–	Negative	[23]
TiO_2 nanparticles in oil/water	Paricle size: 10 × 50 nm Skin: Pig	–	Negative	[79]
ZnO nanoparticles	Skin: Human epidermal	–	0.03%	[102]
	Size: 15–30 nm Duration: 24 h Skin: Human	–	Negative	[64]

are limited. Furthermore, nano-silver is also not listed Cosmetic Directive of the European Union (Annex-5) [70], which is the list of approved preservatives permitted in cosmetic products (Table 1).

3.2 *Automobiles and Transportation*

According to the United Nations, it is estimated that globally vehicle fleet will be double from 750 × 10^8 million today to around 1.5 × 10^9 by 2030 [31]. It raised the question linked to the safety of passengers, traffic regulation system, contaminant reduction and adequate recycling at the end of value added chain to protect scarce resources are turning more crucial. In regard to these concerns, nanotechnology contributes significantly to essential development and fabrication of advanced processes and materials in a different sector, particularly in the automobile industry. For example, new era tyres attained their high durability, mileage as well as grip via nano-scale soot particles and silica [63]. Materials with nano-scale layer and particles have valuable impacts on the interior and exterior surfaces, on the body/engine and drive. Nanomaterials are employed to different body parts, tyre, engines, interiors, emissions, electronics and chassis. The nanomaterials linked with automobiles are briefly described below:

a. **Light-Weight Body Parts**

One of the most known topics associated with automobiles is its weight reduction. The light weight automobiles will increase fuel efficiency, decrease manufacturing cost and cut down the carbon dioxide (CO_2) emissions. It is estimated that through reducing the automobile weight by 10%, there will be 7% fuel economy [17]. Moreover, issues like stability, smooth drive and crash resistance could also be improved by vehicle weight reduction. Considerable number of development had been conducted on this path but failed to manage efficiency and safety. In addition, materials close to engine parts must hold significant thermal resistance, but outer and structural parts must be build of materials that possess excellent mechanical stability. Yet, materials employed such as thermoplastics have shown low mechanical and thermal resistance, and thus it can only be considered after modified via reinforcements.

Carbon nanotubes are light-weight and approximately 150 times stronger than steel. Consequently, CNTs are an appropriate alternative of steel in automobile parts. Incorporation of nano-scale clay in the polymer matrix can produce a nanocomposite that can be used to produce automobile part close to the engines due to high thermal stability [62]. Moreover, clay nanocomposite reinforced with polyamide, polycarbonates and polypropylene are known used polymer nanocomposites. Once these nanomaterials are combined with polymers, their thermal resistance as well as flame retardance will incline [60]. With the use of hybrid solution and nanocomposite, a vehicle that weights typically 900 kg of steel can be reduced to 300 kg. Nanocomposite plastic provides on average 25% weight savings over highly filled plastics and 80% over steel. Volkswagen uses long fiber reinforced thermoplastic (LFRT) that provides better design flexibility, high strength/weight and impact resistance, and corrosion reduction in comparison to metals [59].

b. **Paint-Coating**

The primary concerns related to conventional paints are in-efficient flame retardant, paint transfer efficiency, limited adhesive and surface finish to water and dust particles. Most of these issues can be solved with the incorporation of different nanoparticles into these paints [71]. Nanoparticles like boehmite (AlOOH), silicon dioxide (SiO_2) and zirconium dioxide (ZrO_2) are entrenched in UV curable lacquers which offering in enhanced abrasion resistance. TiO_2 and ZnO nanoparticles will support to enhance UV resistance features and also reflecting those noxious rays [45, 84]. In addition, a combination of nanoparticles with fluoro-methyl group extend its pore volume and surface area that support surface roughness.

c. **Scratch–Resistance**

The brand-new appearance of vehicle body shell must be ensured after number of years of operation and washes. The vehicle coating can be affected due to the scratch/abrasion on the surface. Thus, it is a difficult task for science society to come up with scratch/abrasion resistant coatings without damaging their other features [54, 100]. Nanotechnology plays a vital role in regards to scratch and abrasion resistant coating. Glasel and co-workers produced scratch and abrasion resistant film with the support of siloxane encapsulated silicon dioxide (SiO_2) nanoparticles [7]. Because of the identical distribution of nanoparticles in polymers, scratch resistant feature can be enhanced without affecting other properties. Similarly, nano-alumina have displayed adequate reaction in this aspect.

Compared to conventional paints, nano-varnish have shown good paint brilliance and scratch resistance. The purpose of this technological effect is the reinforced ceramic particles which are combined to varnish layer, nano-scale. Degussa AEROSIL R9200 is the known commercially used nanoparticles in the varnish [3].

d. **Tyres**

Carbon black is the first nanomaterial used as a reinforcing element that was mixed with the tyre. The two main ingredients are soot and silica that are employed as a reinforcing element in the tyre [53]. Substantial fuel efficiency, as well as sustained durability, is attained through the use of soot at nano-scale, as it has coarser surface compared to those employed in normal tyres. Nanoparticles hold extensive surface energy, an association of soot nanoparticles with normal rubber in the tyres is significant, which leads to improved rolling resistance and lowers inner fraction [90]. The combination of nano-silica (10%) to styrene butadiene rubber (SBR)/natural rubber support to enhance the grip and wear resistance of the tyre. Moreover, it has been reported tensile and hardness properties can be improved through multi-walled carbon nanotubes (MWCNTs) (3%) mixing with SBR or natural rubber [56].

e. **Other Uses in Automobiles**

Apart from above-mentioned automobiles parts, nanomaterials have wide-spectrum application in automobile sector such as nano-fluids, nano-filters, fuel cells, cooling systems, nano-electronic mechanical system (NEMS), and damping and braking system [76]. CNTs are employed for the storage of hydrogen fuel in fuel cells, as it can store a significant amount of hydrogen and can be easily retrieved. The efficiency of the batteries can be enhanced via a metal electrode of lithium ion batteries with nanoparticles. Similarly, magnetorheological fluids can support powerful and smooth damping and braking [74]. Magnetorheological fluids contain magnetic nanoparticles dissolved in a solvent and at the time of minor pressure change/magnetic field happens, the particles will align, consequently fluid viscosity increases (Table 2).

3.3 Agriculture

The primary usage of nanomaterials in agriculture is to incline the crops and lands productivity, specifically under sub-optimal situations, started in the early twenty-first century [33]. Nevertheless, nanotechnology stays a moderately under-discovered area in agricultural science [94]. Different nanomaterials with promising potential to modernise the agricultural sector have been introduced, analysed through a significant number of merits and demerits [12]. They normally elevate food safety, crop and quality growth and monitor environmental surroundings, moreover, cracking number of agricultural issues, for instance, soil structure issues, pesticides delivery, plant disease, pollutants detection and fertilisers [75]. Varied nanomaterials are considered in the agricultural sector such as CNTs (single-walled CNTs, multi-walled CNTs),

Table 2 Nanomaterials applications in automobile sector [69]

Nanomaterials	Merits	Demerits	Applications
Long fiber reinforced thermoplastic	Extensive impact resistance, high stability	Production is challenging	Body parts
Silicon dioxide (SiO_2)	Lower paint toxicity, support self-healing	Uniform dispersion is challenging	Paint coating
Copper dioxide (CuO_2), Molybdenum disulphide (MoS_2), Titanium dioxide (TiO_2)	Coefficient of friction can be lowered up to 50%	Life-span is shorter	Engines
Layered double hydroxides (LDH)	Absorbs heat, enhances char growth	Stability concerns	Paint coating
Carbon nanotubes (CNTs)	High mechanical stability and light-weight	Expensive	Body parts
Soot, carbon black, silica	Exntended durability, fuel efficiency is higher	Uniform dispersion is challenging	Vehicle tyres
Copper oxide (nano-fluids)	Upgrades heat transfer coefficient	Agglomeration issues	Engines
Carbon nanotubes (CNTs), copper (Cu), zinc oxide (ZnO), Gold (Au)	Zero-toxicity, good efficiency	Relatively expensive	Vehicle interior

silver (Ag), zinc (Zn), titanium dioxide (TiO_2) and graphene oxide (GO) [33, 47]. Hence, methods of this promising methodology involve safeguard against diseases and pathogens, enhanced efficiency of fungicides, pesticides, herbicides, leading to improved plant growth and maintained discharge [13, 33]. Nanomaterial enhance efficiency, production and agricultural safeguard, as they are valid in nearly every component of the agricultural sector such as transportation, processing, production and storage [19].

Practically, nanomaterials usage covers numerous purposes in agricultural. The primary purpose is that of enhanced fertilisers' delivery, resultant in elevated elements uptake via plants cells and lessened nutrients loss. They offer synchronisation of micro and macronutrients delivery [30]. Nano-structured fertiliser improves the effectiveness of nutrient usages via larger surface areas, mark delivery mechanism, slow and maintained discharge in response to biological demands and environmental triggers. Nano-iron (Fe), nano-phosphorous (P), nano-zinc (Zn) and nano-magnesium (Mg) are few common nano-fertilisers examples. Furthermore, CNTs can be employed as nutrients delivery for micro/macronutrients to reduce their applied quantities with promising consequences in agricultural since they have competitive thermal, chemical, electrical and mechanical properties [47, 86]. The second purpose is that of pest and insect management. Few nanoparticles have excellent capability in control and management of pests because of its slow and active discharge of active-compounds, for instance, zinc oxide, titanium dioxide and gold, considering them

Table 3 Few examples of nanomaterials employed in the agricultural sector

Nanomaterial	Function of nanomaterial	Industrial name	Application
Zinc (Zn)-nanoparticles, copper (Cu)-nanoparticles, boron (B)-anoparticles, Iron (Fe)-nanoparticles, bio humus	Fertiliser	Motawazen, Saula, Solocross, Green Earth NanoPlant	Nano-fertilizers
Copper doped montmorillonite	Detection of propineb fungicide	–	Nano-sensors
Copper (Cu)-nanoparticles, Polymeric-nanoparticles, Zinc (Zn)-nanoparticles	Bactericide, controlled release and fungicide	Bio-nano technology	Nano-pesticides
Poly-epsilon caprolactone	Controlled release	–	Nano-herbicides

as economical and reliable substitutes for synthetic chemicals with negative-side effects. Nano-pesticides decline organic solvent discharge and undesired pesticides movement. Moreover, nanomaterials can locate, detect as well as report pathogens in crops as intelligent field systems and speedy analytic tools for pathogens detection alerting growers to use the required materials earlier to on-set of symptoms [19, 83].

Another purpose is that of nano-fungicides. Chemical fungicides have a harmful impact on plants with health and environmental concerns. Nano-fungicides offer a worthful solution to afore-mentioned issues based on the size and shape of nanoparticles. Ag-nanoparticles are well-known nano-fungicides because of its antimicrobial characteristics. Similarly, ZnO-nanoparticles also show anti-fungal, anti-bacterial and antimicrobial properties, whereas, TiO_2-nanoparticles possess anti-bacterial and photo-catalytic activities that follow to plant safeguard [23, 46].

The last purpose comes in nano-sensors forms, vastly considered in the agricultural sector because of its ability for pollutions monitoring in environment, aquifer and soil. Monitoring crop health, soil conditions and plant pathogens detection can be performed conveniently and precisely through nano-sensors [97]. The well-known nanomaterial employed as a sensor is Ag-nanoparticles. Table 3 lists the common nanomaterials used in the agricultural sector [27] (Table 4).

4 Risks of Nanotechnology

Despite the remarkable properties of nanomaterials, it is essential to evaluate their toxicity and potential adverse impacts on human health, as well as the environment. The toxicity of nanomaterials is dependent on the particle sizes, nature, shapes, solubility, surface properties and chemical composition of the nanoparticles. The possible exposures of nanomaterials with the human system are via inhalation, ingestion, and skin penetration [80]. There are several researchers reported the

Table 4 Few other application of nanomaterials in various fields

Applications	Examples	References
Biomedical and health care	• Gold nanoparticles for targeted delivery of drugs directly to cancer tumours • Liposomes nanoparticles for controlled and targeted drug delivery • Quantum dots used for detection and diagnosis of cancer • Graphene and carbon-based nanomaterial as wearable multifunctional sensor, for monitoring skin temperature, wrist pulse, breathing, and pressure mapping	[22, 51, 89]
Construction Materials	• CNT, GO and SiO_2 nanomaterials are used as reinforced cement composites • CNT as reinforced metal matrix composite to produce stronger, ultralight and corrosion-resistant nanocomposite materials • ZnO nanoparticles are used as coatings to reduce ultraviolet radiation exposure • Silver nanoparticles are embedded in paint to provide antimicrobial properties to the surface • TiO_2 nanoparticles are used as glaze coating to provide anti-fouling properties	[2, 15, 52, 66, 85]
Electronical and electronics	• Carbon-based nanomaterials as electrodes in supercapacitors, which has large storage capacities and fast charged/discharged rates	[14, 40, 98]
Environmental remediation	• Nanomaterials, such as CNT, graphene oxide, as adsorbents to remove heavy metals and pollutants from water bodies • Incorporation of functional nanomaterials, such as TiO_2, Al_2O_3, CNT and zeolites, into filtration membranes to improve the mechanical and thermal stabilities, permeability, fouling resistance, as well as adsorption capability of pollutants	[26, 41, [57, 95]
Food packaging	• Nanoclay for food packaging due to their strong mechanical and thermal properties • Nanoemulsion as edible coating for preserving and storing food • Gold nanoparticles as food sensors to monitor the food quality and safety	[77, 81]

(continued)

potential human health risks of nanomaterials, such as carcinogenicity, genotoxicity, pulmonary inflammation and circulatory effects [6, 73]. Table 5 summarises the possible toxic effects and risks of nanomaterials.

Table 4 (continued)

Applications	Examples	References
Food processing	• SiO_2 nanomaterial as anti-caking agent to maintain the flow properties in powdered products • Titanium dioxide (TiO_2) used as food colour additives and flavour enhancers • SiO_2-gallic acid nanoparticles as antioxidants • Silver (Ag) and zinc oxide (ZnO) nanoparticles as antimicrobial agents in food packaging industry	[10, 32, 72]
Food agriculture	• Fertilisers coated with nanocapsules or nanoparticles for delivery of agri-chemicals and fertilisers to enhance the absorption of the nutrients by the plants • Nanomaterial as a pest control agent to increase the effectiveness towards pests • Zinc oxide (ZnO) quantum dots used as pesticide detection • Magnetic nanoparticles for the smart agrochemical and herbicide delivery system • MWCNT nanomaterials to increase the plant production rate and • TiO_2 nanomaterial as sensor for monitoring soil condition, pesticides and crop growth	[28, 47, 72]
Renewable Energy	• Carbon nanomaterial for direct conversion of sunlight into electricity • Graphene used as metal-free catalyst in fuel cells • Metal–organic frameworks for hydrogen storage	[1, 37, 101]
Sensor	• Metal oxide nanomaterials for various sensor applications, such as gas sensor, chemical oxygen demand (COD) sensor, and biosensor	[5, 11, 49]

5 Conclusion

Today, nanomaterials have become one of the most focused research areas due to their rapid development and their great potential for commercial purposes. The recent advances in engineered nanomaterials have opened up new opportunities in diverse industrial applications, ranging from biotechnology and biomedical to electronics and energy storage applications. Despite the superior properties of nanomaterials, the toxicity and potential associated risks of nanomaterials need to be addressed and regulated satisfactorily before the complete realisation of nanotechnology in industrial applications. Until now, there is a lack of standardised procedures to evaluate the safety and impact of nanomaterials on the environment. More research and

Table 5 Possible risks and toxic effects of nanomaterials

Nanomaterials	Possible risks and toxic effects	References
CNT Fullerene	• Pulmonary inflammation • Granulomas • Fibrosis • Genotoxicity • Lung cancer	[9, 82]
Al_2O_3 CuO TiO_2 ZnO	• Oxidative stress • Genotoxicity • Skin penetration • Malfunction of organ, such as liver, kidney, spleen, lung and brain	[4, 24]
Gold	• Distribution into other organs, such as central nervous system	[16, 58]

efforts are required to minimise the potential adverse impact of nanomaterials on the environment and human health.

References

1. Abdin, Z., Alim, M., Saidur, R., Islam, M., Rashmi, W., Mekhilef, S., Wadi, A.: Solar energy harvesting with the application of nanotechnology. Renew. Sustain. Energy Rev. **26**, 837–852 (2013)
2. Agarwal, A., Bakshi, S.R., Lahiri, D.: Carbon Nanotubes: Reinforced Metal Matrix Composites. CRC Press (2018)
3. Anis, M., AlTaher, G., Sarhan, W., Elsemary, M.: Automotive Applications, Nanovate, pp. 113–130. Springer (2017)
4. Antisari, L.V., Carbone, S., Gatti, A., Vianello, G., Nannipieri, P.: Toxicity of metal oxide (CeO_2, Fe_3O_4, SnO_2) engineered nanoparticles on soil microbial biomass and their distribution in soil. Soil Biol. Biochem. **60**, 87–94 (2013)
5. Bai, J., Zhou, B.: Titanium dioxide nanomaterials for sensor applications. Chem. Rev. **114**, 10131–10176 (2014)
6. Bakand, S., Hayes, A., Dechsakulthorn, F.: Nanoparticles: a review of particle toxicology following inhalation exposure. Inhalation Toxicol. **24**, 125–135 (2012)
7. Bauer, F., Ernst, H., Decker, U., Findeisen, M., Gläsel, H.J., Langguth, H., Hartmann, E., Mehnert, R., Peuker, C.: Preparation of scratch and abrasion resistant polymeric nanocomposites by monomer grafting onto nanoparticles, 1 FTIR and multi-nuclear NMR spectroscopy to the characterisation of methacryl grafting. Macromol. Chem. Phys. **201**, 2654–2659 (2000)
8. Biswas, A., Bayer, I.S., Biris, A.S., Wang, T., Dervishi, E., Faupel, F.: Advances in top–down and bottom–up surface nanofabrication: techniques, applications & future prospects. Adv. Coll. Interface. Sci. **170**, 2–27 (2012)
9. Braakhuis, H.M., Oomen, A.G., Cassee, F.R.: Grouping nanomaterials to predict their potential to induce pulmonary inflammation. Toxicol. Appl. Pharmacol. **299**, 3–7 (2016)
10. Bumbudsanpharoke, N., Ko, S.: Nano-food packaging: an overview of market, migration research, and safety regulations. J. Food Sci. **80**, R910–R923 (2015)
11. Carpenter, M.A., Mathur, S., Kolmakov, A.: Metal Oxide Nanomaterials for Chemical Sensors. Springer Science & Business Media (2012)

12. Chaudhry, N., Dwivedi, S., Chaudhry, V., Singh, A., Saquib, Q., Azam, A., Musarrat, J.: Bio-inspired nanomaterials in agriculture and food: current status, foreseen applications and challenges. Microb. Pathog. **123**, 196–200 (2018)
13. Chhipa, H.: Nanofertilizers and nanopesticides for agriculture. Environ. Chem. Lett. **15**, 15–22 (2017)
14. Choi, H., Yoon, H.: Nanostructured electrode materials for electrochemical capacitor applications. Nanomaterials **5**, 906–936 (2015)
15. Chuah, S., Pan, Z., Sanjayan, J.G., Wang, C.M., Duan, W.H.: Nano reinforced cement and concrete composites and new perspective from graphene oxide. Constr. Build. Mater. **73**, 113–124 (2014)
16. Chuang, S.-M., Lee, Y.-H., Liang, R.-Y., Roam, G.-D., Zeng, Z.-M., Tu, H.-F., Wang, S.-K., Chueh, P.J.: Extensive evaluations of the cytotoxic effects of gold nanoparticles. Biochim. Biophys. Acta (BBA) Gener. Subj. **1830**, 4960–4973 (2013)
17. Coelho, M.C., Torrão, G., Emami, N.: Nanotechnology in automotive industry: research strategy and trends for the future—small objects, big impacts. J. Nanosci. Nanotechnol. **12**, 6621–6630 (2012)
18. Donaldson, K., Stone, V., Tran, C., Kreyling, W., Borm, P.J.: Nanotoxicology. BMJ Publishing Group Ltd. (2004)
19. Duhan, J.S., Kumar, R., Kumar, N., Kaur, P., Nehra, K., Duhan, S.: Nanotechnology: the new perspective in precision agriculture. Biotechnol. Rep. **15**, 11–23 (2017)
20. Erol, O., Uyan, I., Hatip, M., Yilmaz, C., Tekinay, A.B., Guler, M.O.: Recent advances in bioactive 1D and 2D carbon nanomaterials for biomedical applications. Nanomed. Nanotechnol. Biol. Med. **14**, 2433–2454 (2018)
21. Feynman, R.P.: Plenty of Room at the Bottom, APS Annual Meeting (1959)
22. Galvin, P., Thompson, D., Ryan, K.B., McCarthy, A., Moore, A.C., Burke, C.S., Dyson, M., MacCraith, B.D., Gun'ko, Y.K., Byrne, M.T.: Nanoparticle-based drug delivery: case studies for cancer and cardiovascular applications. Cell. Mol. Life Sci. **69**, 389–404 (2012)
23. Gamer, A., Ev, L., Van Ravenzwaay, B.: The in vitro absorption of microfine zinc oxide and titanium dioxide through porcine skin. Toxicol. In Vitro **20**, 301–307 (2006)
24. García, A., Espinosa, R., Delgado, L., Casals, E., González, E., Puntes, V., Barata, C., Font, X., Sánchez, A.: Acute toxicity of cerium oxide, titanium oxide and iron oxide nanoparticles using standardized tests. Desalination **269**, 136–141 (2011)
25. Ghanashyam, G., Jeong, H.K.: Synthesis of nitrogen-doped plasma treated graphite for supercapacitor applications. Chem. Phys. Lett. **725**, 31–37 (2019)
26. Ghasemzadeh, G., Momenpour, M., Omidi, F., Hosseini, M.R., Ahani, M., Barzegari, A.: Applications of nanomaterials in water treatment and environmental remediation. Front. Environ. Sci. Eng. **8**, 471–482 (2014)
27. Gill, S.S., Tuteja, N.: Reactive oxygen species and antioxidant machinery in abiotic stress tolerance in crop plants. Plant Physiol. Biochem. **48**, 909–930 (2010)
28. Gogos, A., Knauer, K., Bucheli, T.D.: Nanomaterials in plant protection and fertilisation: current state, foreseen applications, and research priorities. J. Agric. Food Chem. **60**, 9781–9792 (2012)
29. Gregorczyk, K., Knez, M.: Hybrid nanomaterials through molecular and atomic layer deposition: top down, bottom up, and in-between approaches to new materials. Prog. Mater Sci. **75**, 1–37 (2016)
30. Guo, H., White, J.C., Wang, Z., Xing, B.: Nano-enabled fertilisers to control the release and use efficiency of nutrients. Current Opin. Environ. Sci. Health **6**, 77–83 (2018)
31. Gurjar, B., Tyagi, P.: Applications of nanotechnology in automobile industry for efficiency enhancement and energy saving—a review. Int. J. Interdiscip. Res. **2**, 1–7 (2015)
32. He, X., Hwang, H.M.: Nanotechnology in food science: functionality, applicability, and safety assessment. J. Food Drug Anal. **24**, 671–681 (2016)
33. He, X., Deng, H., Hwang, H.M.: The current application of nanotechnology in food and agriculture. J. Food Drug Anal. **27**, 1–21 (2019)

34. Hola, K., Markova, Z., Zoppellaro, G., Tucek, J., Zboril, R.: Tailored functionalisation of iron oxide nanoparticles for MRI, drug delivery, magnetic separation and immobilisation of biosubstances. Biotechnol. Adv. **33**, 1162–1176 (2015)
35. Honeywell-Nguyen, P.L., de Graaff, A.M., Groenink, H.W., Bouwstra, J.A.: The in vivo and in vitro interactions of elastic and rigid vesicles with human skin. Biochim. Biophys. Acta (BBA) Gener. Subj. **1573**, 130–140 (2002)
36. Hosseinzadeh, A., Bidmeshkipour, S., Abdi, Y., Arzi, E., Mohajerzadeh, S.: Graphene based strain sensors: a comparative study on graphene and its derivatives. Appl. Surf. Sci. **448**, 71–77 (2018)
37. Hussein, A.K.: Applications of nanotechnology in renewable energies—a comprehensive overview and understanding. Renew. Sustain. Energy Rev. **42**, 460–476 (2015)
38. Iqbal, P., Preece, J.A., Mendes, P.M.: Nanotechnology: the "Top-Down" and "Bottom-Up" approaches. Supramolecular chemistry: from molecules to nanomaterials (2012)
39. Jara, A.D., Betemariam, A., Woldetinsae, G., Kim, J.Y.: Purification, application and current market trend of natural graphite: a review. Int. J. Min. Sci. Technol. **29**, 671–689 (2019)
40. Jiang, H., Ma, J., Li, C.: Mesoporous carbon incorporated metal oxide nanomaterials as supercapacitor electrodes. Adv. Mater. **24**, 4197–4202 (2012)
41. Jun, L.Y., Mubarak, N.M., Yee, M.J., Yon, L.S., Bing, C.H., Khalid, M., Abdullah, E.C.: An overview of functionalised carbon nanomaterial for organic pollutant removal. J. Ind. Eng. Chem. **67**, 175–186 (2018)
42. Jun, L.Y., Mubarak, N.M., Yon, L.S., Bing, C.H., Khalid, M., Abdullah, E.C.: Comparative study of acid functionalisation of carbon nanotube via ultrasonic and reflux mechanism. J. Environ. Chem. Eng. **6**, 5889–5896 (2018)
43. Jun, L.Y., Mubarak, N.M., Yon, L.S., Bing, C.H., Khalid, M., Jagadish, P., Abdullah, E.C.: Immobilisation of peroxidase on functionalized MWCNTs-buckypaper/polyvinyl alcohol nanocomposite membrane. Sci. Rep. **9**, 2215 (2019)
44. Karunadasa, K.S.P., Manoratne, C.H., Pitawala, H.M.T.G.A., Rajapakse, R.M.G.: A potential working electrode based on graphite and montmorillonite for electrochemical applications in both aqueous and molten salt electrolytes. Electrochem. Commun. **108**, 106562 (2019)
45. Khanna, A.: Nanotechnology in high performance paint coatings. Asian J. Exp. Sci **21**, 25–32 (2008)
46. Khodakovskaya, M.V., De Silva, K., Biris, A.S., Dervishi, E., Villagarcia, H.: Carbon nanotubes induce growth enhancement of tobacco cells. ACS Nano **6**, 2128–2135 (2012)
47. Khot, L.R., Sankaran, S., Maja, J.M., Ehsani, R., Schuster, E.W.: Applications of nanomaterials in agricultural production and crop protection: a review. Crop Prot. **35**, 64–70 (2012)
48. Kroto, H.W., Heath, J.R., O'Brien, S.C., Curl, R.F., Smalley, R.E.: C60: Buckminsterfullerene. Nature **318**, 162–163 (1985)
49. Kumar, R., Al-Dossary, O., Kumar, G., Umar, A.: Zinc oxide nanostructures for NO_2 gas–sensor applications: a review. Nano-Micro Lett. **7**, 97–120 (2015)
50. Lee, X.J., Hiew, B.Y.Z., Lai, K.C., Lee, L.Y., Gan, S., Thangalazhy-Gopakumar, S., Rigby, S.: Review on graphene and its derivatives: synthesis methods and potential industrial implementation. J. Taiwan Inst. Chem. Eng. **98**, 163–180 (2019)
51. Lehner, R., Wang, X., Marsch, S., Hunziker, P.: Intelligent nanomaterials for medicine: carrier platforms and targeting strategies in the context of clinical application. Nanomed. Nanotechnol. Biol. Med. **9**, 742–757 (2013)
52. Lewicka, Z.A., Benedetto, A.F., Benoit, D.N., William, W.Y., Fortner, J.D., Colvin, V.L.: The structure, composition, and dimensions of TiO_2 and ZnO nanomaterials in commercial sunscreens. J. Nanopart. Res. **13**, 3607 (2011)
53. Li, C., Fan, Z., Wu, S., Li, Y., Gan, Y., Zhang, A.: Effect of carbon black nanoparticles from the pyrolysis of discarded tires on the performance of asphalt and its mixtures. Appl. Sci. **8**, 624 (2018)
54. Liang, K., Richardson, J.J., Cui, J., Caruso, F., Doonan, C.J., Falcaro, P.: Metal–organic framework coatings as cytoprotective exoskeletons for living cells. Adv. Mater. **28**, 7910–7914 (2016)

55. Lin, P.C., Lin, S., Wang, P.C., Sridhar, R.: Techniques for physicochemical characterisation of nanomaterials. Biotechnol. Adv. **32**, 711–726 (2014)
56. Liu, J., Zheng, Z., Li, F., Lei, W., Gao, Y., Wu, Y., Zhang, L., Wang, Z.L.: Nanoparticle chemically end-linking elastomer network with super-low hysteresis loss for fuel-saving automobile. Nano Energy **28**, 87–96 (2016)
57. Liu, X., Wang, M., Zhang, S., Pan, B.: Application potential of carbon nanotubes in water treatment: a review. J. Environ. Sci. **25**, 1263–1280 (2013)
58. López-Serrano, A., Olivas, R.M., Landaluze, J.S., Cámara, C.: Nanoparticles: a global vision. Characterisation, separation, and quantification methods. Potential environmental and health impact. Anal. Methods **6**, 38–56 (2014)
59. Luo, T., Wei, X., Huang, X., Huang, L., Yang, F.: Tribological properties of Al_2O_3 nanoparticles as lubricating oil additives. Ceram. Int. **40**, 7143–7149 (2014)
60. Lyu, M.Y., Choi, T.G.: Research trends in polymer materials for use in light-weight vehicles. Int. J. Precis. Eng. Manuf. **16**, 213–220 (2015)
61. Machado, F.M., Lima, É.C., Jauris, I.M., Adebayo, M.A.: Carbon nanomaterials for environmental applications. In: Bergmann, C.P., Machado, F.M. (eds.) Carbon Nanomaterials as Adsorbents for Environmental and Biological Applications, pp. 85–105. Springer International Publishing, Cham (2015)
62. Malani, A.S., Chaudhari, A.D., Sambhe, R.U.: A review on applications of nanotechnology in automotive industry. Int. J. Mech. Aerosp. Ind. Mechatron. Manuf. Eng. **10** (2016)
63. Mathew, J., Joy, J., George, S.C.: Potential applications of nanotechnology in transportation: a review. J. King Saud Univ. Sci. **31**, 586–594 (2019)
64. Monteiro-Riviere, N.A., Wiench, K., Landsiedel, R., Schulte, S., Inman, A.O., Riviere, J.E.: Safety evaluation of sunscreen formulations containing titanium dioxide and zinc oxide nanoparticles in UVB sunburned skin: an in vitro and in vivo study. Toxicol. Sci. **123**, 264–280 (2011)
65. Mu, L., Sprando, R.L.: Application of nanotechnology in cosmetics. Pharm. Res. **27**, 1746–1749 (2010)
66. Norhasri, M.M., Hamidah, M., Fadzil, A.M.: Applications of using nano material in concrete: a review. Constr. Build. Mater. **133**, 91–97 (2017)
67. Novoselov, K.S., Geim, A.K., Morozov, S.V., Jiang, D., Zhang, Y., Dubonos, S.V., Grigorieva, I.V., Firsov, A.A.: Electric field effect in atomically thin carbon films. Science **306**, 666–669 (2004)
68. Ourique, A., Pohlmann, A., Guterres, S., Beck, R.: Tretinoin-loaded nanocapsules: preparation, physicochemical characterisation, and photostability study. Int. J. Pharm. **352**, 1–4 (2008)
69. Pandiyan, G.K., Prabaharan, T.: Implementation of nanotechnology in fuel cells. Mater. Today Proc. (2020)
70. Pauwels, M., Rogiers, V.: EU legislations affecting safety data availability of cosmetic ingredients. Regul. Toxicol. Pharmacol. **49**, 308–315 (2007)
71. Peters, C.A., Nichols, M.E., Ellwood, K.R.: The evolution of surface texture in automotive coatings. J. Coat. Technol. Res. **8**, 469–480 (2011)
72. Peters, R.J., Bouwmeester, H., Gottardo, S., Amenta, V., Arena, M., Brandhoff, P., Marvin, H.J., Mech, A., Moniz, F.B., Pesudo, L.Q.: Nanomaterials for products and application in agriculture, feed and food. Trends Food Sci. Technol. **54**, 155–164 (2016)
73. Pietroiusti, A., Stockmann-Juvala, H., Lucaroni, F., Savolainen, K.: Nanomaterial exposure, toxicity, and impact on human health. Wiley Interdiscip. Rev. Nanomed. Nanobiotechnol. **10**, e1513 (2018)
74. Poznić, A., Zelić, A., Szabó, L.: Magnetorheological fluid brake–basic performances testing with magnetic field efficiency improvement proposal. Hung. J. Ind. Chem. **40**, 107–111 (2012)
75. Prasad, R., Kumar, V., Prasad, K.S.: Nanotechnology in sustainable agriculture: present concerns and future aspects. Afr. J. Biotech. **13**, 705–713 (2014)
76. Presting, H., König, U.: Future nanotechnology developments for automotive applications. Mater. Sci. Eng. C **23**, 737–741 (2003)

77. Robledo, N., López, L., Bunger, A., Tapia, C., Abugoch, L.: Effects of antimicrobial edible coating of thymol nanoemulsion/quinoa protein/chitosan on the safety, sensorial properties, and quality of refrigerated strawberries (Fragaria × ananassa) under commercial storage environment. Food Bioprocess Technol. **11**, 1566–1574 (2018)
78. Ryman-Rasmussen, J.P., Riviere, J.E., Monteiro-Riviere, N.A.: Penetration of intact skin by quantum dots with diverse physicochemical properties. Toxicol. Sci. **91**, 159–165 (2006)
79. Sadrieh, N., Wokovich, A.M., Gopee, N.V., Zheng, J., Haines, D., Parmiter, D., Siitonen, P.H., Cozart, C.R., Patri, A.K., McNeil, S.E.: Lack of significant dermal penetration of titanium dioxide from sunscreen formulations containing nano-and submicron-size TiO_2 particles. Toxicol. Sci. **115**, 156–166 (2010)
80. Sajid, M., Ilyas, M., Basheer, C., Tariq, M., Daud, M., Baig, N., Shehzad, F.: Impact of nanoparticles on human and environment: review of toxicity factors, exposures, control strategies, and future prospects. Environ. Sci. Pollut. Res. **22**, 4122–4143 (2015)
81. Salvia-Trujillo, L., Rojas-Graü, M.A., Soliva-Fortuny, R., Martín-Belloso, O.: Use of antimicrobial nanoemulsions as edible coatings: Impact on safety and quality attributes of fresh-cut Fuji apples. Postharvest Biol. Technol. **105**, 8–16 (2015)
82. Savolainen, K., Pylkkänen, L., Norppa, H., Falck, G., Lindberg, H., Tuomi, T., Vippola, M., Alenius, H., Hämeri, K., Koivisto, J.: Nanotechnologies, engineered nanomaterials and occupational health and safety–a review. Saf. Sci. **48**, 957–963 (2010)
83. Servin, A., Elmer, W., Mukherjee, A., De la Torre-Roche, R., Hamdi, H., White, J.C., Bindraban, P., Dimkpa, C.: A review of the use of engineered nanomaterials to suppress plant disease and enhance crop yield. J. Nanopart. Res. **17**, 92 (2015)
84. Seubert, C., Nietering, K., Nichols, M., Wykoff, R., Bollin, S.: An overview of the scratch resistance of automotive coatings: exterior clearcoats and polycarbonate hardcoats. Coatings **2**, 221–234 (2012)
85. Sundararaj, U., Gelves, G., Al-Saleh, M.: Nanomaterial composites and methods of making. Google Patents (2011)
86. Taha, R.: Nano carbon applications for plant. Adv. Plants Agric. Res. **5**, 00172 (2016)
87. Taniguchi, N., Arakawa, C., Kobayashi, T.: On the basic concept of'nano-technology. In: Proceedings of the International Conference on Production Engineering, 1974–8. 一般社団法人日本機械学会, pp. 18–23 (1974)
88. Teow, Y., Asharani, P., Hande, M.P., Valiyaveettil, S.: Health impact and safety of engineered nanomaterials. Chem. Commun. **47**, 7025–7038 (2011)
89. Tietze, R., Lyer, S., Dürr, S., Struffert, T., Engelhorn, T., Schwarz, M., Eckert, E., Göen, T., Vasylyev, S., Peukert, W.: Efficient drug-delivery using magnetic nanoparticles—biodistribution and therapeutic effects in tumour bearing rabbits. Nanomed. Nanotechnol. Biol. Med. **9**, 961–971 (2013)
90. Tomar, S.: Innovative nanotechnology applications in automobiles. Int. J. Eng. Res. Technol. **1** (2012)
91. Verma, S.K., Das, A.K., Gantait, S., Kumar, V., Gurel, E.: Applications of carbon nanomaterials in the plant system: a perspective view on the pros and cons. Sci. Total Environ. **667**, 485–499 (2019)
92. Wan, J., Van Aken, D.C., Qing, J., Yaniak, T.J., Clements, T.E., Xu, M.: Developing a graphitic white iron for abrasive wear application: thermal and wear properties. Wear **436–437**, 202967 (2019)
93. Wang, J., Jin, X., Li, C., Wang, W., Wu, H., Guo, S.: Graphene and graphene derivatives toughening polymers: toward high toughness and strength. Chem. Eng. J. **370**, 831–854 (2019)
94. Wu, H., Tito, N., Giraldo, J.P.: Anionic cerium oxide nanoparticles protect plant photosynthesis from abiotic stress by scavenging reactive oxygen species. ACS Nano **11**, 11283–11297 (2017)
95. Wu, Y., Pang, H., Liu, Y., Wang, X., Yu, S., Fu, D., Chen, J., Wang, X.: Environmental remediation of heavy metal ions by novel-nanomaterials: a review. Environ. Pollut. (2018)
96. Yang, Z., Tian, J., Yin, Z., Cui, C., Qian, W., Wei, F.: Carbon nanotube- and graphene-based nanomaterials and applications in high-voltage supercapacitor: a review. Carbon **141**, 467–480 (2019)

97. Yao, K.S., Li, S., Tzeng, K., Cheng, T.C., Chang, C.Y., Chiu, C., Liao, C., Hsu, J., Lin, Z.: Fluorescence silica nanoprobe as a biomarker for rapid detection of plant pathogens, advanced materials research. Trans. Tech. Publ. 513–516 (2009)
98. Yun, J., Kim, D., Lee, G., Ha, J.S.: All-solid-state flexible micro-supercapacitor arrays with patterned graphene/MWNT electrodes. Carbon **79**, 156–164 (2014)
99. Zhang, Y., Gao, Z., Song, N., He, J., Li, X.: Graphene and its derivatives in lithium–sulfur batteries. Mater. Today Energy **9**, 319–335 (2018)
100. Zheng, S., Li, J.: Inorganic–organic sol gel hybrid coatings for corrosion protection of metals. J. Sol-Gel. Sci. Technol. **54**, 174–187 (2010)
101. Zhu, C., Liu, T., Qian, F., Chen, W., Chandrasekaran, S., Yao, B., Song, Y., Duoss, E.B., Kuntz, J.D., Spadaccini, C.M.: 3D printed functional nanomaterials for electrochemical energy storage. Nano Today **15**, 107–120 (2017)
102. Zvyagin, A.V., Zhao, X., Gierden, A., Sanchez, W., Ross, J., Roberts, M.S.: Imaging of zinc oxide nanoparticle penetration in human skin in vitro and in vivo. J. Biomed. Opt. **13**, 064031 (2008)

Graphene and Its Composites

Marlinda Ab Rahman, Suresh Sagadevan, and Mohd Rafie Johan

Abstract Recent years have seen many innovations of graphene in various fields such as physics, chemistry, biology, and materials science. Graphene-based materials and their composites have a wide range of promising applications including electronics, biomedical devices, membranes, wearable sensors, and actuators. Graphene is a 2-dimensional (2D) array of carbon atoms in planar and hexagonal forms. Each carbon is sp^2-hybridized and connects three stable bonds C–C–μ at 120° apart. The unhybridized p-orbital, together with the same p-orbitals across the entire 2-D plane, is perpendicular to the sp^2-hybridization plane on other carbon atoms by Ś interaction. Graphene-based nanocomposites have drawn a great deal of attention in scientific communities, due to its extraordinary magnetic, mechanical, thermal and optical properties, and a large surface region. This chapter, therefore, presents different techniques in the graphene synthesis method and its excellent physical and chemical properties. Various manufacturing processes of graphene-based composites are introduced. In conclusion, the remaining challenges and perspectives in functional science and engineering for graphene nanocomposites are discussed.

Keywords Graphene · Composites · Synthesis methods · Physics · Chemical properties

1 Introduction

The graphite was discovered in a 16th-century mine close to Borrowdale in Cumbria, England. At the beginning of its finding, the graphite used for marking and descriptive purposes for example, nearby farmers used mine graphite blocks to mark their sheep. Due to its softness and dark colour, graphite has long been used as a particular type of lead. Graphene is a single carbon layer with a graphical structure in the chemical literature. Graphene is also classified as single carbon layers that appeared

M. A. Rahman · S. Sagadevan (✉) · M. R. Johan
Nanotechnology and Catalysis Research Centre, University of Malaya, 50603 Kuala Lumpur, Malaysia
e-mail: sureshsagadevan@gmail.com

N. M. Mubarak et al. (eds.), *Contemporary Nanomaterials in Material Engineering Applications*, Engineering Materials,
https://doi.org/10.1007/978-3-030-62761-4_2

in the inter-compound of graphite. For example, it exists in a sequence of naphthalene, anthracenes, coronene, etc. For standard carbon jargon, the term "graphene layer" is also considered [1]. Since its discovery [2], the atomic structure with one layer has revolutionized the nanotechnology platform. After 2004, with its exponentially growing uses, graphene has invaded the nanotechnology market. Graphene is a "magic bullet" for the composite world due to its extraordinary properties. Graphene research achieved an unusually high elevation and was established in 2009 by Geim as a champion of the applied sciences.

Graphene is a single flat sheet from graphite and has the ideal two-dimensional (2D) structure with a monolayer of carbon atoms packed into a honeycomb crystal plane. Figure 1 shows the two-dimensional honeycomb structure of carbon atoms in graphene along with the high-resolution transmission electron microscopic (TEM) image. The TEM image clearly shows the hexagonal lattice with carbon–carbon distance 0.14 nm. Graphene is considered as the fundamental building block for graphitic materials of all other dimensions. It can be wrapped up into zero-dimensional (0D) fullerenes, rolled into one-dimensional (1D) nanotubes and stacked into three-dimensional (3D) graphite. Therefore, graphene is called the mother of all graphitic carbon-based nanomaterials. Figure 2 illustrates the schematic diagram demonstrating that graphene is the mother of all graphitic forms. Graphene is a crucial structural element in carbon allotropic, including graphite, carbon nanotubes, and fullerenes.

Therefore, graphene is a unique example of a layer of atomic membranes. Like any other membrane, two-dimensional structure distortions cost very little energy in a three-dimensional environment, so graphene has excellent potential for folding, distorting, forming scrolls, wrinkles, blisters, creases, etc. Nevertheless, graphene

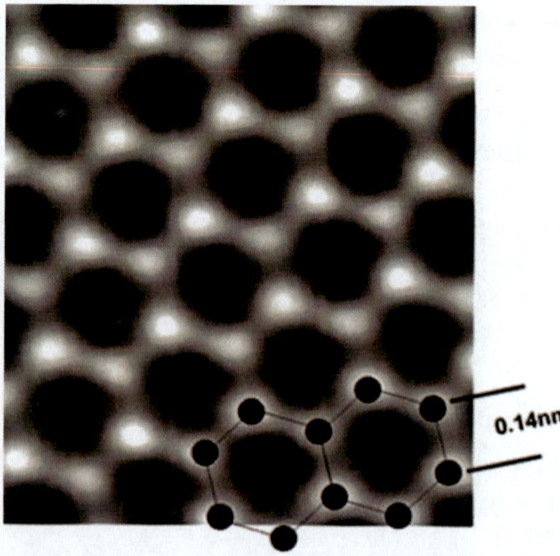

Fig. 1 TEM image of a graphene sheet illustrating the crystalline lattice (bond length ~0.14 nm) (Dato et al. 2009) [3]. It can be wrapped in 0D fullerenes, rolled in 1D nanotubes, or stacked in 3D graphite (see Fig. 2) [4]

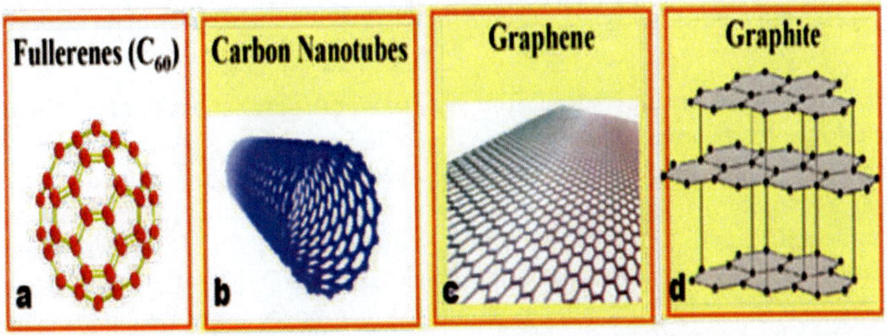

Fig. 2 Graphitic allotrope structure of **a** 0 D-fullerenes, **b** 1D-carbon, **c** 2D-graphene and **d** 3D-graphite (stacking of graphene layer)

can be stabilized to measure its properties through a substratum or scaffold experimentally. The high exfoliation potential of graphite-stacked strata is the most challenging aspect of graphite preparation [5]. Graphite is defined as a stack of graphene layers that rely on the stacking arrangement. This stacking relatively depends on the graphene orientation plane. Due to the weak interaction of each layer of graphite, the graphite constituent can easily be cleavaged. Graphite is a heap of graphene sheets that are tied together by van der Waals (see Fig. 2d), which is less fragile in-plane bonds than the covalent. As mentioned, the weak van der Waals interaction between the graphene sheets may make it easy to exfoliate a thin stack of graphene sheets from a bulk graphite, i.e., when it is scratched over a rough surface thin layers of graphite sticks to that surface.

There have been several attempts to synthesize graphene on a large scale in response to the demands of different industries, in particular, the composite sector, where graphene use has changed the state-of-the-art global market radically for the manufacturers. Graphene has drawn enormous interest in the scientific community because of its extraordinary physical, electrical, and mechanical properties [6, 7]. Some of the unique properties of graphene including high electron mobility at 2.5×10^5 cm^2 V^{-1} s^{-1} [8], high mechanical rigidity with Young's modulus of 1 TPa and intrinsic strength of 130 GPa [9], extremely high thermal efficiency over 3,000 W m K^{-1} [10], α-optical absorption is about 2.3% (in the network constant α [11], entirely permeable for any form of gas [12] and capable of extremely high densities of electrical current (a million times higher than copper) [13]. Graphene is applied in many sectors to a host matrix with a range of improved properties, including aeronautical technologies [14], electronics [15], long-lasting electricity [16], environmental [17], medicines [18] and food.

2 Synthesis Methods

Since Geim and Novoselov's first discovery of stable graphene monolayer in 2004, tremendous efforts have been made to prepare the graphene. Two approaches, classified as top-down and bottom-up, can be used to prepare graphene, as shown in Fig. 3. There are six significant techniques, including mechanical exfoliation [2, 19], epitaxial growth [20], CVD [21, 22], solvothermal [23, 24], graphite thermal expansion and chemical reduction of graphene oxide (GO) [25, 26] and microwave-assisted exfoliation [27] have been employed to prepare these two approaches to produce graphene. However, thermal [28, 29], hydrothermal [30], chemical [31, 26] and solvothermal reduction can be obtained in bulk quantities [23, 32]. These approaches are also versatile, scalable, and suitable for wide-ranging applications.

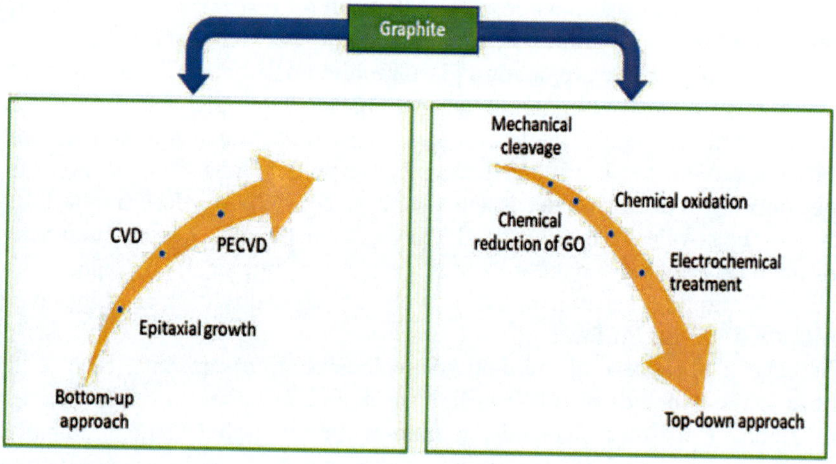

Fig. 3 Synthesis method of graphene from graphite

Fig. 4 shows the electronic band structure of single-layer graphene [33]

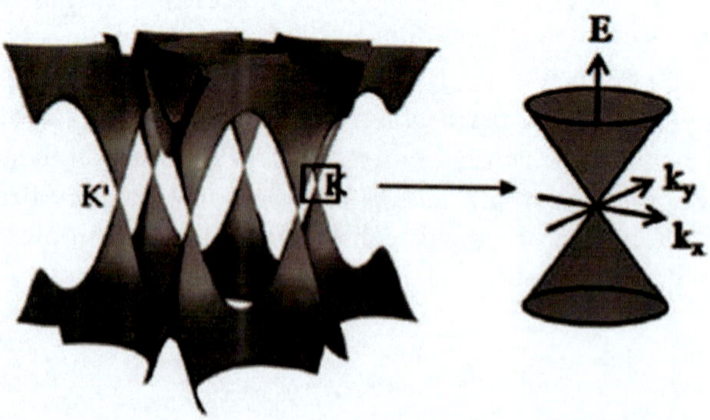

2.1 Top-Down Approach

In the top-down approach, graphene is made from graphite or graphite derivatives physical exfoliation, including mechanical cleavage, sonication, chemical oxidation, and graphite electrochemical exfoliation. The micromechanical technique of cleavage exfoliated graphene, which highly oriented via a scotch tape technique within graphite flakes. In 2004, Geim and Novaselov pioneered this technique. The consistency of single sheets of graphs [4] has benefited from this method. Stankovich et al. later introduced the ionizing process for the preparation of graphite oxide. This method involves the dispersion of graphite oxide n by acoustic hydrazine in a fluid medium [26]. Centrifugation eventually is an isolation process of non-exfoliated graphite from graphene. Unfortunately, incomplete removal of different functional groups during reduction processes causes the low quality of graphene as compared to graphene that is produced by the scotch-tape method. The two primary forms of liquid exfoliation methods, i.e., surfactant-free and surfactant-assisted liquid phase exfoliation have been successfully identified in Ciesielski and Samorì [34]. At present, chemical oxidation is the most promising technique with a high yield of graphene products. Graphene oxide sheets are obtained from graphite oxidation flakes subjected to three main methods formulated by Brodie [35], Hummers (1958), and Staudenmaier [36]. These methods produced GO layer structures with a brighter colour than graphite due to the loss of electronic conjugation brought during the oxidation of GO.

The GO layers are constructed of oxidized graphene sheets adorned with epoxy and hydroxy groups and with edge carboxy (Lerf—Klinowski model). These layers are composed of the basal planes. Such oxygen functionalities make hydrophone oxide galleries simple to interpose with GO hydrophilic graphene and water molecules. This oxygen versatility can also be called a graphite-type of intercalation complex between covalently binding oxygen-bonding, non-covalent, water-binding layers of carbon graphite oxide, and water molecules. In comparison, electrochemical graphite preparation was used to create the colloidal suspension of chemical sheets of Graphene [37].

The graphite sheets also can be prepared by using commercial graphite rods. This technique can be achieved by submerged graphite rods in an imidazolium-based phase-separated mixture of water and ionic liquids with an applied potential of 10–20 V. This technique was claimed to create ionic-liquid graphite functional sheets deriving from the precipitated anode of graphite (Liu et al.) [37]. In this DMF suspension, AFM calculated the average CMG box thickness of ~1,1 nm. Diverse forms of ionic fluids, as well as various ratios of ionic liquid to water, can affect the properties of graphene nanosheets.

2.2 *Bottom-Up Approach*

Walt de Heer and Claire Berger from Atlanta group [38], has developed another alternative method to obtain graphene consists of exposing to high temperature (>1100 °C) known as an epitaxial growth on SiC crystal (4H or 6H–) [39]. This method produces graphene epitaxial growth that the dimensions are depending on the size of the selected SiC substratum. The SiC surface shapes graphene, silicone, or finished carbon and has excellent effects on the thickness of the graphene, its stability, and its carrier density. This approach helped to define the critical characteristics of graphene and attracted more researchers, especially in the semiconductor industry.

The synthesis of graphene in a single layer (0001) ruthenium crystal plane using the chemical vapour deposition (CVD) process was reported in 2008 by Vazquez de Parga et al. [40]. The CVD process for the production of few-layer graphic image films on nickel layers documented by Kim et al. [21]. The template films were also successfully transferred without extreme mechanical or chemical processing onto random substrates to maintain the high crystalline consistency of the generated graph. Graphene may also be synthesized by using plasma-enhanced chemical vapour (PECVD) deposition to reduce costs at lower reaction temperatures [41]. Wang et al. have reported the first development of mono and some layers of graphite by radiofrequency PECVD on multitudes by injecting a gas mix of CH_4 and H_2 at 900 W at a temperature of 680 °C [42]. and by introducing a gas mixture of CH_4 and H_2 at a temperature of 680 °C [42].

3 Physical and Chemical Properties

This section aims to show the exciting features of a single, two, or few layers of graphene sheets. Until implementing the properties of the single graphene, it is crucial to understand the graphene structure clearly. Graphene in one layer may be described as a single two-dimensional material that is densely packed into a honeycomb crystal grid with carbon atom sheets. Bilayer and few graphene layers have two-dimensional carbon atom sheets of 2 and 3–10 layers, respectively. The graphene structures composed of over 10 two dimensional layers are used as a dense graphene board. The best way to determine the thickness of the graph sheet is to use microscopic and spectroscopic techniques.

Information on morphology and graph layer numbers can be investigated by using transmission electron microscopy (TEM). Single-layer Graphene displays a dark line and two or a few layers of graphene showing a folded edge with two faint lines. Raman spectroscopy, on the other hand, is another way to investigate graphene layer thickness. The first graphene spectra order of the Raman spectra of D, G, and 2D bands that match plane vibrations [43]. The number of graphene layers of G and 2D bands, their form, and the section strength ratio can be verified. G-band sensitivity decreases as the number of layers increases. The G-band downshift also demonstrates the increasing proportion of graphene layers. The change in layers shown by changing the shape, width, and position of the 2D band. The 2D bandwidth will distinguish between the single, bi, and few layers of graphene. The graphene of a single layer shows a peculiar electrical structure, showing an alternating band in the Brillouin region (Fig. 4), in two sharp points (K and K′). SLG is, therefore, expected to possess specific unique characteristics concerning metals, semiconductors, and semimetals. The optimistic features of the space are portrayed in single-layer graphene, while there is no energy difference in the few layers graphene. The more layers the graphene structure contains, the more polished the stuff. Graphene has an excellent gas adsorption structure and hydrogen absorption.

4 Graphene with Composite Materials

Graphene can increase the conductivity and strength of bulk materials. This may also help to establish superior composite consistency. To create composites that are conductive and highly resistant to heat and pressure, graphene can be added to metals, polymers and ceramics. Graphene composites have many potential applications, creating unique and innovative materials with a lot of research going on. Recently, there are many applications based on coupling materials like graphene-polymer, which is widely used in medical implants, automotive and recycling, etc. The composite materials are structures formed to create unique finished content, incorporating two or more elements with different properties. They are also known as combination materials or direct composites. They're not combined or merged, but in the final organic form, they remain separate. Thanks to their properties, composite materials can be made stronger, lighter, or softer than traditional materials by incorporating their different components. The bulk of the composites contain two materials-matrices (or binder) composed of a bundle of more durable fibres or fragments. Fibre glass, produced as the first plastic composite in the 1940s, is a typical example of this structure and is still commonly used. Fine glass fibres, tissued- into a fabric, are used as reinforcing fibre glass in a plastic or resin matrix. Although composites are not a new concept (for example, for thousands of years, dry mud bricks embedded in stroke have existed), many new and exciting composites have been produced using the latest technology. Through carefully choosing and updating the matrix, as well as the best manufacturing process to implement it, substantially superior products can be produced with tailored features for specific needs. Composite building materials, for example, concrete or asphalt, various metal composites, plastic blends, and ceramic composites, are usually constructed.

4.1 Composites with Polymers, Metals, Metal Oxides

It has attracted excellent graphene properties to be used as potential composite materials to achieve more exciting performance in electronics, optics, and photonic applications. In various matrix substances such as polymers, epoxy metals, and metal oxides, graphene and its derivatives have been incorporated. In 2006, Ruoff et al. synthesized the early introduction of graphene into the polymer matrix by solution-phase mixing with polystyrene. The polystyrene—graphene composite that formed exhibits a low percolation threshold of nearly 0.1 vol. percent at room temperature. Most of the previous composite graphene/polymer studies used graphene derivatives such as graphene oxide (GO) and reduced GO (RGO) as polymer matrix fillers combined with polyvinyl alcohol (PVA) [44], poly(methyl methacrylate) (PMMA) [45], poly(arylene sulfide) [46] and epoxy [47, 48].

However, the technique of ex-situ hybridization is likely much more straightforward than the method of in-situ hybridization. The graphene and pre-synthesized or commercially available nanocrystals are generally mixed in the medium of the solution. In this process, stacking through some linking molecule is one of the mechanisms involved in absorbing nanoparticles such as several examples, gold nanoparticles [49] and CdS nanoparticles [50] with surface modification on initial nanocrystal preparation and dispersion of graphene. Also reported was the manufacture of other metals such as silver, palladium, and platinum nanoparticles/graphene composites synthesized through covalent interaction [51]. Whereas graphene hybridization using semiconductor oxides was typically applied to TiO_2 or ZnO as well as to nanocomposite materials based on graphene synthesis. This method can be used to produce a high-quality graph-based nanocomposite without the use of any stabilizing agent. Many approaches have been used for the acquisition and analysis of these composites: electrochemical path, covalent mixture, thermal breakdown, and a sol–gel phase.

With support from UV and ethanol solution-photocatalytic reduction, William et al. [52], as well as Akhavan et al. [53], developed graphene/ZnO or graphene/TiO_2 composites. In all the above synthesis, carbon material/ZnO systems have outstanding characteristics. Graphene ZnO is therefore vital for environmental applications employing a quick, reliable, and cost-effective method.

5 Civil Engineering Applications

The unique mechanical, thermal, optical, and electrical properties of graphene will soon benefit civil engineering, says the National Physical Laboratory's Andrew Pollard. Graphene has many potential applications in civil engineering, but consistency raw material issues need to be overcome. Scientists at Manchester University first isolated graphene flakes in 2004 and winning a Nobel Prize in Physics 6 years later. Every year in this new and exciting area of research, the many excellent properties of the material have led to thousands of scientific publications and patents. It has already been used for flexible electronics, energy storage, nanocomposites, sensors, thermal management, and liquid filtration. Carbon is one of the universe's most common chemical elements and Life's essential elements. Furthermore, graphene is one of the universe's most straightforward molecular structures. For this 2D carbon material, though, while the future is bright, it is not without challenges. Lack of trust in the supply chain is one of the crucial barriers to real-world applications.

Most companies that want to develop graphene-enabled products are commonly struggling to find a reliable source of graphene suppliers' material. It is due to batch-to-batch variability. This problem can be worsened by the fact that the supplier nor the end-user uncertain about the physical or chemical properties of the graphene. The issue of commercialized graphene is the black powder or liquid containing flakes of one or several layers of graphene, as well as graphite, from the hundreds of suppliers worldwide. It is challenging to measure a statistically representative amount of nanoscale flakes for each piece of material that can be produced on the ton-scale. Similarly, there are no standardized measurement procedures to do this, and therefore there are no well-defined quality control techniques.

Graphene, another type of carbon nanoparticle, has recently revealed its excellence in enhancing various advanced properties as well as the desired mechanical properties, showing a path to the formation of a super concrete structure. Graphene and derivatives such as graphic nanoplatelets, graphene oxide, and lower graphene oxide provide increased stress sensing, temperature sensing and EMI shielding, wave absorption. Graphene has some remarkable and excellent properties such as large unique surfaces (2630 m^2 g^{-1}), high intrinsic strength (130 GPa). Young's firm package (~1.0 TPa). There is no improvement only in the rheology characteristics, but in the electrical properties of the material by integrating graphene into the cement composite. The electrical properties of the GNP-cement composite are critical. They can be used to track damage in a concrete structure so that the civil infrastructure is protected, secure, and reliable. Non-destructive testing has been shown to provide quick and effective inspection of these systems [54, 55].

Nevertheless, it is the concrete self-sensor that can track its stress that is important for this age. Through measuring shifts in the electrical resistivity factor [54], the pressure of cement-based composite reinforced conduction fillers can be detected. Automotive sensing is associated with the breakdown of conduction fibres when cracks are initiated in the cement composite, which enhances the sample's resistivity. As cracks open due to tensile loading or cracking, the resistivity values will be positive while the compressive load will be negative. A multifunction, high-performance composite based on types of cement, capable of perceiving damages in real-time automatically, will be generated in newly developed Nano-carbon engineering [56–59]. CNT consists of sp^2 derivatives, graphene sheets rolled up into a cylindrical tube, of a single layer of carbon [60]. The two-dimensional structure will be formed if these rolled CNT sheets are opened in one direction. Such two-dimensional sheets have an even higher surface area and aspect ratio (GNP). The effect of the improved GNP mortar crack on the adjustment of electrical resistance [61]. Samples (cube and prism forms) have been found to have responded to an increase in electrical strength with an increase in crack size. For the same relative crack size, the adjustment of the electrical resistance of the cube was more critical than that of the prism. According to the authors' knowledge, however, the response of the GNP-cement composite to different damage levels was not investigated. Therefore, in this analysis, Bingham, Revised Bingham, Herschel—Bulkley, and Casson models were used to test the fluvial properties of the GNP cement paste. In the flow curves of cement paste, a discrepancy was measured for different percentages of graphene nanoplatelets. The rheological properties of graphene cement paste with varying periods of rest (time between sample casting preparations) and shear rate cycles also were evaluated. The self-sensing properties of cemented concrete based on GNP were later determined. For calculating the electrical resistance, the four-sample procedure was used. Fractional improvements to strain sensitivity and intensity have been observed and used to determine self-sensing characteristics. Ultimately, the use of the GNP-cement composite sample was tested with the reinforced plate.

6 Mechanical Engineering Applications

The global automotive industry is facing major challenges, including CO_2 emission reductions, health risk, and energy efficiency. A new direction for greener and healthier cars is thus more than necessary. This situation opens up such a chance that graphene-based innovation is the right solution. With our commercial secret manufacturing process, the final cost of innovating the car industry by graphene is no longer the slow down factor. Graphene has significant uses in composites and coatings for producing light, sturdy and healthy vehicles. Graphene is more than 200-fold stronger than steel and can absorb carbon fibre, concrete, and aluminium components which able to increase the physical stability of the car. Heated car seats graphene-enhanced can be realized now without high-energy waste. In the years to come, graphene will strengthen or replace existing lithium batteries.

About seven or eight years, lightweight vehicles with graphene will be on the market and could lead to a much lighter and more energy-efficient production of automobiles. That is the expectation of Sunderland University, leading a consortium promoting revolutionary material. There is currently a pilot plant operating in Japan. University scientists worked on bumper made of graphene that can absorb 40% more energy than the standard item when added to carbon-reinforced plastic in some way. The result could be much lighter cars and more compact vehicles. Graphene is developed to produce a single-atomic substance or 1 million times thinner than a human hair, by breaking down graphite as the same material used in pencils. When dissolved powder in fluid, small amounts can be combined with the solvent used for impregnating carbon fibre. In the same process, a thermoplastic material can be moulded when it is heated and cooled actively. In this case, when molten, the graphene is combined with the rubber.

7 Electronics Engineering Applications

7.1 Energy Storage

Graphene has been looked at as an alternative to the current materials used in storing ions on the electrodes of supercapacitors. The greater the surface area, the more ions can be stored on it. Graphene has a theoretical surface area of around 2600 square meters per gram. Unfortunately, that theoretical surface area can be translated into a real device because that surface area is only achievable with a standalone sheet of graphene. In order to get it to work in a practical device that will provide a decent volumetric capacitance, you need to stack several sheets on top of each other. At this point, the theoretical surface area is lost, and you get about the same surface area you can get with the activated carbon used today.

Graphene thus makes it a rare commodity for batteries and supercapacitors to use. Graphene can enable faster charging and more energy-saving devices. It is also possible to use graphene to improve fuel cells. Graphene has a range of other practical applications: anti-corrosion coatings and lacquers, reliable, precise sensors, faster and better electronics, versatile screens, powerful solar panels, quicker DNA sequencing, and more. Applications of graphene are so broad and received attention in every company to gain benefits from this new material. The impact of these new materials will be more fitting in the soon future.

7.2 Electronics

Graphene is an important and broad building block that seems to gain every industry's attention for this new material. Moreover, this entire carbon atom structure has unique quantum properties. These graphene properties will help us study Majorana Fermions' properties and achieve superconductors at a high critical temperature. Based on such research, new types of flexible solar cells and new types of quantum computers can be created. In new energy and quantum computing, graphene technology has excellent potential. Graphene edges have become the best material for making supercapacitors in the field of new energy because they can carry a large amount of charge. In addition, a large number of studies have been conducted using graphene or derivative structures for thin-film solar cell electrodes. It is possible to make the most efficient all-carbon-based solar cells with 3D structures and organic chemical modification based on the newly discovered graphene quantum properties. Carbon nanotubes are excellent microelectrode materials as graphene derivatives and have great potential in brain-computer interfaces. It is biocompatible compared to the metal electrode material while having excellent electrical conductivity, and it is also easier to perform molecular-level modifications to induce synapse growth. Also, these electrodes have been tested by some studies. Smaller and more efficient computers can be manufactured using graphene-based materials. This computer can improve the efficiency of the EEG signal analysis, thus achieving a more individualized two - way human and computer adaptation. Thus, it can help users to learn more quickly on how to operate the brain-computer interface.

8 Conclusions and Future Work

Graphene has excellent mechanical, thermal, electrical properties and is a better substitute for traditional nanofillers for polymer matrices. We addressed recent advances in the materials, processing and future applications of graphene nanocomposites. It also defined various techniques and the excellent physical and chemical properties of the graphene synthesis process. Therefore, it is necessary to make further contributions to the identification of biocompatibility, to improve the stability of nanocarriers, to their size and also to provide more studies on the toxicity of this new material. The physical and chemical features of graphene have appeared with graphene electronics, including biosensors, organic solar cells, actuators, and touch screens. Graphene is used in future electronics via lightweight, stretchable structures as a possible part of these properties. Although many technically feasible solutions are currently being established, there are still several practical difficulties. The advantages of graphics-based materials are primarily mentioned in this analysis, as described and listed.

Nonetheless, the world of graphene is feasible with less costly pilot-scale production technologies. Rising GO can achieve this to graphene, one of the major challenges shortly. After the reduction process successfully reaches the pure graphene, its outstanding electronic and mechanical properties will be recovered. In short, it has unique features for various engineering uses. Graphene and its composites have unique properties.

References

1. Boehm, H.P., Setton, R., Stumpp, E.: Nomenclature and terminology of graphite intercalation compounds (IUPAC Recommendations 1994). Pure Appl. Chem. **66**, 1893 (1994)
2. Novoselov, K.S., Geim, A.K., Morozov, S.V., Jiang, D., Zhang, Y., Dubonos, S.V., Grigorieva, I.V., Firsov, A.A.: Electric field effect in atomically thin carbon films. Science **306**(5696), 666–669 (2004)
3. Dato, A., Lee, Z., Jeon, K.J., Erni, R., Radmilovic, V., Richardson, T.J., Frenklach, M.: Clean and highly ordered graphene synthesized in the gas phase. Chem. Commun. **40**, 6095 (2009)
4. Geim, A.K., Novoselov, K.S.: The rise of graphene. In: Nanoscience and Technology, pp. 11–19 (2007)
5. Yang, X., Dou, X., Rouhanipour, A., Zhi, L., Räder, H.J., Müllen, K.: Two-dimensional graphene nanoribbons. J. Am. Chem. Soc. **130**(13), 4216–4217 (2008)
6. Liang, H.W., Zhuang, X., Brüller, S., Feng, X., Müllen, K.: Hierarchically porous carbons with optimized nitrogen doping as highly active electrocatalysts for oxygen reduction. Nat. Commun. **5**, 4973 (2014)

7. Yi, M., Shen, Z.: A review on mechanical exfoliation for the scalable production of graphene. J. Mater. Chem. A **3**(22), 11700–11715 (2015)
8. Mayorov, A.S., Gorbachev, R.V., Morozov, S.V., Britnell, L., Jalil, R., Ponomarenko, L.A., Blake, P., Novoselov, K.S., Watanabe, K., Taniguchi, T., Geim, A.K.: Micrometer-scale ballistic transport in encapsulated graphene at room temperature. Nano Lett. **11**(6), 2396–2399 (2011)
9. Liu, F., Ming, P., Li, J.: Ab initio calculation of ideal strength and phonon instability of graphene under tension. Phys. Rev. B **76**(6), 064120 (2007)
10. Balandin, A.A.: Thermal properties of graphene and nanostructured carbon materials. Nat. Mater. **10**, 569 (2011)
11. Nair, R.R., Blake, P., Grigorenko, A.N., Novoselov, K.S., Booth, T.J., Stauber, T., Peres, N.M., Geim, A.K.: Fine structure constant defines visual transparency of graphene. Science **320**(5881), 1308–1308 (2008)
12. Bunch, J.S., Verbridge, S.S., Alden, J.S., van der Zande, A.M., Parpia, J.M., Craighead, H.G., McEuen, P.L.: Impermeable atomic membranes from graphene sheets. Nano Lett. **8**(8), 2458–2462 (2008)
13. Moser, J., Barreiro, A., Bachtold, A.: Current-induced cleaning of graphene. Appl. Phys. Lett. **91**(16), 163513 (2007)
14. Manta, A., Gresil, M., Soutis, C.: Graphene in aerospace composites: characterizing thermal response. AIP Conf. Proc. **1932**(1), 020001 (2018)
15. Eda, G., Chhowalla, M.: Graphene-based composite thin films for electronics. Nano Lett. **9**(2), 814–818 (2009)
16. Kumar, A., Sharma, K., Dixit, A.R.: A review of the mechanical and thermal properties of graphene and its hybrid polymer nanocomposites for structural applications. J. Mater. Sci. **54**(8), 5992–6026 (2019)
17. Shen, Y., Chen, B.: Sulfonated graphene nanosheets as a superb adsorbent for various environmental pollutants in water. Environ. Sci. Technol. **49**(12), 7364–7372 (2015)
18. Bonanni, A., Pumera, M.: Graphene platform for hairpin-DNA-based impedimetric genosensing. ACS Nano **5**(3), 2356–2361 (2011)
19. Geim, A.K.: Graphene: status and prospects. Science **324**(5934), 1530–1534 (2009)
20. Sutter, P.W., Flege, J.I., Sutter, E.A.: Epitaxial graphene on ruthenium. Nat. Mater. **7**, 406 (2008)
21. Grande, C.D., Mangadlao, J., Fan, J., De Leon, A., Delgado-Ospina, J., Rojas, J.G., Rodrigues, D.F., Advincula, R.: Chitosan cross-linked graphene oxide nanocomposite films with antimicrobial activity for application in food industry. Macromol. Symp. **374**(1), 1600114 (2017)
22. Yan, Z., Peng, Z., Tour, J.M.: Chemical vapor deposition of graphene single crystals. Acc. Chem. Res. **47**(4), 1327–1337 (2014)
23. Wang, H., Robinson, J.T., Li, X., Dai, H.: Solvothermal reduction of chemically exfoliated graphene sheets. J. Am. Chem. Soc. **131**(29), 9910–9911 (2009)
24. Ai, L., Zhang, C., Chen, Z.: Removal of methylene blue from aqueous solution by a solvothermal-synthesized graphene/magnetite composite. J. Hazard. Mater. **192**(3), 1515–1524 (2011)
25. Gómez-Navarro, C., Weitz, R.T., Bittner, A.M., Scolari, M., Mews, A., Burghard, M., Kern, K.: Electronic transport properties of individual chemically reduced graphene oxide sheets. Nano Lett. **9**(5), 2206–2206 (2009)
26. Stankovich, S., Dikin, D.A., Piner, R.D., Kohlhaas, K.A., Kleinhammes, A., Jia, Y., Wu, Y., Nguyen, S.T., Ruoff, R.S.: Synthesis of graphene-based nanosheets via chemical reduction of exfoliated graphite oxide. Carbon **45**(7), 1558–1565 (2007)
27. Hu, H., Zhao, Z., Zhou, Q., Gogotsi, Y., Qiu, J.: The role of microwave absorption on formation of graphene from graphite oxide. Carbon **50**(9), 3267–3273 (2012)
28. Lee, C., Wei, X., Kysar, J.W., Hone, J.: Measurement of the elastic properties and intrinsic strength of monolayer graphene. Science **321**(5887), 385–388 (2008)
29. Zhao, B., Liu, P., Jiang, Y., Pan, D., Tao, H., Song, J., Fang, T., Xu, W.: Supercapacitor performances of thermally reduced graphene oxide. J. Power Sources **198**, 423–427 (2012)

30. Marlinda, A.R., Huang, N.M., Muhamad, M.R., An'amt, M.N., Chang, B.Y.S., Yusoff, N., Harrison, I., Lim, H.N., Chia, C.H., Kumar, S.V.: Highly efficient preparation of ZnO nanorods decorated reduced graphene oxide nanocomposites. Mater. Lett. **80**, 9–12 (2012)
31. Lomeda, J.R., Doyle, C.D., Kosynkin, D.V., Hwang, W.F., Tour, J.M.: Diazonium functionalization of surfactant-wrapped chemically converted graphene sheets. J. Am. Chem. Soc. **130**(48), 16201–16206 (2008)
32. Nethravathi, C., Rajamathi, M.: Chemically modified graphene sheets produced by the solvothermal reduction of colloidal dispersions of graphite oxide. Carbon **46**(14), 1994–1998 (2008)
33. Rao, C.N.R., Biswas, K., Subrahmanyam, K.S., Govindaraj, A.: Graphene, the new nanocarbon. J. Mater. Chem. **19**, 2457–2469 (2009)
34. Ciesielski, A., Samorì, P.: Graphene via sonication assisted liquid-phase exfoliation. Chem. Soc. Rev. **43**(1), 381–398 (2014)
35. Brodie, B.C.: Sur le poids atomique du graphite. Ann. Chim. Phys. **59**, 466–472 (1860)
36. Staudenmaier, L.: Verfahren zur Darstellung der Graphitsäure. Ber. Dtsch. Chem. Ges. **31**(2), 1481–1487 (1898)
37. Liu, N., Luo, F., Wu, H., Liu, Y., Zhang, C., Chen, J.: One-step ionic-liquid-assisted electrochemical synthesis of ionic-liquid-functionalized graphene sheets directly from graphite. Adv. Func. Mater. **18**(10), 1518–1525 (2008)
38. de Heer, W.A., Berger, C., Wu, X., First, P.N., Conrad, E.H., Li, X., Li, T., Sprinkle, M., Hass, J., Sadowski, M.L., Potemski, M., Martinez, G.: Epitaxial graphene. Solid State Commun. **143**(1), 92–100 (2007)
39. Berger, C., Song, Z., Li, T., Li, X., Ogbazghi, A.Y., Feng, R., Dai, Z., Marchenkov, A.N., Conrad, E.H., First, P.N., de Heer, W.A.: Ultrathin epitaxial graphite: 2D electron gas properties and a route toward graphene-based nanoelectronics. The J. Phys. Chem. B **108**(52), 19912–19916 (2004)
40. Vázquez de Parga, A.L., Calleja, F., Borca, B., Passeggi, M.C.G., Hinarejos, J.J., Guinea, F., Miranda, R.: Periodically rippled graphene: growth and spatially resolved electronic structure. Phys. Rev. Lett. **100**(5), 056807 (2008)
41. Liu, W.W., Chai, S.P., Mohamed, A.R., Hashim, U.: Synthesis and characterization of graphene and carbon nanotubes: a review on the past and recent developments. J. Ind. Eng. Chem. **20**(4), 1171–1185 (2014)
42. Wang, J., Zhu, M., Outlaw, R.A., Zhao, X., Manos, D.M., Holloway, B.C.: Synthesis of carbon nanosheets by inductively coupled radio-frequency plasma enhanced chemical vapor deposition. Carbon **42**(14), 2867–2872 (2004)
43. Ferrari Andrea, C., Robertson, J.: Raman spectroscopy of amorphous, nanostructured, diamond–like carbon, and nanodiamond. Philos. Trans. Roy. Soc. Lond. Series A: Math., Phys. Eng. Sci. **362**(1824), 2477–2512 (2004)
44. Liu, P., Gong, K., Xiao, P., Xiao, M.: Preparation and characterization of poly(vinyl acetate)-intercalated graphite oxide nanocomposite. J. Mater. Chem. **10**(4), 933–935 (2000)
45. Jang, J.Y., Kim, M.S., Jeong, H.M., Shin, C.M.: Graphite oxide/poly(methyl methacrylate) nanocomposites prepared by a novel method utilizing macroazoinitiator. Compos. Sci. Technol. **69**(2), 186–191 (2009)
46. Du, X.S., Xiao, M., Meng, Y.Z., Hay, A.S.: Direct synthesis of poly(arylenedisulfide)/carbon nanosheet composites via the oxidation with graphite oxide. Carbon **43**(1), 195–197 (2005)
47. Wang, S., Tambraparni, M., Qiu, J., Tipton, J., Dean, D.: Thermal expansion of graphene composites. Macromolecules **42**(14), 5251–5255 (2009)
48. Bortz, D.R., Heras, E.G., Martin-Gullon, I.: Impressive fatigue life and fracture toughness improvements in graphene oxide/epoxy composites. Macromolecules **45**(1), 238–245 (2012)
49. Huang, J., Zhang, L., Chen, B., Ji, N., Chen, F., Zhang, Y., Zhang, Z.: Nanocomposites of size-controlled gold nanoparticles and graphene oxide: formation and applications in SERS and catalysis. Nanoscale **2**(12), 2733–2738 (2010)
50. Feng, M., Sun, R., Zhan, H., Chen, Y.: Lossless synthesis of graphene nanosheets decorated with tiny cadmium sulfide quantum dots with excellent nonlinear optical properties. Nanotechnology **21**(7), 075601 (2010)

51. Liu, J., Fu, S., Yuan, B., Li, Y., Deng, Z.: Toward a universal "Adhesive Nanosheet" for the assembly of multiple nanoparticles based on a protein-induced reduction/decoration of graphene oxide. J. Am. Chem. Soc. **132**(21), 7279–7281 (2010)
52. Williams, G., Kamat, P.V.: Graphene–semiconductor nanocomposites: excited-state interactions between ZnO nanoparticles and graphene oxide. Langmuir **25**(24), 13869–13873 (2009)
53. Akhavan, O.: Photocatalytic reduction of graphene oxides hybridized by ZnO nanoparticles in ethanol. Carbon **49**(1), 11–18 (2011)
54. Chung, D.D.L.: Piezoresistive cement-based materials for strain sensing. J. Intell. Mater. Syst. Struct. **13**(9), 599–609 (2002)
55. Kashif Ur Rehman, S., Ibrahim, Z., Memon, S.A., Jameel, M.: Non-destructive test methods for concrete bridges: a review. Constr. Build. Mater. **107**, 58–86 (2016)
56. Raki, L., Beaudoin, J., Alizadeh, R., Makar, J., Sato, T.: Cement and concrete nanoscience and nanotechnology. Materials **3**(2), 918–942 (2010)
57. Sanchez, F., Sobolev, K.: Nanotechnology in concrete—a review. Constr. Build. Mater. **24**(11), 2060–2071 (2010)
58. Sobolev, K., Gutierrez, M.F.: How nanotechnology can change the concrete world. Am. Ceram. Soc. Bull. **84**(10), 14–17 (2005)
59. Mukhopadhyay, A.K.: Next-generation nano-based concrete construction products: a review. In: Nanotechnology in Civil Infrastructure: A Paradigm Shift, pp. 207–223 (2011)
60. Vajtai, R.: Springer Handbook of Nanomaterials. Springer, Berlin (2013)
61. Sixuan, H.: Multifunctional Graphite Nanoplatelets (GNP) reinforced cementitious composites. Master's thesis, Tsinghua University, Beijing, China (2012). https://scholarbank.nus.edu.sg/handle/10635/34752

Carbon Nanotubes and Their Composites: From Synthesis to Applications

Mahesh Vaka, Rashmi Walvekar, and Swarnalatha Yanamadala

Abstract Carbon nanotubes (CNTs), accidentally discovered wonder material from fullerene, is still a material of researcher's interest because of its superior properties and ability to transform/bind into different material via different hybridisations (sp, sp^2, sp^3). After the structural identification of carbon nanotubes in 1991by Iijima, the different types of carbon nanotubes, such as single, double and multi-walled have become one of the potential candidates of the study. In the 1990s and 2000s, the physics and chemistry behind the carbon nanotubes have been developed from both conceptual and experimental studies. After that, the introduction of different synthesis methods and the evolution of properties for different applications are isolated and intensively studied. Besides, CNT based composites have grabbed a lot of attention in the scientific world due to their superior properties that are beneficial for various commercial purposes. This chapter aims to provide different properties of CNTs, synthesis techniques of CNTs, CNT-based composites and their applications in different fields. In conclusion, current challenges and future perspectives for the development of low-cost synthesis and production for commercialisation and material/device development.

Keywords CNT synthesis and properties · CNT-based composites · CNT applications

M. Vaka (✉)
School of Computer Science and Engineering, Taylor's University Lakeside Campus, 47500 Subang Jaya, Selangor, Malaysia
e-mail: maheshsusan@gmail.com

R. Walvekar
Department of Chemical Engineering, School of Energy and Chemical Engineering, Xiamen University Malaysia, Jalan Sunsuria, 43900 Bandar SunsuriaSelangor, Malaysia

S. Yanamadala
Department of Biotechnology, Sathyabama Institute of Science and Technology, Chennai 600119, India

N. M. Mubarak et al. (eds.), *Contemporary Nanomaterials in Material Engineering Applications*, Engineering Materials,
https://doi.org/10.1007/978-3-030-62761-4_3

1 Introduction

The evolution and identification of a potential candidate within the area of nanotechnology called carbon nanotubes (CNTs) from fullerene came into existence in 1991 [1, 2]. Iijima first observed a cylindrical tube-like graphitic carbon structure known as multi-walled carbon nanotubes (MWCNTs) visually by using a transmission electron microscope (TEM) was only 2–20 layers, followed by synthesis of MWCNTs in large quantities [3–5]. Later, Iijima and Bethune come up with the synthesis of single-walled carbon nanotubes (SWCNTs) in 1993 [6, 7]. SWCNTs is nothing but an elongated form of fullerene or graphene rolled into a cylindrical shape, as shown in Fig. 3.1. Moreover, the single-walled carbon nanotubes (SWCNTs) and multi-walled carbon nanotubes (MWCNTs) differ in properties; the structure differs length in microns, diameter in nanometers, their L/D ratio also varies depending upon the number of walls present as shown in Fig. 3.2 and a careful selection of SWCNTs and MWCNTs is required for application studies.

In addition to this, double-walled carbon nanotubes (DWCNTs) were also discovered during 1991, but less attention was paid due to lack of synthesis and purity. However, the first synthesis of DWCNTs was reported in 1998 and came into existence only in late 2000 [8, 9]. Thess et al. [10] first reported the Vertically-aligned carbon nanotubes (VACNTs) in 1996, where 70% of nanotubes were rolled into rope-like structures. Later, in the same year, Li et al. revealed the large-scale synthesis of vertically-aligned carbon nanotubes, which is of the 50-μm thick film grown using chemical vapour deposition (CVD) method [11]. In 2006, Qian [12] reported a new kind of few walled carbon nanotubes (FWCNTs) with 2–5 layers of sidewalls and diameter less than 10 nm. Thermal chemical vapour deposition technique was used for the synthesis of FWCNTs, which shows extraordinary mechanical and electrical properties.

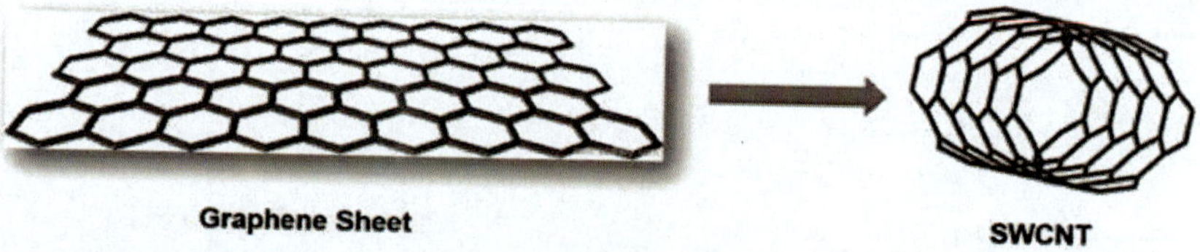

Fig. 3.1 Graphene rolled into a cylindrical form of SWCNT

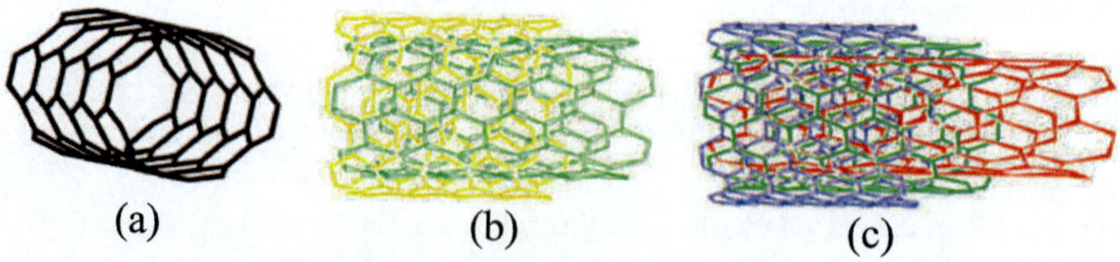

Fig. 3.2 **a** Single-walled **b** double-walled and **c** multi-walled carbon nanotube structures

1.1 Structure

Iijima first reported and discovered that the carbon nanotubes were cylindrically rolled graphene-like form/sheet with different diameters and orientations (helicities and chiralities). The structural pattern of the carbon nanotube comes from the rolled graphene sheet that can be labelled by a pair of integers (n, m). For suppose if we consider the rectangle with corner equivalent points OABB′ in the graphene sheet as depicted in Fig. 3.3, where $\underset{C_h}{\rightarrow}$ indicates the boundary of the nanotube connecting the points O and A, while $\underset{T}{\rightarrow}$ represents the axis of the tube. The cross-section of a tube to high symmetry direction in the sheet is described by angle θ. The integers (n, m) represents the unit vectors along with bi-directional in the lattice structure of the graphene sheet, which determines the shape and chirality of the nanotube [13, 14].

For example, if ($m = 0$, $\theta = 0°$) the nanotube structure is named as zigzag nanotubes, whereas, if ($n = m$, $\theta = 30°$) the nanotube structure is categorised as armchair nanotubes. If ($0 < |m| < n$, $0 < |\theta| < 30°$) then the nanotubes are classified into chiral nanotubes as shown in Fig. 3.4. The change in diameter and chiral angle of nanotubes affect the electronic properties/characteristics of nanotubes during the synthesis process, as shown in Fig. 3.5. For instance, if $n - m/3 = i$ with a finite bandgap makes zigzag and chiral nanotubes metallic or else in most of the cases, nanotubes can be semiconductors [15]. The formation of such metallic and semi-conducting single-walled carbon nanotubes can be isolated in a particular region by type, i.e., density differentiation and this density gradient can be due to the use of chemicals.

The comparison between the SWCNTs, DWCNTs, and MWCNTs characteristics in different aspects are tabulated in Table 3.1. Moreover, the thermal and mechanical properties of CNTs depend on the type of CNTs, defects, diameter, length, and finally, on different techniques used for the synthesis of different CNTs.

Based on the above superior properties, the usage of CNTs is categorised for different kinds of applications. Moreover, the use of different types of CNTs for certain applications requires large quantities of CNTs with high purity and reliable synthesis procedures. Additionally, the synthesis process determines the type of

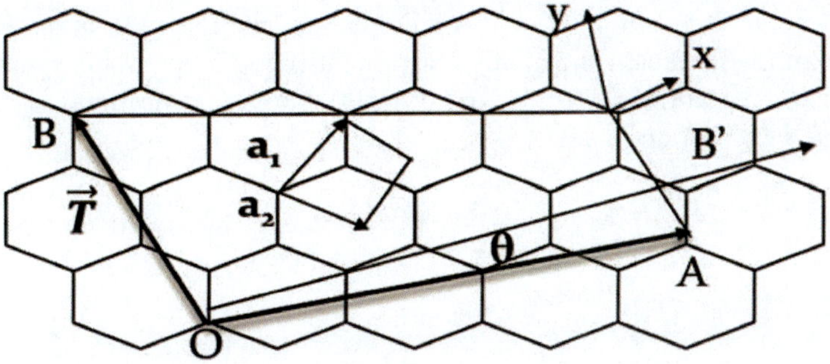

Fig. 3.3 Carbon nanotubes construction from a graphene sheet

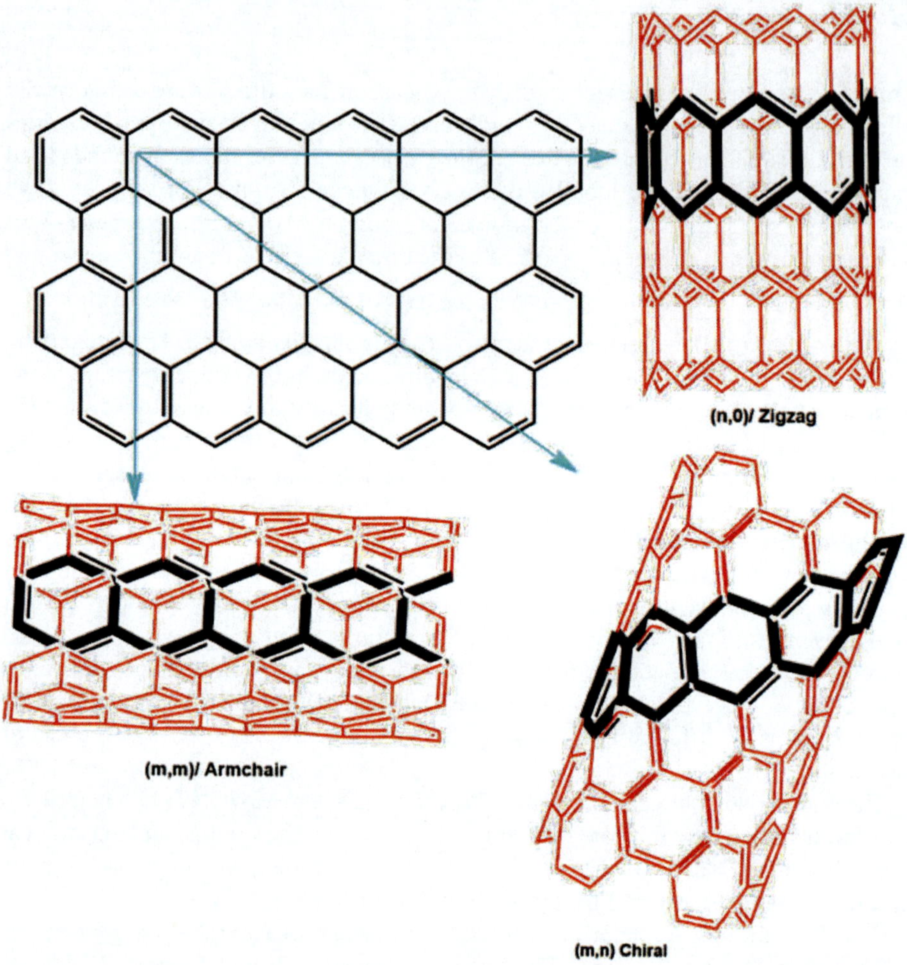

Fig. 3.4 Different structures formed from a graphene sheet such as Armchair, Zigzag, and Chiral

CNTs, and some of the most commonly used synthesis techniques of CNTs are Chemical Vapour Deposition (CVD), Arc Discharge, and Laser Ablation. However, CVD is the most significant and highly recommended method for bulk synthesis and direct device integration. Different type of CNT synthesis procedures is discussed below in the next section in detail.

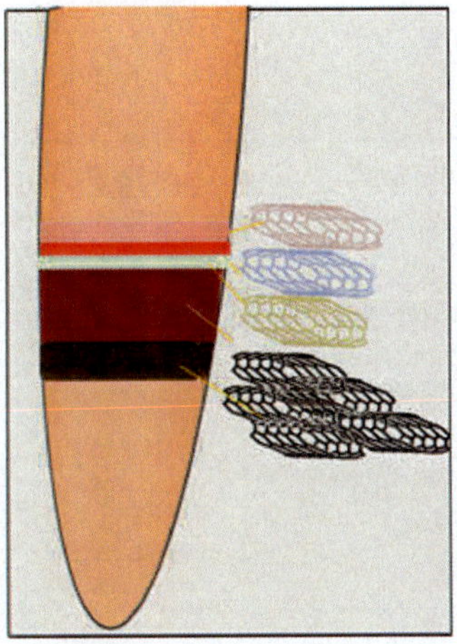

Fig. 3.5 A metallic and semiconductor single-walled carbon nanotube isolated from density gradient

Table 3.1 Comparison of SWCNTs, DWCNTs, and MWCNTs features in different aspects [16]

Characteristics	SWCNTs	DWCNTs	MWCNTs
Size	0.4 to few nm in diameter	0.33 to 0.42 nm inter-wall distance	˜ 0.33 nm inter-wall distance
Layers	Single-layer	Double layer	Multiple layers
Purity	Poor (70%)	Good (90%)	Good (95%)
Synthesis techniques	Laser ablation, Arc-discharge, CVD	Arc-discharge, CVD, Peapod method	Laser ablation, Arc-discharge, CVD, Peapod method
Characterisation	Simple	A bit complex	Very complex
Bulk synthesis production	Difficult	easy	Very easy
Electrical behaviour	Metallic/semiconductor	Metallic/semiconductor	Metallic
Bandgap	Varies from 0.4 to 2 eV	Very few bandgaps	Zero bandgap
Thermal properties	Thermally stable up to ˜ 1800 °C in Ar, 750 °C in air Thermal conductivity > 3500 Wm^{-1} K^{-1}	Thermally stable up to ˜ 2000 °C in Ar, 800 °C in air Thermal conductivity ˜ 600 Wm^{-1} K^{-1}	Thermal stability differs with length and diameter Thermal conductivity > 3000 Wm^{-1} K^{-1}
Mechanical properties	Young's modulus: ˜ 1.8 TPa (E + T) Tensile strength: ˜ 1 TPa (E)	Young's modulus: 0.73 to 1.33 TPa (E) Tensile strength: 13 to 46 GPa (E)	Young's modulus: 0.27 to 0.95 TPa (E) Tensile strength: 11 to 63 GPa (E)

Table 3.2 Comparison of different techniques used for CNTs synthesis [3, 64]

Factors	Arc-discharge	Laser ablation	Chemical Vapour Deposition (CVD)
Process	Difficult	Difficult	Easy
Availability of raw materials	Difficult	Difficult	Easy and abundantly available
Conditions used	Ar/N_2 gas at 500 torr	Temp range from 500–1000 °C at atmospheric conditions	Ar gas (low pressure)
Production	Low	Low	High
Yield	70%	80–85%	95–99%
Purity	Low	Medium–High	High
Working temp range	˜4000 °C	30–1000 °C	500–1200 °C
Product synthesis	SWCNTs, MWCNTs, DWCNTs, FWCNTs	SWCNTs, MWCNTs	SWCNTs, MWCNTs, DWCNTs, vertically-aligned MWCNTs
Carbon sources	Graphite	Graphite	Hydrocarbons
Cost per unit	High	High	Low

Table 3.3 Comparison of mechanical properties between different materials [68, 73–75]

Materials	Young's modulus (GPa)	Tensile Strength (GPa)	Density (g/cm^3)
SWCNTs	1054	150	N/A
DWCNTs (zigzag)	1260	–	–
MWCNTs	1200	150	2.6
Steel	208	0.4	7.8
Epoxy	3.5	0.005	1.25
Wood	16	0.008	0.6

2 Synthesis Techniques of CNTs

The most commonly used synthesis techniques are arc discharge and laser ablation to produce CNT at a higher temperature. However, recently chemical vapour deposition is the promising preparation method that replaces the other existing techniques because of its low-temperature conditions (< 800 °C). Another important aspect of being considered during CNT production is nanotube length, diameter, orientation, alignment, and purity. Moreover, the synthesis of CNT in bulk production requires catalysts that will be one of the major concerns. The other problems while producing in large quantities are different types of impurities such as graphite, fullerene, amorphous carbon, and different types of metals (Fe, Co, Mo or Ni) and quantity depends on the type of techniques used [17]. Thus, techniques with high precision and high purity are considered to produce CNT in bulk production with low cost is preferable.

2.1 CNT Synthesis Using Arc Discharge

Arc discharge method is the first known technique used for the assembly of SWCNTs and MWCNTs and further optimised to produce in higher quantities later. Iijima first time recognised MWCNTs using arc discharge technique [3]. They used similar conditions of the Kratschmer-Huffman technique for fullerene production and carbon whiskers production technique described by Bacon [18]. This method uses two graphite electrodes placed 1-mm distance from each other with a diameter of 5–20 mm and used an inert gas (He or Ar) under atmosphere conditions at a pressure of 500 torrs [19]. The setup uses the very high current (50–120 A), and voltage (12–25 V) produces an arc between the two electrodes, which raises the temperature inside and leads to carbon evaporation from the electrodes. The evaporated carbon then again re-condenses and forms CNT on the cathode, but the main drawback with this method is impurities [14]. Anyhow, the synthesis of SWCNTs and MWCNTs became easy and cost production need to be considered.

MWCNTs synthesis is very easy, and the most commonly used technique is electric arc discharge (DC arc discharge). The experimental setup consists of two water-cooled graphite electrodes placed inside a chamber with helium gas at atmospheric pressure. Ebbesen and Ajayan reported the large-scale synthesis of CNTs using a modified arc discharge method under He atmospheric pressure [4]. Few studies have been published on the bulk synthesis of CNTs using the arc-discharge technique. Shi et al. [20, 21] evidenced the production of high yield of SWCNTs (10 g per day) under very high He pressure of 500–700 torr. Ando et al. [22] demonstrated bulk production of SWCNTs (1.24 g/min) with high purity obtained when the highest current of 100 A used with plasma jet-assisted DC arc-discharge method.

The synthesis of DWCNTs in large-scale is a bit complex compared to SWCNTs and MWCNTs production; still, few researchers achieved the synthesis of DWCNTs using an arc-discharge technique. Hutchison et al. [23] used a graphite rod of 8.2 mm in diameter as an anode with the help of catalysts (Ni, Co, Fe, and S) for the production of DWCNTs under Ar and hydrogen mixture using arc discharge technique. Sugai et al. [24] used Y/Ni catalyst for the production of high-quality DWCNTs using high-temperature pulsed arc discharge techniques. Huang et al. [25] developed DWCNTs super bundles etched on the bowl-like cathode under a hydrogen-free atmosphere using an arc discharge method.

Su et al. [26] first time reported the synthesis of few-walled carbon nanotubes (FWCNTs) in large quantity under low-pressure air condition using the arc discharge method with iron and sulfur as catalyst and promoter. The effective synthesis of FWCNTs was found between 6–12 kPa under air pressure. However, when low pressure and high pressure of 3 kPa and 15 kPa restricts the low-synthesis of FWCNTs.

2.2 *CNT Synthesis Using Laser Ablation*

The other significant method used for CNT synthesis is pulsed laser deposition (PLD) technique, where the properties of CNTs rely on different factors such as peak power, continuous wave versus pulse, energy fluence, the oscillation wavelength of the laser, the pressure inside the chamber, the chemical composition of the material, pressure and flow rate of the gas, temperature, different substrates. Laser ablation is considered as one of the most critical ones of PLD, that produces high purity SWCNTs with high quality. Guo et al. [27] first demonstrated this technique for the preparation of SWCNTs using the same principle and mechanism of arc discharge. Still, the only difference is the usage of laser on to the target graphite material using catalyst materials such as Ni and Co.

Laser ablation process works similarly as the arc discharge technique. In both the techniques, carbon is formed from the graphite vapourisation. SWCNTs are formed when metals such as Ni, Co and Pt are doped with the graphite target material. In this method, the target material is inserted into a quartz tube and placed inside a furnace further heat to 800–1500 °C temperature. A black powdery substance of carbon is formed near the water-cooled Cu collector after passing 500 torrs of Ar gas through the tube. Thess et al. [10] reported high-yield (70–90%) of CNT from graphite using this technique. The main drawback of this technique is production cost is very high, and the cost of the laser is also expensive. The formed CNTs using this method has impurities due to catalysts, carbon impurities and byproducts from carbon material require purification that again increases the total cost of the process. Additionally, the purification step might damage the CNT structure and dope the CNTs (Fig. 3.6) [28].

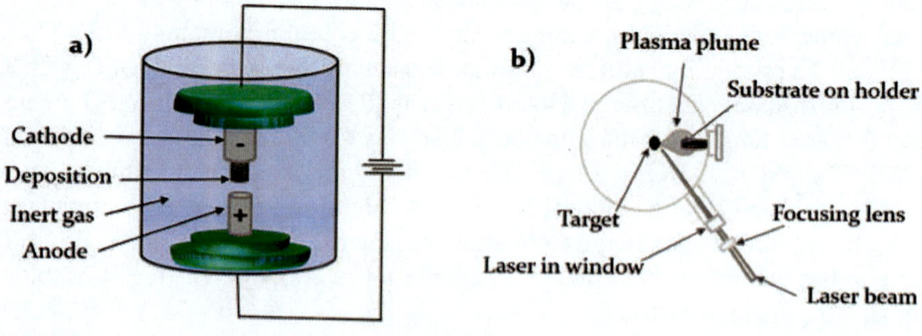

Fig. 3.6 **a** Schematic representation of Arc discharge method for the production of CNT, **b** Pictorial diagram of Laser ablation technique for the production SWCNTs

Bolshakov et al. demonstrated the synthesis of SWCNTs without any additional heat supply to the graphite material by using a 2 kW continuous-wave CO_2 laser ablation technique in Ar condition [29]. The results showed that the SWCNTs achieved by CO_2 laser with increased laser power was about 1.2–1.3 nm in diameter. Moreover, introducing UV laser opens a novel production of CNTs mechanism because of its high quality compared to infrared laser in the photochemical ablation that is advantageous for photothermal ablation technique. Few studies used different laser sources or types for the preparation of SWCNTs using metal catalysts and reinforced agents such as polyurethane and polystyrene. Lebel et al. [30] prepared SWCNTs incorporating into the polymer matrix of polyurethane using UV-laser (KrF excimer) ablation technique. The results revealed that the mechanical strength and mechanical modulus increased by 15 times compared to the infrared laser preparation method. Kusaba and Tsunawaki [31] reported the preparation of SWCNTs using XeCl excimer laser with the oscillation wavelength of 308 nm under Ar gas using Co and Ni catalyst at 1273 K, 1373 K, 1473 K and 1623 K temperature. The results indicated that the formation of high quality and yield of SWCNTs with a mean diameter of 1.2–1.7 nm and length of 2 mm.

Hu et al. [32] demonstrated the preparation of thin-walled carbon nanocages with different shapes in liquid using pulsed-laser ablation (PLA) technique without any presence of a catalyst. In this method, the formation of carbon nanocages is not due to the condensation of gas but because of bubble blowing mechanism. Compagnini et al. [33] successfully deposited carbon nanowalls at the electrode surface using a laser ablation technique in liquid under the electric field. The negatively charged particles generated a flower-like structure at the electrode (anode) surface using plasma and deposition mechanisms. Schauerman et al. [34] synthesised high purity, high yield, and purification efficiency of SWCNTs. They investigated the impact of catalyst on the particle size of SWCNTs by using pulsed laser vaporisation technique. The results indicated that the use of nanometal catalysts such as Ni and Co produces SWCNT of high purity, yield and offers thermal oxidation of SWCNTs by removing the carbon impurities without the use of any acid treatments.

Recently, Stramel et al. [35] prepared MWCNTs and MWCNT-based thin films using pulsed laser deposition with pulsed; diode-pumped, Tm:Ho:LuLF laser. The results showed that the high-quality MWCNTs formed with very less degradation and these thin films when formed with the combination of polystyrene targets (PSNTs) onto the silicon substrate. The results revealed that the thin films formed from pure MWCNTs showed higher quality than the PSNTs targets. Bonaccorso et al. [36] reported a similar study for the preparation of MWCNTs thin films using the PLD technique that commercially uses polystyrene-nanotubes pellets on alumina substrates.

2.3 *Chemical Vapour Deposition (CVD)*

Chemical vapour deposition using thermal [37] or plasma-enhanced (PE) is one of the most popular and commonly used methods for the production of CNTs. The most economic feasibility and large scale production of high purity CNTs is possible with other CVD methods such as water-assisted CVD [38, 39], oxygen assisted CVD [40], hot-filament (HFCVD) [41], microwave plasma (MPECVD) [42] or radiofrequency CVD [43] are some of the currently used techniques compared to laser ablation. The main advantage of the CVD technique is to obtain high purity CNTs and easy control of the reaction rate.

Fotopoulos and Xanthakis [44] produced SWCNTs by base growth in thermal CVD and formation of a subsurface carbon layer after the deformation of Ni catalyst nanoparticle. These effects are due to the adsorption on the catalyst nanoparticles during C_2H_2 pyrolysis confirmed by the recent in situ video-rate TEM studies. The growth of CNTs using catalysts in the CVD method through plasma irradiance or heat (thermal CVD) by decomposition of carbon source. For the efficient growth of CNTs, the most commonly used catalysts are metals such as Fe, Co, and Ni [45] and by improving the growth of CNT, catalysts are doped sometimes with metals such as Au [46]. The most common hydrocarbon sources such as methane [47], ethane [48], ethylene [49], acetylene [50], xylene [51], their mixtures [52], isobutene [53] and ethanol [54] are used in CVD for the synthesis of CNTs [55].

Zhu et al. [56] prepared DWCNTs using catalytic chemical vapour deposition (CCVD) method using metal catalysts such as Fe and Co decomposed on to the mesoporous silica substrate. The results showed a uniform shaped high percentage of DWCNTs with crystal-like alignment. Ramesh et al. [57] imitated a similar method to produce high-yield DWCNTs using Fe/Co catalyst on mesoporous silica by CCVD of alcohol. Hiraoka et al. synthesised DWCNTs using zeolites as a substrate in CCVD of acetylene on the well-dispersed metal particles fixed in heat-resistant zeolites at 900 °C and above [58].

Lyu et al. used Fe-Mo/Al_2O_3 as a catalyst at 900 °C by decomposition of benzene as a perfect feedstock to synthesise high quality and pure DWCNTs [59]. Zhang et al. used nickel oxide-silica aerogels as a catalyst for the preparation of MWCNTs with 40–60 nm diameters by the catalytic decomposition of methane at 680–700 °C for 120 min [60]. Jiang et al. [61] investigated the CNT growth in-situ on the pretreated graphite electrode (GE) using CCVD technique and $Ni(NO_3)_2$ as a catalyst. The prepared CNTs has a length of 200 to 1000 nm due to shorter growing time with the outer diameter of 80 nm and an inner diameter of 20 nm. Feng et al. prepared high-quality DWCNTs in one-step CCVD technique under argon gas using acetone as a carbon source, ferrocene as a source of Fe catalyst and thiophene as a promoter at a synthesis temperature of 1170 °C.

Li demonstrated the synthesis of well-aligned MWCNTs on a large area of SiO_2/Si substrate coated with Ni through pyrolysis of C_2H_2 using thermal CVD method at 900 °C. Because of the higher density of Ni particles and steric hindrance effect between the neighbouring CNTs, a better arrangement of CNTs can be obtained.

The crystallisation of CNTs increased with the increasing NH_3 pretreatment, which plays a critical role in controlling the surface morphology of metals and getting the alignment of MWCNTs. Moreover, CNTs grow by a tip growth mechanism that was displayed using Energy-dispersive X-ray spectrum (EDX) analysis [62]. Kim et al. [63] proposed a new technique for the production of CNTs using three various iron-containing proteins such as myoglobin, cytochrome c and haemoglobin to control the structure and size of CNTs accurately. These proteins are firmly bound to the amine-terminated self-assembled monolayer surfaces (SAM) by peptide bonds between the amine group of SAM and carboxyl group of the proteins for the CNT production. This technique is strictly constraint to SWCNTs synthesis (Table 3.2).

3 Properties of CNTs

CNTs are one of the promising material among the carbon family with larger surface area, electrical, mechanical and high aspect ratios. CNTs are tubular structures with different chirality, degree of graphitisation and different diameters show some versatile and superior properties. This section particularly highlights the physical, mechanical, electrical and optical properties of CNTs.

3.1 Mechanical Properties of CNT

Carbon nanotubes are predominantly dominant in mechanical properties such as tensile strength and Young's modulus. Several studies in the last few years [65–67] have explained the mechanical properties of CNTs, both theoretically and experimentally. The tensile strength of CNTs is due to the covalent sp^2 bond between the neighbouring existing carbon atoms. Due to the C–C bonds in the lattice, CNTs are expected to achieve high Young's modulus in the axial direction. Moreover, when SWCNTs are formed from graphite sheets, the σ bonding between individual carbon atoms increases along their axis [68]. This bonding makes CNTs stronger and stiffer with the highest tensile strength and Young's modulus. Young's modulus is generally related to the material stiffness and it is defined as the stress applied to the material with the resulting strain. For practical applications such as nanoelectromechanical systems (NEMS) resonators [69], bioengineering [70] and electrochemical actuators [71], etc., Young's modulus of the material must be known before using it.

The Young's modulus for SWCNTs reported being very high about 1 to 1.8 TPa due to its diameter is between 1–2 nm, which makes them a suitable material for probe tips of the scanning electron microscope. The modulus of SWCNTs mainly depends on the diameter and chirality. However, in the case of MWCNTs, it depends on the disorder in the sidewalls [72]. Besides, Young's modulus of MWCNTs is 1.2 TPa higher than the SWCNTs, and this is due to the difference in diameter of the tubes and also the van der Waals forces acting between the tubes [68]. Due to minute structural defects of CNTs, their elasticity function is tremendous and resumes back to the original shape when subjected to severe twists and bends (Table 3.3).

3.2 *Electrical Properties of CNT*

CNTs exhibit excellent electrical properties and remain a material of interest for most of the researchers until date. The structural defects on the CNTs make it electrically more conductive than the existing materials. CNTs possess low electrical resistivity of 10^{-6} Ω-m. However, changes in structure, chirality, diameter, and other functional changes might significantly alter the electrical properties of CNTs. In a few cases, the chiral forms of CNTs also exhibit different electrical properties [76–79]. The electronic structure of CNTs changes when different materials bind to the surface of CNT, resulting in electrical conductivity variation. More interestingly, the addition of different materials occupy and bind with different functional groups on CNT surface via different interactions such as covalent, non-covalent, electrostatic, hydrogen, and finally Π- Π interactions that affects the electrical properties. On the other hand, based on the chirality and diameter, SWCNTs and MWCNTs can be classified as either metallic or semiconducting.

3.3 *Thermal Properties of CNT*

Since graphite and diamond show low heat capacity and high thermal conductivity, CNTs are rolled form of graphite structure are expected to exhibit the same thermal properties due to the photon band structure [80]. In addition to excellent electrical and mechanical properties, CNTs also displays exceptional thermal properties. Moreover, the inter-tube coupling in SWCNTs and inter-shell coupling in MWCNTs indicates low specific heat that is similar to three-dimensional graphite [81, 82]. Pop et al. developed a thermal conductivity model as a function of CNT diameter and temperature and found the thermal conductivity of 1.7 diameters and 2.6 mm long SWCNTs at room temperature was nearly 3500 Wm^{-1} K^{-1}. The thermal properties of SWCNTs were achieved by Joule self-heating over 300–800 K temperature range produced from high-bias (I-V) electrical characteristics [83]. Kim et al. [84] measured the thermal conductivity of MWCNTs at room temperature is above 3000 Wm^{-1} K^{-1} that is two orders higher magnitude than the bulk MWCNTs. Few other parameters such as photon-active modes, length of the free path for the photons, and boundary surface scattering also affects the thermal properties of CNTs [85–87]. Moreover, the diameter, length of the tube, morphology, structural defects, atomic arrangements, and the presence of impurities also influence the properties of CNTs [88–90]. Some of the physical, electrical, thermal, elastic, and optical properties are summarised in Table 3.4.

Table 3.4 Properties of CNTs [91]

Physical properties		**Values**
Equilibrium structure	SWCNTs (average diameter)	1.2–1.4 nm
	Distance between opposite carbon atoms	2.83 Å
	Parallel carbon atoms separation	2.456 Å
	Analogous carbon bonds separation	2.45 Å
	Carbon bond length	1.42 Å
	C–C tight bonding overlap energy	~2.5 eV
	Group symmetry (10,10)	C5V
Lattice constant		17 Å
Lattice parameter	(10,10) Armchair	16.78 Å
	(17,0) Zigzag	16.52 Å
	(12,6) Chiral	16.52 Å
Density	(10,10) Armchair	1.33 g/cm^3
	(17,0) Zigzag	1.34 g/cm^3
	(12,6) Chiral	1.40 g/cm^3
Interlayer spacing	(n, n) Armchair	3.38 Å
	(n, 0) Zigzag	3.41 Å
	(2n, n) Chiral	3.39 Å
Optical properties		
Fundamental gap	Metallic (n-m/3)	0 eV
	Semiconducting (n-m not divided by 3)	~~0.5 eV
Electrical transport	Conductance	12.9 KΩ^{-1}
	Resistivity	10^{-4} Ω.cm
	Current density (maximum)	1,013 A/m^2
Thermal transport	Conductivity	~3,000 W/m/K
	Phonon mean free path	~100 nm
	Relaxation time	~10 to 11 s
Elastic behaviour	SWCNTs (Young's modulus)	1 TPa
	MWCNTs (Young's modulus)	1.28 TPa
	Maximum tensile strength	~100 GPa

4 CNT Based Composites/hybrids

Composites are formed by combining two or more materials to enhance the electrical, mechanical or specific properties [92, 93]. In the case of composites, one material acts as a filler (nanoparticle, nanowires, nanotubes, etc.) and the other behaves as supporting material or matrix (polymer, ceramics). Usually, random mixing of different materials results in unique properties and, in some cases, non-uniform properties are observed [94]. Hybrids nanoparticles are traditionally formed via the assembly of different materials/particles on the carbon material. In this situation, one acts as an active agent, and the other behaves as inactive. Different types of materials such as metals, metal oxides, polymers, ceramics, nanowires, and peptides/proteins formed interactions with carbon materials for various applications [95, 96]. These interactions are of two types: (1) Strong covalent interactions, and (2) Electrostatic interactions, hydrogen, Vander Waals, and Π-Π interactions [97].

Due to high surface area, high conductivity, excellent chemical and electrochemical stability, and very good mechanical strength, CNTs have emerged as a new class of material used for a wide range of applications. However, the hydrophobic nature of these materials such as CNTs, MWCNTs, and SWCNTs makes it hard to dissolve in polar fluids such as water compared to non-polar fluids such as silicon oil etc. Different materials such as metals (Ag, Au, Pt, Pd, Ru and their alloys) [98, 99], metal oxides (SiO_2, TiO_2, Al_2O_3, Fe_2O_3, Fe_3O_4, ZnO, CuO, CeO_2 etc.) [100], polymers [101], ceramics [102] bind to CNTs either by covalent or non-covalent interactions. These composites show tremendous improvement in different properties compared to single nanoparticles/nanomaterials.

4.1 CNT-Based Metal Nanoparticle (CNT/MNPs) Composites

The interaction of a new kind of materials such as metal nanoparticles and CNTs (MNPs/CNTs) hybrids that combine the properties of both the materials and exhibit unique properties due to the interaction between MNPs/CNTs. Planeix et al. [103] first reported the synthesis and application of hybrid MNPs/CNTs in 1994 for heterogeneous catalysis. Taking advantage of this discovery of hybrids/composites, several researchers started exploring different hybrids in different applications.

Interestingly, the composites of metal or semiconducting NPs with CNTs are obtained via two different ways. The first approach is using the bare NPs, i.e., with the help of precursors NPs are deposited or grown on the surface of the CNTs. The second approach where NPs are formed before and later bind to CNTs through oxygen functional groups either by covalent or non-covalent interactions. Generally, in the case of the first scenario, CNTs do not react with the metal salts but behaves as a conducting wire and supports for the deposition of NPs. Mostly, these MNPs grow via the process of nucleation on the CNTs surface via van der Waals interactions, and in few cases, the adhesion of MNPs is, even more, stronger than this. This process is carried out by performing the metal cation reduction by applying heat, light, and reducing agents [99]. Researchers have explored and extensively studied some metals such as Au, Ag, Pt, Pd, Rh, Ru and Ni [104–110]. Xue et al. [105] used different kinds of metals such as Pt, Ag, Au, Pd and Cu NPs were grown on MWCNTs by thermal decomposition of metal salts under hydrogen stream. Additionally, CNTs not only act as a supporting material but also act as a template in determining the metal nanoparticles size.

Ang et al. [111] used Pd and Sn catalytic nuclei to activate CNT using a single-step approach. The formation of Ni and Pd NPs on CNTs by electroless plating

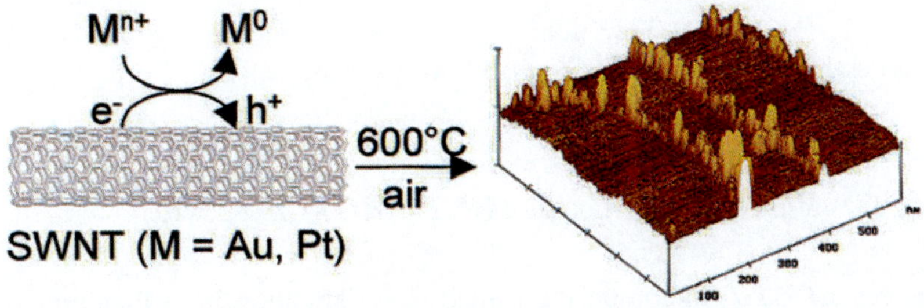

Fig. 3.7 Electroless deposition of metal nanoparticles on CNTs (Preprinted with the permission of the publisher) [108]

technique using activated carbon nanotubes as a precursor. Choi et al. [108] reported MNPs/CNTs hybrid material by forming Pt and Au NPs on the sidewalls of SWCNTs. They claimed that the reduction of metal ions on the CNT surface did not involve any catalyst or reducing agent. The author noticed that the reaction was spontaneous and observed the formation of MNPs on CNTs after the dispersion of CNT into $HAuCl_4$ (Au^{+3}) or Na_2PtCl_2 (Pt^{+2}) for 3 min duration. CNTs act as a template for the deposition of MNPs on CNT structure and was successfully examined by atomic force microscopy (AFM) as shown in Fig. 3.7.

Few studies [112–114] reported the formation of MNPs on CNTs using the electrodeposition method. The deposition of Au, Pt and Pd NPs on CNTs formed under control potential and concentration of salt precursors in the electrochemical cell. Moreover, the CNTs surface fully occupied by the metal NPs, and the presence of hydroxyl, carboxylic and other functional groups did not behave as activators during electrodeposition.

Coleman et al. [115] used the Bingel reaction to modify the CNT surface with thioether groups that act as a bridge between Au NPs and CNTs binding. Ellis et al. [116] demonstrated the hydrophobic interactions between the monolayer coated Au NPs on to the sidewalls of CNTs. Octanethiols act as a capping agent to coat gold NPs that are anchoring to the acetone-activated CNTs used for device applications.

Researchers carried out the functionalisation of CNTs with pyrene derivatives, other aromatic compounds via Π-Π interactions [117–120]. The pyrene molecule chain ended with thiol group is coated on CNTs, where Au NPs are introduced to bind on to the CNTs surface through Π-Π stacking. The fluorescence of pyrene is quenched fully by CNTs, and total quenching after binding with Au NPs is observed although, the enhancement in the signal is observed using Raman spectroscopy and other spectroscopic studies. The electron transfer between the electron donor, Zn porphyrin and acceptor SWCNTs is due to Π-Π, and Coulombic interactions between CNTs bound with pyrene and Zn porphyrin [119]. The result indicates the charge transport and charge transfer between CNTs, pyrene, and Zn porphyrin.

The metal NPs attached to the oxidised CNTs through electrostatic interaction [121–129] have been reported in the literature. The cationic polyelectrolyte bind to the carboxylic groups on the oxidised surface of CNT via electrostatic interactions. Thus, the negatively charged Au NPs anchor to the polyelectrolyte chain attached to the surface of CNTs [121]. Few studies also reported the assembly of MNPs/CNTs nanohybrids through electrostatic layer-by-layer technique [130, 131].

4.2 CNTs-Based Metal Oxide (CNTs/MO) Composite

In recent years, CNTs/MO composites gained a lot of attention due to their remarkable superior properties and significant impact on the performance of the composites. CNTs/MO are broadly divided into two categories based on the application and the material composition in the composite. Class I: oxidised functional groups on CNT bind to the MO either by covalent or non-covalent binding. Class II: CNTs are strongly fixed or embedded within the MO matrix [132–134]. The synthesis of CNTs/MO composites have been reported as (1) in-situ approach and (2) ex-situ. In the case of in-situ technique, the nanoparticles are prepared and attached on CNTs through functional groups either covalently or non-covalently, whereas, in the case of ex-situ, the nanoparticles are decorated on the CNTs using a different technique [135].

The most commonly used in-situ techniques are an electrochemical reduction, electrodeposition, sol–gel, hydrothermal and CVD [136–138]. The main advantage of using this method is the formation of different types of NPs, nanorods, or nanobeads with controlled thickness, size, structure and also morphology. The main drawback with this kind of approach is the structural damage to CNTs, and sometimes it led to decomposition of the structure/nanoparticles during the harsh treatments that eventually effects the overall performance of the composite. The advantage of the ex-situ method was easy controlling of the size and morphology of the nanoparticles that were already prepared using different techniques. As the nanoparticles bind on the surface of the CNTs through various interactions and oxygen functionalisation, the distribution and concentration of MO NPs on the CNT surface can be easily controlled [139, 140].

The important factor was the mechanical properties of CNTs/MO composite, where the quality, uniformity and low quantities of CNTs do not show any effect on the mechanical properties of CNTs/MO composite. Moreover, the higher the number of CNTs in MO suppress grain growth and restricts the densification. The decrease in the toughness of MWCNTs in the matrix is due to a more defective structure compared to SWCNTs. The load transfer is only on the outer layer of MWCNTs, and the inner layers do not have any interaction with MO. Therefore, MWCNTs have a very less load transfer effect than the SWCNTs [141, 142].

The other parameters taken into consideration are the ratio of nanoparticle on the CNTs surface by determining the initial concentration and type of precursor used. Moreover, the increase in MO ratio results in the formation of agglomeration on the surface of CNTs. The size and morphology of the nanoparticles on CNTs depend on synthesis treatment, temperature and time. Finally, the uniform distribution of nanoparticles on CNTs to form a good composite based on the parameters and conditions [143].

4.3 CNTs-Based Polymer Composite

The rapid development of CNTs-polymer composite due to extremely superior properties such as mechanical, electrical, and thermal properties used as a potential candidate for different applications, where CNTs replaces other materials to form CNTs-polymer composite with improved effectiveness and performance. This improved performance of the composite is due to the strong covalent interaction and weak non-covalent interaction between the CNT-polymer. The preparation of CNTs-polymer composite via acid functionalisation enhances the strong bonding properties between CNTs and polymer matrix. Thus, the carboxylic end groups present on CNTs effectively interact with the polymer to improve the mechanical strength and Young's modulus [144–147]. Additionally, the presence of carboxylic functional groups on the CNTs actively form a covalent type of interaction with the polymer molecule. The interaction between CNTs-polymer classified into "grating from and grating to" [148–150].

The initiators are fixed onto the CNTs surface initially in the grating-from technique, followed by the development of polymer molecules bound to the CNTs through in-situ polymerisation. The advantage of this method is the synthesis of a high grafting density polymer-functionalised CNTs. However, the amount of initiator and substrate need to be considered strictly during this technique due to the limited availability of the polymer. Moreover, the higher the amount of CNTs restricts the matrix grain growth and thus hinder the densification. The latter method is based on the binding of the polymer molecule to the CNTs functional group with the existing functionalised end polymer through a favourable chemical reaction. The benefit of this technique is that the polymers with functional groups available in the market can also be used for interaction. However, the main disadvantage of this approach is that the binding of polymer chains sterically on the CNTs surface initially restricts the scattering of any other additional macromolecules to the CNTs surface hence resulting in low polymer loading [151].

The non-covalent functionalisation of CNTs and polymer molecules involves van der Waals, Π-Π, CH-Π or electrostatic interactions. The main benefit of this type of functionalisation is that the CNTs structure does not change while retaining the electrical and mechanical properties. However, the limitation with this non-covalent functionalisation is the ability of the load transfer reduce because of the weak interactions between the wrapping molecules and the CNTs surface [150, 152, 153].

The other significant factor in developing a better physical binding between CNTs and polymer matrix is designing a proper manufacturing technique of the composites apart from enhancing the mechanical, electrical, thermal and stress transfer properties of CNTs-polymers [154].

4.4 CNTs-Based Ceramic Composites

Roy, Komarnenei and colleagues first coined the term 'nanocomposite' to ceramics. They used the sol–gel technique to synthesise a hybrid ceramic composite with the metal nanoparticle leads to monophasic or multiphasic composite [155]. Niihara [156] first proposed the concept of structural ceramic composite in 1991 and later, An et al. [157] developed a structural ceramic composite on Si_3N_4/SiC and Al_2O_3/SiC system by adopting the same technique to study the effect of CNTs on mechanical and tribological properties. However, ceramics face some serious issues such as poor resistance to creep, degradation of mechanical properties at high temperatures, low fracture toughness and strength, fatigue and thermal shock. Several attempts have been made to overcome all these issues by introducing the micron-sized particulates such as platelets, whiskers, and fibres at the matrix grain boundaries. However, these micron-sized fillers do not resolve the problems successfully. The incorporation of nanometer-sized particles into ceramics has opened a new window to control the microstructures passively [158].

Due to strong mechanical and electrical properties, CNTs act as an alternative material than others to form ceramic-based composites. In few cases, the structure of the CNTs causes damage by high-temperature sintering. Still, the CNTs structure can be retained during spark-plasma-sintering to form the composite material [159, 160]. In some cases, CNTs form poor interactions with solvents or matrix components because of their large surface areas and high aspect ratios [161]. Due to van der Waals forces acting between the neighbouring nanotubes, SWCNTs turns into ropes and bundles and form an agglomeration.

Moreover, the high loading fractions of material occupies the available matrix material and favours agglomeration because of the large area of filler and shortage of the matrix material. It is known that the nanocomposites show improved properties up to certain loading fractions and beyond the limit tends to the agglomeration of CNTs. However, when considering the electrical conductivity of CNTs, mesh-like structures connecting is desired. To obtain better results, the initial dispersion of CNTs is significant and allowed the CNTs network to spread [162, 163].

5 Applications of CNTs

CNTs evolved as an emerging material in different applications due to its superior properties such as mechanical, electrical, optical and chemical properties. Due to the large surface area to the aspect ratio of CNTs, various methods were employed and used in several applications such as biomedicine, energy-related applications, environment, and sensors [164]. Several researchers have explored various types of materials to integrate with CNTs to improve the properties further or discover novel properties in other diverse areas of nanotechnology to fulfil the current demands. The addition of CNTs to other materials tune their properties and enhance the properties that apply to the desired application. Few of the CNT applications are discussed under this section:

5.1 *Transparent Electrodes*

Organic solar cells and organic light-emitting diodes (OLED) are made up of optically transparent electrodes that are strong, flexible and conductive. Thin layers of CNTs provide additional benefits by acting as a flexible electrode compared to the existing metal oxide like indium tin oxides [165]. The sheet resistance of CNTs with OLEDs remains the same under bending because of its flexible and superior nature. In contrast, the sheet resistance increases drastically for ITO-based electrodes due to cracks on the surface [166].

5.2 *Supercapacitors*

Supercapacitors are the perfect replacement for lithium-ion batteries due to the presence of the charged double layer to store energy. CNTs act as potential material for electrochemical double-layer capacitors (EDLC) because of the high surface area. Signorelli et al. [167] formed CNTs on top of the conductive substrate, which performed seven times higher than the existing activated carbon-based devices in terms of energy densities. Xie et al. [168] reported that the CNTs/metal composite can further improve the performance with increased capacitance from 29.8 F g^{-1} to 250.5 F g^{-1}.

5.3 *Fuel Cells*

Proton exchange membrane fuel cells (PEMFCs) and direct methanol fuel cells (DMFCs) have grabbed a lot of attention as a sustainable and clean energy source [114, 169–171]. The drawback with these fuel cells was the increased cost of Pt catalyst, and the cost can be minimised by depositing the metal catalyst on the CNTs surface.

5.4 *Sensors*

CNTs used as a natural sensing element with high sensitivity to detect specific molecules were reported by Swager et al. [172]. CNTs with a large surface area and adsorptive properties acts as a carrier to bind with metals, metal oxides, polymers to form a sensory system with high specificity. Jiang et al. developed a novel CuO/MWCNTs electrode for the detection of glucose in human serum samples, and the composite showed high sensitivity of 2596 $\mu A\ mM^{-1}\ cm^{-2}$ and response time of about 1 s with the addition of 0.10 mM glucose than the normal MWCNTs electrode [173].

5.5 *Hydrogen Storage*

CNTs showed a significant effect on hydrogen storage research due to its high aspect ratio. Few studies reported that CNTs as a promising material for hydrogen storage capacity but most of the experimental data was controversial, and some of them are non-reproducible as well. Dillion et al. [174] reported 10 mass % of hydrogen storage capacity for SWCNTs. Still, this topic is debatable, and it is unclear whether CNTs can store hydrogen. More intensive research needed to explore the significance of CNTs in hydrogen storage applications [175]. Silambarasan et al. [176] outlined the reversible and reproducible hydrogen storage using functionalised SWCNTs with borane. Absorption of hydrogen by the functionalised SWCNTs for different time durations has been designed and employed using hydrogenation setup. The results showed maximum storage of hydrogen at 50 °C was about 4.77 wt% and at a higher temperature ranging from 90–125 °C, the samples become dehydrogenated. Besides, the amount of hydrogen storage in functionalised SWCNTs increases with the increasing hydrogenation duration.

5.6 *Heat Transfer and Thermal Energy Storage Applications*

The addition of nanoparticles into the base fluids can enhance the thermophysical properties such as thermal conductivity, specific heat capacity and thermal stability, etc. Several studies reported that the addition of CNTs, MWCNTs and CNTs hybrids improved the heat transfer performance [177–184]. Additionally, surface wettability plays a significant role in fluid flow behaviour and heat transfer performance. The potential use of microstructured materials in heat transfer applications has been highlighted by a few researchers [185–188]. Few other studies also reported the addition of CNTs showed increase heat transfer rate or performance [189–192].

5.7 *Biomedical*

The attachment of biomolecules to CNT by covalent and non-covalent interactions makes the CNTs development possible in biomedical applications [193–195]. CNTs for bioelectronic applications have become an interesting field, such as neural prosthetic devices [165]. Another important field of research that CNTs widely used is therapeutics. CNTs have been reported as molecular transporters for human T-cells and as an anticancer drug [196, 197].

6 Challenges

CNTs due to their high surface area, aspect ratio, and superior properties widely used in different applications. The synthesis, purification, processing and fabrication of CNTs are still facing problems and remain a key challenge. However, during the synthesis process of metal/MWCNTs composite, the traces of metal nanoparticles are most commonly seen impurities in CNTs. These impurities play a significant role in changing the behaviour or properties of CNTs reported by Sljukic et al. [198].

Another crucial question associated with CNTs is solubility and dispersion, and several efforts have been made to overcome this by forming non-covalent interactions such as diazonium chemistry, by oxidation of defect sites, and using solubilising agents but the pristine CNTs observed to be insoluble in polar solvents such as water in few cases [199–202]. This allows a different type of solution methods such as spin coating, drop-casting to form CNT-based composites for future applications. However, many studies reported that the covalently functionalised CNTs showed higher solubility but still impact the electronic properties of the materials [165].

7 Conclusions

We conclude that SWCNTs, DWCNTs, MWCNTs, and the nanocomposites have emerged as a potential candidate for different applications. The synthesis, purification, processing and fabrication of CNTs-based devices exhibit superior properties such as mechanical, electrical, high aspect ratios and large surface area for wide-range of applications. The structural changes may vary the properties of nanocomposites after covalent and non-covalent interactions and further increase the quality of composites that currently restricts the applications of CNTs. With the continued improvement of research and new technologies, tremendous progress can be expected in the field of CNTs.

References

1. Andzelm, J., Govind, N., Maiti, A.: Nanotube-based gas sensors–role of structural defects. Chem. Phys. Lett. **421**(1–3), 58–62 (2006)

2. Yu, B., Meyyappan, M.: Nanotechnology: role in emerging nanoelectronics. Solid-State Electr. **50**(4), 536–544 (2006)
3. Iijima, S.: Helical Microtubules of Graphitic Carbon. Nature **354**(6348), 56 (1991)
4. Ebbesen, T., Ajayan, P.: Large-scale synthesis of carbon nanotubes. Nature **358**(6383), 220 (1992)
5. Ebbesen, T., Hiura, H., Fujita, J., Ochiai, Y., Matsui, S., Tanigaki, K.: Patterns in the bulk growth of carbon nanotubes. Chem. Phys. Lett. **209**(1–2), 83–90 (1993)
6. Bethune, D., Kiang, C.H., De Vries, M., Gorman, G., Savoy, R., Vazquez, J., et al.: Cobalt-catalysed growth of carbon nanotubes with single-atomic-layer walls. Nature **363**(6430), 605 (1993)
7. Iijima, S., Ichihashi, T.: Single-shell carbon nanotubes of 1-nm diameter. Nature **363**(6430), 603 (1993)
8. Endo, M., Muramatsu, H., Hayashi, T., Kim, Y., Terrones, M., Dresselhaus, M.: Nanotechnology: 'buckypaper' from coaxial nanotubes. Nature **433**(7025), 476 (2005)
9. Green, A.A., Hersam, M.C.: Processing and properties of highly enriched double-wall carbon nanotubes. Nat. Nanotechnol. **4**(1), 64 (2009)
10. Thess, A., Lee, R., Nikolaev, P., Dai, H., Petit, P., Robert, J., et al.: Crystalline ropes of metallic carbon nanotubes. Science **273**(5274), 483–487 (1996)
11. Li, W., Xie, S., Qian, L., Chang, B., Zou, B., Zhou, W., et al.: Large-scale synthesis of aligned carbon nanotubes. Science **274**(5293), 1701–1703 (1996)
12. Qian, C., Qi, H., Gao, B., Cheng, Y., Qiu, Q., Qin, L.C., et al.: Fabrication of small diameter few-walled carbon nanotubes with enhanced field emission property. J. Nanosci. Nanotechnol. **6**(5), 1346–1349 (2006)
13. Dresselhaus, MS., Avouris, P.: Introduction to carbon materials research. Carbon Nanotubes, pp. 1–9. Springer (2001)
14. Dresselhaus, G., Riichiro, S.: Physical properties of carbon nanotubes. World scientific (1998)
15. O'connell, MJ.: Carbon nanotubes: properties and applications. CRC press (2006)
16. Shen, C., Brozena, A.H., Wang, Y.: Double-walled carbon nanotubes: challenges and opportunities. Nanoscale. **3**(2), 503–518 (2011)
17. Prasek, J., Drbohlavova, J., Chomoucka, J., Hubalek, J., Jasek, O., Adam, V., et al.: Methods for carbon nanotubes synthesis. J. Mater. Chem. **21**(40), 15872–15884 (2011)
18. Bacon, R.: Growth, structure, and properties of graphite whiskers. J. Appl. Phys. **31**(2), 283–290 (1960)
19. Harris, P.: Carbon Nanotubes and Related Structures. Cambridge University Press, Cambridge, UK (1999)
20. Shi, Z., Lian, Y., Liao, F.H., Zhou, X., Gu, Z., Zhang, Y., et al.: Large scale synthesis of single-wall carbon nanotubes by arc-discharge method. J. Phys. Chem. Solids **61**(7), 1031–1036 (2000)
21. Shi, Z., Lian, Y., Zhou, X., Gu, Z., Zhang, Y., Iijima, S., et al.: Mass-production of single-wall carbon nanotubes by arc discharge method. Carbon **37**(9), 1449–1453 (1999)
22. Ando, Y., Zhao, X., Hirahara, K., Suenaga, K., Bandow, S., Iijima, S.: Mass production of single-wall carbon nanotubes by the arc plasma jet method. Chem. Phys. Lett. **323**(5–6), 580–585 (2000)
23. Hutchison, J., Kiselev, N., Krinichnaya, E., Krestinin, A., Loutfy, R., Morawsky, A., et al.: Double-walled carbon nanotubes fabricated by a hydrogen arc discharge method. Carbon **39**(5), 761–770 (2001)
24. Sugai, T., Yoshida, H., Shimada, T., Okazaki, T., Shinohara, H., Bandow, S.: New synthesis of high-quality double-walled carbon nanotubes by high-temperature pulsed arc discharge. Nano Lett. **3**(6), 769–773 (2003)
25. Huang, H., Kajiura, H., Tsutsui, S., Murakami, Y., Ata, M.: High-quality double-walled carbon nanotube super bundles grown in a hydrogen-free atmosphere. J. Phys. Chem. B **107**(34), 8794–8798 (2003)
26. Su, Y., Zhou, P., Zhao, J., Yang, Z., Zhang, Y.: Large-scale synthesis of few-walled carbon nanotubes by DC arc discharge in low-pressure flowing air. Mater. Res. Bull. **48**(9), 3232–3235 (2013)

27. Guo, T., Nikolaev, P., Thess, A., Colbert, D.T., Smalley, R.E.: Catalytic growth of single-walled manotubes by laser vaporization. Chem. Phys. Lett. **243**(1–2), 49–54 (1995)
28. Liu, W.W., Chai, S.P., Mohamed, A.R., Hashim, U.: Synthesis and characterization of graphene and carbon nanotubes: a review on the past and recent developments. J. Ind. Eng. Chem. **20**(4), 1171–1185 (2014)
29. Bolshakov, A., Uglov, S., Saveliev, A., Konov, V., Gorbunov, A., Pompe, W., et al.: A novel CW laser–powder method of carbon single-wall nanotubes production. Diam. Relat. Mater. **11**(3–6), 927–930 (2002)
30. Lebel, L.L., Aissa, B., El Khakani, M.A., Therriault, D.: Preparation and mechanical characterization of laser ablated single-walled carbon-nanotubes/polyurethane nanocomposite microbeams. Compos. Sci. Technol. **70**(3), 518–524 (2010)
31. Kusaba, M., Tsunawaki, Y.: Production of single-wall carbon nanotubes by a XeCl excimer laser ablation. Thin Solid Films **506**, 255–258 (2006)
32. Hu, S., Yang, J., Liu, W., Dong, Y., Cao, S.: Carbon nanocage bubbles produced by pulsed-laser ablation of carbon in water. Carbon **49**(4), 1505–1507 (2011)
33. Compagnini, G., Sinatra, M., Russo, P., Messina, G.C., Puglisi, O., Scalese, S.: Deposition of few layer graphene nanowalls at the electrodes during electric field-assisted laser ablation of carbon in water. Carbon **50**(6), 2362–2365 (2012)
34. Schauerman, C.M., Alvarenga, J., Landi, B.J., Cress, C.D., Raffaelle, R.P.: Impact of nanometal catalysts on the laser vaporization synthesis of single wall carbon nanotubes. Carbon **47**(10), 2431–2435 (2009)
35. Stramel, A., Gupta, M., Lee, H., Yu, J., Edwards, W.: Pulsed laser deposition of carbon nanotube and polystyrene–carbon nanotube composite thin films. Opt. Lasers Eng. **48**(12), 1291–1295 (2010)
36. Bonaccorso, F., Bongiorno, C., Fazio, B., Gucciardi, P., Marago, O., Morone, A., et al.: Pulsed laser deposition of multiwalled carbon nanotubes thin films. Appl. Surf. Sci. **254**(4), 1260–1263 (2007)
37. Steiner, S.A., III., Baumann, T.F., Bayer, B.C., Blume, R., Worsley, M.A., MoberlyChan, W.J., et al.: Nanoscale zirconia as a nonmetallic catalyst for graphitization of carbon and growth of single-and multiwall carbon nanotubes. J. Am. Chem. Soc. **131**(34), 12144–12154 (2009)
38. Smajda, R., Andresen, J., Duchamp, M., Meunier, R., Casimirius, S., Hernadi, K., et al.: Synthesis and mechanical properties of carbon nanotubes produced by the water assisted CVD process. Physica Status Solidi (B). **246**(11–12), 2457–2460 (2009)
39. Patole, S., Alegaonkar, P., Lee, H.C., Yoo, J.B.: Optimization of water assisted chemical vapor deposition parameters for super growth of carbon nanotubes. Carbon **46**(14), 1987–1993 (2008)
40. Byon, H.R., Lim, H.S., Song, H.J., Choi, H.C.: A synthesis of high purity single-walled carbon nanotubes from small diameters of cobalt nanoparticles by using oxygen-assisted chemical vapor deposition process. Bull. Korean Chem. Soc. **28**(11), 2056–2060 (2007)
41. Varshney, D., Weiner, B.R., Morell, G.: Growth and field emission study of a monolithic carbon nanotube/diamond composite. Carbon **48**(12), 3353–3358 (2010)
42. Brown, B., Parker, C.B., Stoner, B.R., Glass, J.T.: Growth of vertically aligned bamboo-like carbon nanotubes from ammonia/methane precursors using a platinum catalyst. Carbon **49**(1), 266–274 (2011)
43. Xu, Y., Dervishi, E., Biris, A.R., Biris, A.S.: Chirality-enriched semiconducting carbon nanotubes synthesized on high surface area MgO-supported catalyst. Mater. Lett. **65**(12), 1878–1881 (2011)
44. Fotopoulos, N., Xanthakis, J.: A molecular level model for the nucleation of a single-wall carbon nanotube cap over a transition metal catalytic particle. Diam. Relat. Mater. **19**(5–6), 557–561 (2010)
45. Lee, O., Jung, J., Doo, S., Kim, S.S., Noh, T.H., Kim, K.I., et al.: Effects of temperature and catalysts on the synthesis of carbon nanotubes by chemical vapor deposition. Met. Mater. Int. **16**(4), 663–667 (2010)

46. Sharma, R., Chee, S.W., Herzing, A., Miranda, R., Rez, P.: Evaluation of the role of Au in improving catalytic activity of Ni nanoparticles for the formation of one-dimensional carbon nanostructures. Nano Lett. **11**(6), 2464–2471 (2011)
47. Palizdar, M., Ahgababazadeh, R., Mirhabibi, A., Brydson, R., Pilehvari, S.: Investigation of Fe/MgO catalyst support precursors for the chemical vapour deposition growth of carbon nanotubes. J. Nanosci. Nanotechnol. **11**(6), 5345–5351 (2011)
48. Tomie, T., Inoue, S., Kohno, M., Matsumura, Y.: Prospective growth region for chemical vapor deposition synthesis of carbon nanotube on C-H–O ternary diagram. Diam. Relat. Mater. **19**(11), 1401–1404 (2010)
49. Narkiewicz, U., Podsiadły, M., Jędrzejewski, R., Pełech, I.: Catalytic decomposition of hydrocarbons on cobalt, nickel and iron catalysts to obtain carbon nanomaterials. Appl. Catal. A **384**(1–2), 27–35 (2010)
50. He, D., Li, H., Li, W., Haghi-Ashtiani, P., Lejay, P., Bai, J.: Growth of carbon nanotubes in six orthogonal directions on spherical alumina microparticles. Carbon **49**(7), 2273–2286 (2011)
51. Shirazi, Y., Tofighy, M.A., Mohammadi, T.: Pak A. Effects of different carbon precursors on synthesis of multiwall carbon nanotubes: Purification and Functionalization. Appl. Surf. Sci. **257**(16), 7359–7367 (2011)
52. Li, H., He, D., Li, T., Genestoux, M., Bai, J.: Chemical kinetics of catalytic chemical vapor deposition of an acetylene/xylene mixture for improved carbon nanotube production. Carbon **48**(15), 4330–4342 (2010)
53. Santangelo, S., Messina, G., Faggio, G., Lanza, M., Pistone, A., Milone, C.: Calibration of reaction parameters for the improvement of thermal stability and crystalline quality of multi-walled carbon nanotubes. J. Mater. Sci. **45**(3), 783–792 (2010)
54. Hou, B., Xiang, R., Inoue, T., Einarsson, E., Chiashi, S., Shiomi, J., et al.: Decomposition of ethanol and dimethyl ether during chemical vapor deposition synthesis of single-walled carbon nanotubes. Jpn. J. Appl. Phys. **50**(6R), 065101 (2011)
55. Li, Q., Yan, H., Zhang, J., Liu, Z.: Effect of hydrocarbons precursors on the formation of carbon nanotubes in chemical vapor deposition. Carbon **42**(4), 829–835 (2004)
56. Zhu, J., Yudasaka, M., Iijima, S.: A catalytic chemical vapor deposition synthesis of double-walled carbon nanotubes over metal catalysts supported on a mesoporous material. Chem. Phys. Lett. **380**(5–6), 496–502 (2003)
57. Ramesh, P., Okazaki, T., Taniguchi, R., Kimura, J., Sugai, T., Sato, K., et al.: Selective chemical vapor deposition synthesis of double-wall carbon nanotubes on mesoporous silica. J. Phys. Chem. B **109**(3), 1141–1147 (2005)
58. Hiraoka, T., Kawakubo, T., Kimura, J., Taniguchi, R., Okamoto, A., Okazaki, T., et al.: Selective synthesis of double-wall carbon nanotubes by CCVD of acetylene using zeolite supports. Chem. Phys. Lett. **382**(5–6), 679–685 (2003)
59. Lyu, S.C., Liu, B.C., Lee, C.J., Kang, H.K., Yang, C.W., Park, C.Y.: High-quality double-walled carbon nanotubes produced by catalytic decomposition of benzene. Chem. Mater. **15**(20), 3951–3954 (2003)
60. Zhang, D., Shi, L., Fang, J., Dai, K., Li, X.: Preparation and desalination performance of multiwall carbon nanotubes. Mater. Chem. Phys. **97**(2–3), 415–419 (2006)
61. Jiang, Q., Song, L., Yang, H., He, Z., Zhao, Y.: Preparation and characterization on the carbon nanotube chemically modified electrode grown in situ. Electrochem. Commun. **10**(3), 424–427 (2008)
62. Li, G.: Synthesis of well-aligned carbon nanotubes on the NH 3 pretreatment Ni catalyst films. Russ. J. Phys. Chem. A **84**(9), 1560–1565 (2010)
63. Kim, H.J., Oh, E., Lee, J., Shim, D.S., Lee, K.H.: Synthesis of carbon nanotubes with catalytic iron-containing proteins. Carbon **49**(12), 3717–3722 (2011)
64. Rafique, M.M.A., Iqbal, J.: Production of carbon nanotubes by different routes-a review. J. Encapsul. Adsorp. Sci. **1**(02), 29 (2011)
65. Salvetat, J.P., Bonard, J.M., Thomson, N., Kulik, A., Forro, L., Benoit, W., et al.: Mechanical properties of carbon nanotubes. Appl. Phys. A **69**(3), 255–260 (1999)

66. Wong, E.W., Sheehan, P.E., Lieber, C.M.: Nanobeam mechanics: elasticity, strength, and toughness of nanorods and nanotubes. Science **277**(5334), 1971–1975 (1997)
67. Poncharal, P., Wang, Z., Ugarte, D., De Heer, W.A.: Electrostatic deflections and electromechanical resonances of carbon nanotubes. Science **283**(5407), 1513–1516 (1999)
68. Meyyappan, M.: Carbon nanotubes: science and applications. CRC press (2004)
69. Peng, H., Chang, C., Aloni, S., Yuzvinsky, T., Zettl, A.: Ultrahigh frequency nanotube resonators. Phys. Rev. Lett. **97**(8), 087203 (2006)
70. PourAkbar Saffar, K., Arshi, A.R., JamilPour, N., Najafi, A.R., Rouhi, G., Sudak, L.: A cross-linking model for estimating Young's modulus of artificial bone tissue grown on carbon nanotube scaffold. J. Biomed. Mater. Res. Part a **94**(2), 594–602 (2010)
71. Yun, Y., Shanov, V., Tu, Y., Schulz, M.J., Yarmolenko, S., Neralla, S., et al.: A multi-wall carbon nanotube tower electrochemical actuator. Nano Lett. **6**(4), 689–693 (2006)
72. Pitroda, J., Jethwa, B., Dave, S.: A critical review on carbon nanotubes. Int J Constr Res Civ Eng. **2**, 36–42 (2016)
73. Krüger, M., Widmer, I., Nussbaumer, T., Buitelaar, M., Schönenberger, C.: Sensitivity of single multiwalled carbon nanotubes to the environment. New J. Phys. **5**(1), 138 (2003)
74. Avouris, P., Dresselhaus, G., Dresselhaus, M.: Carbon nanotubes: synthesis, structure, properties and applications. Topics in Applied Physics (2000)
75. Badehian, H.A., Gharbavi, K.: Transport and mechanical properties of double-walled carbon nanotubes as a function of interwall distance. Mol. Cryst. Liq. Cryst. **650**(1), 138–146 (2017)
76. Saito, R., Fujita, M., Dresselhaus, G., Dresselhaus, M.: Electronic structure of chiral graphene tubules. Appl. Phy. Lett. **60**(18), 2204–2206 (1992)
77. Jin, F.L., Park, S.J.: Recent advances in carbon-nanotube-based epoxy composites. Carbon Lett. **14**(1), 1–13 (2013)
78. Wepasnick, K.A., Smith, B.A., Bitter, J.L., Fairbrother, D.H.: Chemical and structural characterization of carbon nanotube surfaces. Anal. Bioanal. Chem. **396**(3), 1003–1014 (2010)
79. Dai, H., Wong, E.W., Lieber, C.M.: Probing electrical transport in nanomaterials: conductivity of individual carbon nanotubes. Science **272**(5261), 523–526 (1996)
80. Nessim, G.D.: Properties, synthesis, and growth mechanisms of carbon nanotubes with special focus on thermal chemical vapor deposition. Nanoscale. **2**(8), 1306–1323 (2010)
81. Hone, J., Whitney, M., Piskoti, C., Zettl, A.: Thermal conductivity of single-walled carbon nanotubes. Physical Review B. **59**(4), R2514 (1999)
82. Hone, J.: Phonons and thermal properties of carbon nanotubes. Carbon Nanotubes, pp. 273–86. Springer (2001)
83. Pop, E., Mann, D., Wang, Q., Goodson, K., Dai, H.: Thermal conductance of an individual single-wall carbon nanotube above room temperature. Nano Lett. **6**(1), 96–100 (2006)
84. Kim, P., Shi, L., Majumdar, A., McEuen, P.L.: Thermal transport measurements of individual multiwalled nanotubes. Phys. Rev. Lett. **87**(21), 215502 (2001)
85. Yu, C., Shi, L., Yao, Z., Li, D., Majumdar, A.: Thermal conductance and thermopower of an individual single-wall carbon nanotube. Nano Lett. **5**(9), 1842–1846 (2005)
86. Maultzsch, J., Reich, S., Thomsen, C., Dobardžić, E., Milošević, I., Damnjanović, M.: Phonon dispersion of carbon nanotubes. Solid State Commun. **121**(9–10), 471–474 (2002)
87. Ishii, H., Kobayashi, N., Hirose, K.: Electron–phonon coupling effect on quantum transport in carbon nanotubes using time-dependent wave-packet approach. Physica E **40**(2), 249–252 (2007)
88. Maeda, T., Horie, C.: Phonon modes in single-wall nanotubes with a small diameter. Phys. B **263**, 479–481 (1999)
89. Kasuya, A., Saito, Y., Sasaki, Y., Fukushima, M., Maedaa, T., Horie, C., et al.: Size dependent characteristics of single wall carbon nanotubes. Mater. Sci. Eng. **217**, 46–47 (1996)
90. Popov, V.: Theoretical evidence for T1/2 specific heat behavior in carbon nanotube systems. Carbon **42**(5–6), 991–995 (2004)
91. Eatemadi, A., Daraee, H., Karimkhanloo, H., Kouhi, M., Zarghami, N., Akbarzadeh, A., et al.: Carbon nanotubes: properties, synthesis, purification, and medical applications. Nanoscale Res. Lett. **9**(1), 393 (2014)

92. Vilatela, J.J., Eder, D.: Nanocarbon composites and hybrids in sustainability: a review. Chemsuschem **5**(3), 456–478 (2012)
93. Schottner, G.: Hybrid sol– gel-derived polymers: applications of multifunctional materials. Chem. Mater. **13**(10), 3422–3435 (2001)
94. Vaka, M.: Synthesis and characterisation of graphene hybrid nanoarchitechures for potential sensing applications: Deakin University (2015)
95. Rao, R., Pint, C.L., Islam, A.E., Weatherup, R.S., Hofmann, S., Meshot, E.R., et al.: Carbon nanotubes and related nanomaterials: critical advances and challenges for synthesis toward mainstream commercial applications. ACS Nano **12**(12), 11756–11784 (2018)
96. Vaka, M., Walvekar, R., Rasheed, A.K., Khalid, M., Panchal, H.: A review: emphasizing the nanofluids use in PV/T systems
97. Gómez-Romero, P., Sanchez, C.: Hybrid materials. Functional properties. From Maya Blue to 21st century materials. New J. Chem. **29**(1), 57–58 (2005)
98. Wildgoose, G.G., Banks, C.E., Compton, R.G.: Metal nanoparticles and related materials supported on carbon nanotubes: methods and applications. Small. **2**(2), 182–193 (2006)
99. Georgakilas, V., Gournis, D., Tzitzios, V., Pasquato, L., Guldi, D.M., Prato, M.: Decorating carbon nanotubes with metal or semiconductor nanoparticles. J. Mater. Chem. **17**(26), 2679–2694 (2007)
100. Mazloumi, M., Shadmehr, S., Rangom, Y., Nazar, L.F., Tang, X.: Fabrication of three-dimensional carbon nanotube and metal oxide hybrid mesoporous architectures. ACS Nano **7**(5), 4281–4288 (2013)
101. Andrews, R., Weisenberger, M.: Carbon nanotube polymer composites. Curr. Opin. Solid State Mater. Sci. **8**(1), 31–37 (2004)
102. Peigney, A., Laurent, C., Flahaut, E., Rousset, A.: Carbon nanotubes in novel ceramic matrix nanocomposites. Ceram. Int. **26**(6), 677–683 (2000)
103. Planeix, J., Coustel, N., Coq, B., Brotons, V., Kumbhar, P., Dutartre, R., et al.: Application of carbon nanotubes as supports in heterogeneous catalysis. J. Am. Chem. Soc. **116**(17), 7935–7936 (1994)
104. Kim, H.S., Lee, H., Han, K.S., Kim, J.H., Song, M.S., Park, M.S., et al.: Hydrogen storage in Ni nanoparticle-dispersed multiwalled carbon nanotubes. J. Phys. Chem. B **109**(18), 8983–8986 (2005)
105. LeeáTan, K.: Growth of Pd, Pt, Ag and Au nanoparticles on carbon nanotubes. J. Mater. Chem. **11**(9), 2378–2381 (2001)
106. Xu, C., Chen, J., Cui, Y., Han, Q., Choo, H., Liaw, P.K., et al.: Influence of the surface treatment on the deposition of platinum nanoparticles on the carbon nanotubes. Adv. Eng. Mater. **8**(1–2), 73–77 (2006)
107. Ye, J.S., Cui, H.F., Liu, X., Lim, T.M., Zhang, W.D., Sheu, F.S.: Preparation and characterization of aligned carbon nanotube–ruthenium oxide nanocomposites for supercapacitors. Small. **1**(5), 560–565 (2005)
108. Choi, H.C., Shim, M., Bangsaruntip, S., Dai, H.: Spontaneous reduction of metal ions on the sidewalls of carbon nanotubes. J. Am. Chem. Soc. **124**(31), 9058–9059 (2002)
109. Tzitzios, V., Georgakilas, V., Oikonomou, E., Karakassides, M., Petridis, D.: Synthesis and characterization of carbon nanotube/metal nanoparticle composites well dispersed in organic media. Carbon **44**(5), 848–853 (2006)
110. Yoon, B., Wai, C.M.: Microemulsion-templated synthesis of carbon nanotube-supported Pd and Rh nanoparticles for catalytic applications. J. Am. Chem. Soc. **127**(49), 17174–17175 (2005)
111. Ang, L.-M., Hor, T.A., Xu, G.-Q., Tung, C.-H., Zhao, S., Wang, J.L.: Electroless plating of metals onto carbon nanotubes activated by a single-step activation method. Chem. Mater. **11**(8), 2115–2118 (1999)
112. Quinn, B.M., Dekker, C., Lemay, S.G.: Electrodeposition of noble metal nanoparticles on carbon nanotubes. J. Am. Chem. Soc. **127**(17), 6146–6147 (2005)
113. Day, T.M., Unwin, P.R., Wilson, N.R., Macpherson, J.V.: Electrochemical templating of metal nanoparticles and nanowires on single-walled carbon nanotube networks. J. Am. Chem. Soc. **127**(30), 10639–10647 (2005)

114. Wang, C., Waje, M., Wang, X., Tang, J.M., Haddon, R.C., Yan, Y.: Proton exchange membrane fuel cells with carbon nanotube based electrodes. Nano Lett. **4**(2), 345–348 (2004)
115. Coleman, K.S., Bailey, S.R., Fogden, S., Green, M.L.: Functionalization of single-walled carbon nanotubes via the Bingel reaction. J. Am. Chem. Soc. **125**(29), 8722–8723 (2003)
116. Ellis, A.V., Vijayamohanan, K., Goswami, R., Chakrapani, N., Ramanathan, L., Ajayan, P.M., et al.: Hydrophobic anchoring of monolayer-protected gold nanoclusters to carbon nanotubes. Nano Lett. **3**(3), 279–282 (2003)
117. Liu, L., Wang, T., Li, J., Guo, Z.X., Dai, L., Zhang, D., et al.: Self-assembly of gold nanoparticles to carbon nanotubes using a thiol-terminated pyrene as interlinker. Chem. Phys. Lett. **367**(5–6), 747–752 (2003)
118. Yang, D.Q., Hennequin, B., Sacher, E.: XPS demonstration of $\pi-\pi$ interaction between benzyl mercaptan and multiwalled carbon nanotubes and their use in the adhesion of Pt nanoparticles. Chem. Mater. **18**(21), 5033–5038 (2006)
119. Guldi, D.M., Rahman, G., Jux, N., Tagmatarchis, N., Prato, M.: Integrating single-wall carbon nanotubes into donor–acceptor nanohybrids. Angew. Chem. Int. Ed. **43**(41), 5526–5530 (2004)
120. Guldi, D.M., Rahman, G., Prato, M., Jux, N., Qin, S., Ford, W.: Single-wall carbon nanotubes as integrative building blocks for solar-energy conversion. Angew. Chem. Int. Ed. **44**(13), 2015–2018 (2005)
121. Jiang, K., Eitan, A., Schadler, L.S., Ajayan, P.M., Siegel, R.W., Grobert, N., et al.: Selective attachment of gold nanoparticles to nitrogen-doped carbon nanotubes. Nano Lett. **3**(3), 275–277 (2003)
122. Carrillo, A., Swartz, J.A., Gamba, J.M., Kane, R.S., Chakrapani, N., Wei, B., et al.: Noncovalent functionalization of graphite and carbon nanotubes with polymer multilayers and gold nanoparticles. Nano Lett. **3**(10), 1437–1440 (2003)
123. Gao, C., Vo, C.D., Jin, Y.Z., Li, W., Armes, S.P.: Multihydroxy polymer-functionalized carbon nanotubes: synthesis, derivatization, and metal loading. Macromolecules **38**(21), 8634–8648 (2005)
124. Correa-Duarte, M.A., Pérez-Juste, J., Sánchez-Iglesias, A., Giersig, M., Liz-Marzán, L.M.: Aligning Au nanorods by using carbon nanotubes as templates. Angew. Chem. Int. Ed. **44**(28), 4375–4378 (2005)
125. Nikoobakht, B., El-Sayed, M.A.: Preparation and growth mechanism of gold nanorods (NRs) using seed-mediated growth method. Chem. Mater. **15**(10), 1957–1962 (2003)
126. Hu, X., Wang, T., Qu, X., Dong, S.: In situ synthesis and characterization of multiwalled carbon nanotube/Au nanoparticle composite materials. J. Phys. Chem. B **110**(2), 853–857 (2006)
127. Fullam, S., Cottell, D., Rensmo, H., Fitzmaurice, D.: Carbon nanotube templated self-assembly and thermal processing of gold nanowires. Adv. Mater. **12**(19), 1430–1432 (2000)
128. Hrapovic, S., Liu, Y., Male, K.B., Luong, J.H.: Electrochemical biosensing platforms using platinum nanoparticles and carbon nanotubes. Anal. Chem. **76**(4), 1083–1088 (2004)
129. Stoffelbach, F., Aqil, A., Jérôme, C., Jérôme, R., Detrembleur, C.: An easy and economically viable route for the decoration of carbon nanotubes by magnetite nanoparticles, and their orientation in a magnetic field. Chem. Commun. **36**, 4532–4533 (2005)
130. Wang, L., Guo, S., Huang, L., Dong, S.: Alternate assemblies of polyelectrolyte functionalized carbon nanotubes and platinum nanoparticles as tunable electrocatalysts for dioxygen reduction. Electrochem. Commun. **9**(4), 827–832 (2007)
131. Correa-Duarte, M.A., Liz-Marzán, L.M.: Carbon nanotubes as templates for one-dimensional nanoparticle assemblies. J. Mater. Chem. **16**(1), 22–25 (2006)
132. Liu, Y., Chae, H.G., Choi, Y.H., Kumar, S.: Effect of carbon nanotubes on sintering behavior of alumina prepared by sol–gel method. Ceram. Int. **40**(5), 6579–6587 (2014)
133. Leary, R., Westwood, A.: Carbonaceous nanomaterials for the enhancement of TiO2 photocatalysis. Carbon **49**(3), 741–772 (2011)

134. Rajarao, R., Jayanna, R., Sahajwalla, V., Bhat, B.R.: GreenApproach to decorate multi-walled carbon nanotubes by metal/metal oxide nanoparticles. Procedia Materials Science. **5**, 69–75 (2014)
135. Mallakpour, S., Khadem, E.: Carbon nanotube–metal oxide nanocomposites: fabrication, properties and applications. Chem. Eng. J. **302**, 344–367 (2016)
136. Hu, Y., Guo, C.: Carbon nanotubes and carbon nanotubes/metal oxide heterostructures: synthesis, characterization and electrochemical property. Carbon nanotubes—growth and applications InTech, Croatia 3–34 (2011)
137. Yi, J., Xue, W., Xie, Z., Chen, J., Zhu, L.: A novel processing route to develop alumina matrix nanocompositesreinforced with multi-walled carbon nanotubes. Mater. Res. Bull. **64**, 323–326 (2015)
138. Estili, M., Kawasaki, A., Pittini-Yamada, Y., Utke, I., Michler, J.: In situ characterization of tensile-bending load bearing ability of multi-walled carbon nanotubes in alumina-based nanocomposites. J. Mater. Chem. **21**(12), 4272–4278 (2011)
139. Lu, K.: Rheological behavior of carbon nanotube-alumina nanoparticle dispersion systems. Powder Technol. **177**(3), 154–161 (2007)
140. Chu, H., Wei, L., Cui, R., Wang, J., Li, Y.: Carbon nanotubes combined with inorganic nanomaterials: Preparations and applications. Coord. Chem. Rev. **254**(9–10), 1117–1134 (2010)
141. Zhang, T., Kumari, L., Du, G., Li, W., Wang, Q., Balani, K., et al.: Mechanical properties of carbon nanotube–alumina nanocomposites synthesized by chemical vapor deposition and spark plasma sintering. Compos. A Appl. Sci. Manuf. **40**(1), 86–93 (2009)
142. Zarabadi-Poor, P., Badiei, A.: Synthesis of carbon nanotubes using metal-modified nanoporous silicas. Carbon Nanotubes-Growth and Applications, 59–74 (2011)
143. Gupta, V., Saleh, T.A.: Syntheses of carbon nanotube-metal oxides composites; adsorption and photo-degradation. Carbon Nanotubes-From Research to Applications, IntechOpen (2011)
144. Gao, J., Zhao, B., Itkis, M.E., Bekyarova, E., Hu, H., Kranak, V., et al.: Chemical engineering of the single-walled carbon nanotube−nylon 6 interface. J. Am. Chem. Soc. **128**(23), 7492–7496 (2006)
145. Sui, G., Zhong, W., Yang, X., Yu, Y., Zhao, S.: Preparation and properties of natural rubber composites reinforced with pretreated carbon nanotubes. Polym. Adv. Technol. **19**(11), 1543–1549 (2008)
146. Sahoo, N.G., Jung, Y.C., Yoo, H.J., Cho, J.W.: Effect of functionalized carbon nanotubes on molecular interaction and properties of polyurethane composites. Macromol. Chem. Phys. **207**(19), 1773–1780 (2006)
147. Yuen, S.-M., Ma, C.-C.M., Chuang, C.-Y., Yu, K.-C., Wu, S.-Y., Yang, C.-C. et al.: Effect of processing method on the shielding effectiveness of electromagnetic interference of MWCNT/PMMA composites. Comp. Sci. Technol. **68**(3–4), 963–968 (2008)
148. Coleman, J.N., Khan, U., Blau, W.J., Gun'ko, Y.K.: Small but strong: a review of the mechanical properties of carbon nanotube–polymer composites. Carbon **44**(9), 1624–1652 (2006)
149. Coleman, J.N., Khan, U., Gun'ko, Y.K.: Mechanical reinforcement of polymers using carbon nanotubes. Adv. Mater. **18**(6), 689–706 (2006)
150. Liu, P.: Modifications of carbon nanotubes with polymers. Eur. Polymer J. **41**(11), 2693–2703 (2005)
151. Byrne, M.T., Gun'ko, Y.K.: Recent advances in research on carbon nanotube–polymer composites. Adv. Mater. **22**(15), 1672–1688 (2010)
152. Baskaran, D., Mays, J.W., Bratcher, M.S.: Noncovalent and nonspecific molecular interactions of polymers with multiwalled carbon nanotubes. Chem. Mater. **17**(13), 3389–3397 (2005)
153. Hirsch, A.: Functionalization of single-walled carbon nanotubes. Angew. Chem. Int. Ed. **41**(11), 1853–1859 (2002)
154. Lau, K.-T., Gu, C., Hui, D.: A critical review on nanotube and nanotube/nanoclay related polymer composite materials. Compos. Part B: Eng. **37**(6), 425–36 (2006)
155. Nanocomposites, K.S.: J. Mater. Chem. **2**(12), 1219–1230 (1992)

156. Niihara, K.: New design concept of structural ceramics. J. Ceram. Soc. Jpn. **99**(1154), 974–982 (1991)
157. An, J.W., You, D.H., Lim, D.S.: Tribological properties of hot-pressed alumina–CNT composites. Wear **255**(1–6), 677–681 (2003)
158. Samal, S.S., Bal, S.: Carbon nanotube reinforced ceramic matrix composites-a review (2008)
159. Zhan, G.D., Kuntz, J.D., Wan, J., Mukherjee, A.K.: Single-wall carbon nanotubes as attractive toughening agents in alumina-based nanocomposites. Nat. Mater. **2**(1), 38 (2003)
160. Wang, X., Padture, N.P., Tanaka, H.: Contact-damage-resistant ceramic/single-wall carbon nanotubes and ceramic/graphite composites. Nat. Mater. **3**(8), 539 (2004)
161. Harris, P.J.: Carbon nanotube composites. Int. Mater. Rev. **49**(1), 31–43 (2004)
162. Sandler, J., Kirk, J., Kinloch, I., Shaffer, M., Windle, A.: Ultra-low electrical percolation threshold in carbon-nanotube-epoxy composites. Polymer **44**(19), 5893–5899 (2003)
163. Kovacs, J.Z., Velagala, B.S., Schulte, K., Bauhofer, W.: Two percolation thresholds in carbon nanotube epoxy composites. Compos. Sci. Technol. **67**(5), 922–928 (2007)
164. Ibrahim, K.S.: Carbon nanotubes-properties and applications: a review. Carbon Letters. **14**(3), 131–144 (2013)
165. Schnorr, J.M., Swager, T.M.: Emerging applications of carbon nanotubes. Chem. Mater. **23**(3), 646–657 (2010)
166. Hu, L., Li, J., Liu, J., Grüner, G., Marks, T.: Flexible organic light-emitting diodes with transparent carbon nanotube electrodes: problems and solutions. Nanotechnology. **21**(15), 155202 (2010)
167. Signorelli, R., Ku, D.C., Kassakian, J.G., Schindall, J.E.: Electrochemical double-layer capacitors using carbon nanotube electrode structures. Proc. IEEE **97**(11), 1837–1847 (2009)
168. Xie, X., Gao, L.: Characterization of a manganese dioxide/carbon nanotube composite fabricated using an in situ coating method. Carbon **45**(12), 2365–2373 (2007)
169. Li, W., Liang, C., Zhou, W., Qiu, J., Zhou, Z., Sun, G., et al.: Preparation and characterization of multiwalled carbon nanotube-supported platinum for cathode catalysts of direct methanol fuel cells. J. Phys. Chem. B **107**(26), 6292–6299 (2003)
170. Prabhuram, J., Zhao, T., Tang, Z., Chen, R., Liang, Z.: Multiwalled carbon nanotube supported PtRu for the anode of direct methanol fuel cells. J. Phys. Chem. B **110**(11), 5245–5252 (2006)
171. Le Goff, A., Artero, V., Jousselme, B., Tran, P.D., Guillet, N., Métayé, R., et al.: From hydrogenases to noble metal–free catalytic nanomaterials for H2 production and uptake. Science **326**(5958), 1384–1387 (2009)
172. Swager, T.M.: The molecular wire approach to sensory signal amplification. Acc. Chem. Res. **31**(5), 201–207 (1998)
173. Jiang, L.C., Zhang, W.D.: A highly sensitive nonenzymatic glucose sensor based on CuO nanoparticles-modified carbon nanotube electrode. Biosens. Bioelectron. **25**(6), 1402–1407 (2010)
174. Dillon, A.C., Jones, K., Bekkedahl, T., Kiang, C., Bethune, D., Heben, M.: Storage of hydrogen in single-walled carbon nanotubes. Nature **386**(6623), 377 (1997)
175. Dervishi, E., Li, Z., Xu, Y., Saini, V., Biris, A.R., Lupu, D., et al.: Carbon nanotubes: synthesis, properties, and applications. Part. Sci. Technol. **27**(2), 107–125 (2009)
176. Silambarasan, D., Surya, V.J., Vasu, V., Iyakutti, K.: Reversible and reproducible hydrogen storage in single-walled carbon nanotubes functionalized with borane. Carbon Nanotubes-Recent Progress. IntechOpen, pp. 331–348 (2018)
177. Mesgari, S., Taylor, R.A., Hjerrild, N.E., Crisostomo, F., Li, Q., Scott, J.: An investigation of thermal stability of carbon nanofluids for solar thermal applications. Sol. Energy Mater. Sol. Cells **157**, 652–659 (2016)
178. Akhavan-Behabadi, M., Shahidi, M., Aligoodarz, M., Fakoor-Pakdaman, M.: An experimental investigation on rheological properties and heat transfer performance of MWCNT-water nanofluid flow inside vertical tubes. Appl. Therm. Eng. **106**, 916–924 (2016)
179. Manasrah, A.D., Al-Mubaiyedh, U.A., Laui, T., Ben-Mansour, R., Al-Marri, M.J., Almanassra, I.W., et al.: Heat transfer enhancement of nanofluids using iron nanoparticles decorated carbon nanotubes. Appl. Therm. Eng. **107**, 1008–1018 (2016)

180. Esfe, M.H., Arani, A.A.A., Madadi, M.R., Alirezaie, A.: A study on rheological characteristics of hybrid nano-lubricants containing MWCNT-TiO2 nanoparticles. J. Mol. Liq. **260**, 229–236 (2018)
181. Esfe, M.H., Alirezaie, A., Rejvani, M.: An applicable study on the thermal conductivity of SWCNT-MgO hybrid nanofluid and price-performance analysis for energy management. Appl. Therm. Eng. **111**, 1202–1210 (2017)
182. Esfe, M.H., Afrand, M., Yan, W.M., Yarmand, H., Toghraie, D., Dahari, M.: Effects of temperature and concentration on rheological behavior of MWCNTs/SiO2 (20–80)-SAE40 hybrid nano-lubricant. Int. Commun. Heat Mass Transfer **76**, 133–138 (2016)
183. Esfe, M.H., Yan, W.M., Akbari, M., Karimipour, A., Hassani, M.: Experimental study on thermal conductivity of DWCNT-ZnO/water-EG nanofluids. Int. Commun. Heat Mass Transfer **68**, 248–251 (2015)
184. Esfe, M.H., Saedodin, S., Yan, W.M., Afrand, M., Sina, N.: Study on thermal conductivity of water-based nanofluids with hybrid suspensions of CNTs/Al 2 O 3 nanoparticles. J. Therm. Anal. Calorim. **124**(1), 455–460 (2016)
185. Tuteja, A., Choi, W., McKinley, G.H., Cohen, R.E., Rubner, M.F.: Design parameters for superhydrophobicity and superoleophobicity. MRS Bull. **33**(8), 752–758 (2008)
186. Ling, H., Srinivasan, S., Golovin, K., McKinley, G.H., Tuteja, A., Katz, J.: High-resolution velocity measurement in the inner part of turbulent boundary layers over super-hydrophobic surfaces. J. Fluid Mech. **801**, 670–703 (2016)
187. Wei, M., Song, Y., Zhu, Y., Preston, D.J., Tan, C.S., Wang, E.N.: Heat transfer suppression by suspended droplets on microstructured surfaces. Appl. Phys. Lett. **116**(23), 233703 (2020)
188. Lau, K.K., Bico, J., Teo, K.B., Chhowalla, M., Amaratunga, G.A., Milne, W.I., et al.: Superhydrophobic carbon nanotube forests. Nano Lett. **3**(12), 1701–1705 (2003)
189. Ding, Y., Alias, H., Wen, D., Williams, R.A.: Heat transfer of aqueous suspensions of carbon nanotubes (CNT nanofluids). Int. J. Heat Mass Transf. **49**(1–2), 240–250 (2006)
190. Bui, K., Grady, B.P., Papavassiliou, D.V.: Heat transfer in high volume fraction CNT nanocomposites: Effects of inter-nanotube thermal resistance. Chem. Phys. Lett. **508**(4–6), 248–251 (2011)
191. Yazid, M.N.A.W.M., Sidik, N.A.C., Yahya, W.J.: Heat and mass transfer characteristics of carbon nanotube nanofluids: a review. Renew. Sustain. Energy Rev. **80**, 914–941 (2017)
192. Garg, P., Alvarado, J.L., Marsh, C., Carlson, T.A., Kessler, D.A., Annamalai, K.: An experimental study on the effect of ultrasonication on viscosity and heat transfer performance of multi-wall carbon nanotube-based aqueous nanofluids. Int. J. Heat Mass Transf. **52**(21–22), 5090–5101 (2009)
193. Georgakilas, V., Tagmatarchis, N., Pantarotto, D., Bianco, A., Briand, J.P., Prato, M.: Amino acid functionalisation of water soluble carbon nanotubes. Chem. Commun. **24**, 3050–3051 (2002)
194. Star, A., Steuerman, D.W., Heath, J.R., Stoddart, J.F.: Starched carbon nanotubes. Angew. Chem. Int. Ed. **41**(14), 2508–2512 (2002)
195. Yang, W., Thordarson, P., Gooding, J.J., Ringer, S.P., Braet, F.: Carbon nanotubes for biological and biomedical applications. Nanotechnology. **18**(41), 412001 (2007)
196. Feazell, R.P., Nakayama-Ratchford, N., Dai, H., Lippard, S.J.: Soluble single-walled carbon nanotubes as longboat delivery systems for platinum (IV) anticancer drug design. J. Am. Chem. Soc. **129**(27), 8438–8439 (2007)
197. Liu, Z., Winters, M., Holodniy, M., Dai, H.: siRNA delivery into human T cells and primary cells with carbon-nanotube transporters. Angew. Chem. Int. Ed. **46**(12), 2023–2027 (2007)
198. Šljukić, B., Banks, C.E., Compton, R.G.: Iron oxide particles are the active sites for hydrogen peroxide sensing at multiwalled carbon nanotube modified electrodes. Nano Lett. **6**(7), 1556–1558 (2006)
199. Tasis, D., Tagmatarchis, N., Georgakilas, V., Prato, M.: Soluble carbon nanotubes. Chem. Eur. J. **9**(17), 4000–4008 (2003)
200. Backes, C., Schmidt, C.D., Hauke, F., Böttcher, C., Hirsch, A.: High population of individualized SWCNTs through the adsorption of water-soluble perylenes. J. Am. Chem. Soc. **131**(6), 2172–2184 (2009)

201. Asuri, P., Karajanagi, S.S., Sellitto, E., Kim, D.Y., Kane, R.S., Dordick, J.S.: Water-soluble carbon nanotube-enzyme conjugates as functional biocatalytic formulations. Biotechnol. Bioeng. **95**(5), 804–811 (2006)
202. Li, H., Cheng, F., Duft, A.M., Adronov, A.: Functionalization of single-walled carbon nanotubes with well-defined polystyrene by "click" coupling. J. Am. Chem. Soc. **127**(41), 14518–14524 (2005)

Synthesis of Nanoclay Composite Material

Pratap Kumar Deheri and Biswabandita Kar

Abstract The immense research interest in clay sciences is driven by the easy availability in nature, extraordinary properties, wide range of applicability, cheaper and less toxicity. Clay minerals have a huge potential to explore and/or manipulate application specific physical properties in the lab. Nano-structured clay broadly can be classified as aggregated nano-clay, isolated nano-clay particles (tactoid), intercalated clay and exfoliated clay. Clay shows the properties related to nano-structuring to its fullest when exfoliated and many interesting physical, morphological characteristic and improved properties are observed. Clay based nano-composites materials provide significant properties improvements even at low nanoparticles content. Effort is made to strengthen the understanding on effect of size, shape and the chemical compositions to properties relations. The optimization of adsorption properties, swelling behavior, rheological properties optimization, nano-sized clay development and design of polymer–clay composites development opens the new prospect of research and application of clay minerals. To obtain a uniform distribution with strong linking between polymer to particles remains a critical challenge in order to obtain properties like flame retardant, mechanical, barrier and thermal properties, etc. This chapter focuses more in-depth on the synthesis and properties of clay-polymer composite. Processing of Clay-polymer nano-composites such as conventional solution blending, melt blending, in situ polymerization and the use of ultrasounds in enhancement of nanoparticles dispersion has been discussed. To reach the optimal properties that is required for specific applications, selection of composition, clay microstructure and processing is the key and has been elaborated in this section.

Keywords Clay nano-structure · Clay nano-composites · Intercalation · Exfoliation

P. K. Deheri · B. Kar (✉)
School of Applied Sciences, KIIT (Deemed to be University), Bhubaneswar, India
e-mail: bbkarfch@kiit.ac.in

N. M. Mubarak et al. (eds.), *Contemporary Nanomaterials in Material Engineering Applications*, Engineering Materials,
https://doi.org/10.1007/978-3-030-62761-4_4

1 Introduction

To define the term clay or clay mineral is very complicated. The term clay is used to represents the fine particles (less than 2 μm) formed from rock or earth crust by natural erosion and chemical weathering process [1]. In general, clays consisted of mixtures of fine grained clay minerals, quartz, carbonate, and metal oxides. According to the CMS Nomenclature Committee clay is primarily of fine-grained minerals that is generally plastic at appropriate water contents and will loss plasticity when dried or fired [2, 3]. These definitions do not include synthetic clays and clay-like materials. Natural clays are highly heterogeneous in compositions and properties. Structural imperfection, compositions variations, presence of different impurities, variable crystallinity limits the practical applications of natural clay. This leads to more and more research in development of clay minerals in lab with control size, shape, composition and better properties. The diverse application of clay minerals is further corroborated by the wider species of clay present with different compositions, the wide range of physical and chemical properties and the large probability of compositional modifications.

Natural clays are mixture of hydrous metal-phyllosilicates, quartz, carbonate and metal oxides. However, synthetic clays are hydrous metal-phyllosilicates without quartz, carbonate and metal oxides. The metal ions can be aluminum, iron, magnesium, alkali metals, alkaline earth metals etc. in varying atomic percentages. However, the term clay is quite frequently used to represent the clay minerals (sheet metal-silicate structure). The phyllosilicates are broadly divided into two different groups depending on the layer type, structure and numbers. First one is 1:1 layer consisting of one tetrahedral sheet and one octahedral sheet and the second one is a 2:1 layer containing an octahedral sheet between two opposing tetrahedral sheets [4]. The basic building blocks in both the groups are comprised of octahedral and tetrahedral sheets. However, the compositions and arrangements of these layers in different clay minerals are very different. These differences in compositions and layer stacking account for most of the different in their physical and chemical properties. In this chapter, the six groups of clay minerals (Kaolinite, Smectite, Illite, Vermiculite, Chlorite and Allophanes), their chemical and structural properties are discussed. Clay nano-structuring such as clay nano tubes, nano particles and nano sheets is presented. Clay surface modifications and nano clay-polymer composites synthesis with their mechanical and physical properties are discussed.

1.1 Clay: Structure and Compositions

Clay or clay minerals are distinguished based on their different crystal structure [2, 5]. There is a close relation between the crystal structure to the bulk physical and chemical properties of clay. Here we are discussing the clay structure starting from the basic building unit tetrahedron and octahedron. The clay minerals classes and types are so vast it is out of scope of this chapter to discuss each in details. Here, the important and more frequently observed clay structures will be discussed.

Tetrahedral sheet (phyllosilicates): Clay minerals are called phyllosilicates (derived from Greek word phyllus meaning leaf). "Phyllosilicates are characterized by their layered structures composed of polymeric sheets of silica tetrahedra". The tetrahedron is formed by cations surrounded by four oxygen anions. In clay minerals the predominant central cation is silicon (Si^{4+}) (Fig. 1a) [6]. In the SiO_4^{4-} tetrahedral; central Si with a 4+ charges is bonded to surrounding four O with 2− charges. This resulted in a net negative charge of −4 in SiO_4^{4-} tetrahedral. In phyllosilicates, three basal oxygen of the SiO_4^{4-} tetrahedra are shared with other tetrahedra and form a six member rings. As the three basal oxygen of the tetrahedron are shared only the apical oxygen is free and retains a charge of −1. This leads to formation of basic structural unit of $Si_2O_5^{-2}$ (Fig. 1b) [6, 7]. However, in some clay the central silicon cation in a tetrahedron is partially substituted by aluminum and magnesium resulted in extra negative charges.

Octahedral sheet: The second structural unit in a clay mineral is the octahedral sheet. The octahedral layers are very much similar to the structure of either Brucite [$Mg(OH)_2$] or Gibbsite [$Al(OH)_3$]. In brucite octahedra, the cations are +2 ions like Mg^{+2} or Fe^{+2} while in gibbsite [$Al(OH)_3$] octahedra the cations are +3 like Al^{+3} or Fe^{+3}. The octahedral sheet is formed by sharing of the hydroxyl group at the corner of each octahedron (edge sharing) Fig. 2 [6, 7]. In the brucite structure, all the three octahedral sites are occupied. In the Gibbsite structure every 3rd cation site is unoccupied to maintain charge neutrality. When the entire octahedral site is filled it is called tri-octahedral sheet while the other is termed as di-octahedral sheet.

The Clay Layer: These basic building block (tetrahedron and octahedron) are linked together to form the layer clay structures. Most phyllosilicates contain hydroxyl ion (OH^-), with the OH^- located at the centre of the six member ring (Fig. 3) [6]. Thus, the structural unit becomes $Si_2O_4(OH)^{-3}$. In the octahedral sheet

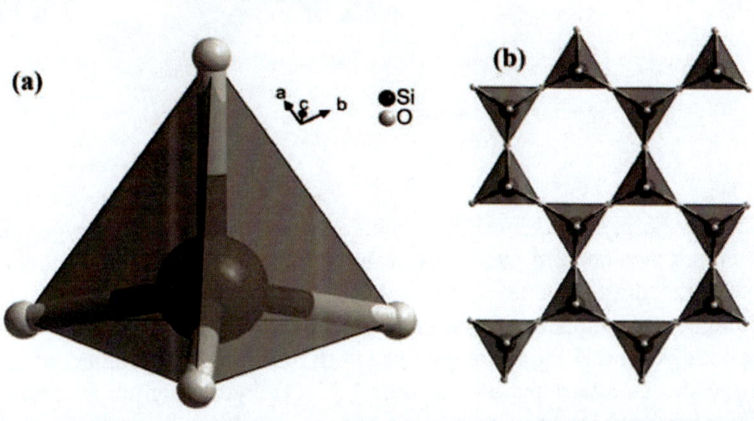

Fig. 1 **a** Silica tetrahedron; dark black color atom is silicon bonded to four oxygen tetrahedral, **b** phyllosilicates; silicate tetrahedra bonded through corner sharing oxygen to other three silicate tetrahedron forming a six member ring [6, 7]

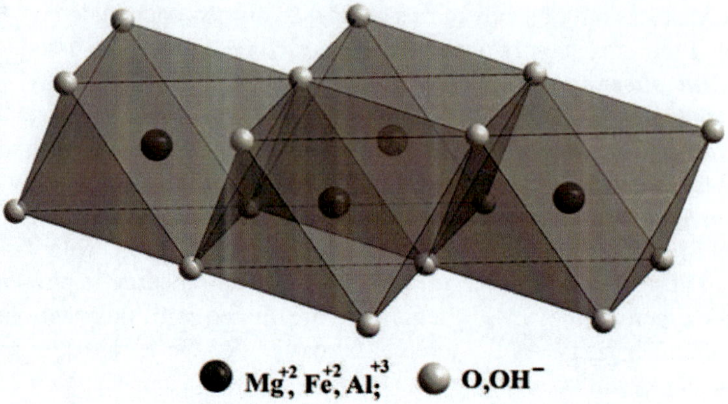

Fig. 2 **a** Octahedral unit; the central octahedral formed by oxygen/hydroxyl ions site is occupied by metal ions, **b** Brucite; if central atom is +2 metal ion and **c** Gibbsite structure; if central atom is +3 metal ion [6, 7]

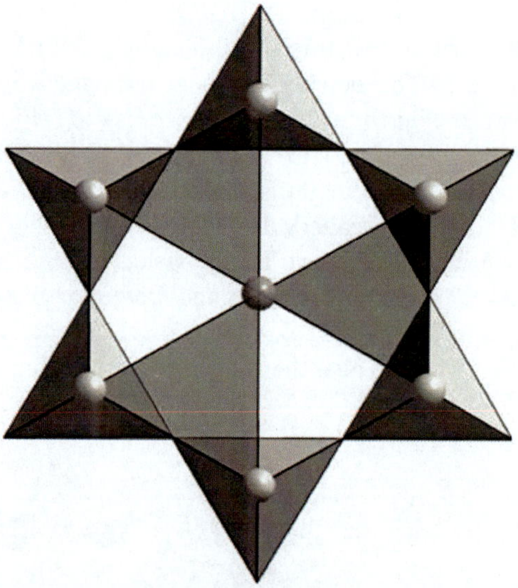

Fig. 3 $Si_2O_4(OH)^{-3}$, the octahedral building block, four of the apical oxygen from $Si_2O_5{}^{-2}$ unit is shared to form the octahedral [6, 7]

cations, usually Fe^{+2}, Mg^{+2}, or Al^{+3} are bonded to four of the apex oxygen of tetrahedral sheet and the centered OH^- ions of the phyllosilicate. The horizontal linkage of multiple octahedra forms the octahedral sheet.

Now imagine the octahedral layer is a tri-octahedral containing Mg^{+2} or Fe^{+2}. If two of the octahedral OH^- ions are replaced with the apical oxygen O^{2-} of the tetrahedral Si_2O_4 sheets, a tetrahedral–octahedral (T–O) layer structure is formed. This kind of layer structure is observed in serpentine mineral. However, if the octahedral layer is di-octahedral containing Al^{+3}, the obtained (T–O) layer is termed as Kaolinite clay (Fig. 4). The formed T–O clay layers are bonded to the top (or bottom) of

Fig. 4 Tetrahedral—octahedral T–O and T–O–T layer structure [6, 7]

another T–O layer by Van der Waals bonds, electrostatic force or hydrogen bonding between the layers. Now, suppose 2 more of the OH ions in the octahedral layer are replaced by O^{2-} of two different Si_2O_4 tetrahedral layers. This leads to a T–O–T layer formation. The T–O–T layers are boned either by weak Van der Waals force, electrostatic force or hydrogen bonding between the layers in a clay system. If an Al^{+3} is substituted for every 4th Si^{+4} in the tetrahedral layer, this causes an excess -1 charge in each T–O–T layer. Positive metal ions like K^{+1} or Na^{+1} is bonded between the two T–O–T layers to maintain the charge neutrality.

Clay minerals classification: As discussed; clay minerals can be divided into two groups depending on the number of tetrahedral and octahedral sheets in the layer structure. In a 1:1 structure one tetrahedra is covalently attached to one octahedra and in a 2:1-type structure one octahedra is sandwich between two tetrahedra. These layers are connected by week Vander Waals force, electrostatic force or hydrogen bonding between the layers to form clay crystallites. These inter-layers are conveniently characterized by x-ray diffraction peak at low 2θ values compared to other characteristic peaks. Interlayer interactions are weak and can easily be disrupt. Hence, interlayer can easily be accessed by water, organic cations or polar organic liquids. This phenomenon is termed as intercalation and causes the lattice to expand. This again can be monitored by x-ray diffraction peak shifting to a lower diffraction angle.

***1:1 (T–O) type minerals* (Kaolinite and Halloysite)**: The 1:1 structure is made up of one Al-octahedra layer and one silicate tetrahedral layer. The common examples of 1:1 layered structures are Kaolinite and Halloysite (Fig. 5). Kaolinite is the most common clay in this group and the common chemical formula is $Al_2Si_2O_5(OH)_4$. It contains Al^{3+} in octahedral site and Si^{4+} in tetrahedral site. The adjacent layer is connected by hydrogen bonding between the H of OH of octahedral sheet and oxygen of tetrahedral sheet. There are three planes of anions; one plane consists of the basal

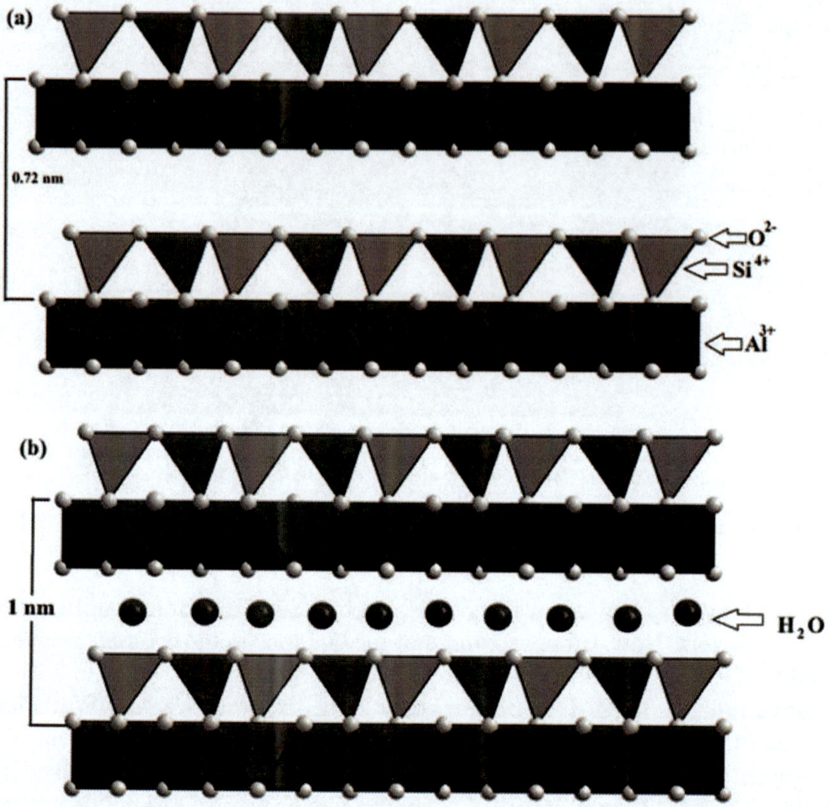

Fig. 5 Kaolinite and Halloysite clay structure [7]

O^{2-} ions of the tetrahedral sheet, the second consists of O^{2-} ions common to both the tetrahedral and octahedral sheets and the third is O^{2-} of octahedral basal plane. It has very little isomorphous substitution at both octahedral or tetrahedral site and hence no permanent charge. As it is charge neutral, the cation exchange capacity is also very low. Halloysite is very similar to kaolinite except the interlayer water. The water is present as a hydration sphere to the cations present between the inter-layers. The representative chemical formula for halloysite is $Al_2Si_2O_5(OH)_{4.}nH_2O$. When n = 2, halloysite is in its fully hydrated form and has a single layer of water between the 1:1 layers and is named halloysite-(10 Å). But when n = 0, halloysite in a less hydrous form and is called halloysite-(7 Å). The name is derived from the differences in the interlayer spacing as a result of the degree of hydration. In general, clays occur as thin plate like structure but halloysite often occur as tubular, fibrous or spherical particles.

2:1 (T–O–T) type minerals: Unlike 1:1 clay mineral, 2:1 structure is more diverse. The 2:1 structure is made up of two silicate tetrahedral layer, one each side of the octahedral layer Fig. 4 [6]. 2:1 structure has four planes of anions; two are the basal oxygen of the two tetrahedral sheets while the other two are the inner planes consist of oxygen and hydroxyls commonly shared between octahedral and the two tetrahedral sheets.

Pyrophyllite: To start with, let's consider the simplest 2:1 pyrophyllite structures. It consists of one Al^{3+} octahedral layer sandwich between two Si^{4+} tetrahedral layer Fig. 6 [7]. As the tetrahedral sheets contain only Si^{4+} and the octahedral sheet contains only Al^{3+} the idealistic chemical formula of pyrophyllite can be written as $Al_2(Si_4)O_{10}(OH)_2$. The charge is balanced making the pyrophyllite clay layer electrically neutral. Hence, the pyrophyllite clay layers are held together only by weak Van-der Waals forces to form the clay crystallites.

Micas: Mica minerals structure is almost similar as pyrophyllite and have the 2:1 layer structure. However, the chemical formula is different. The important difference is; one fourth of tetrahedral site Si is replaced by Al (Fig. 7) [7]. As Si^{+4} is replaced by

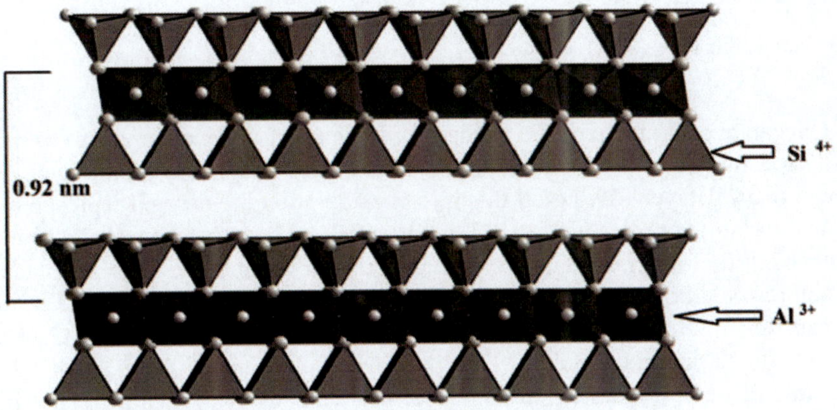

Fig. 6 Pyrophyllite clay structure [7]

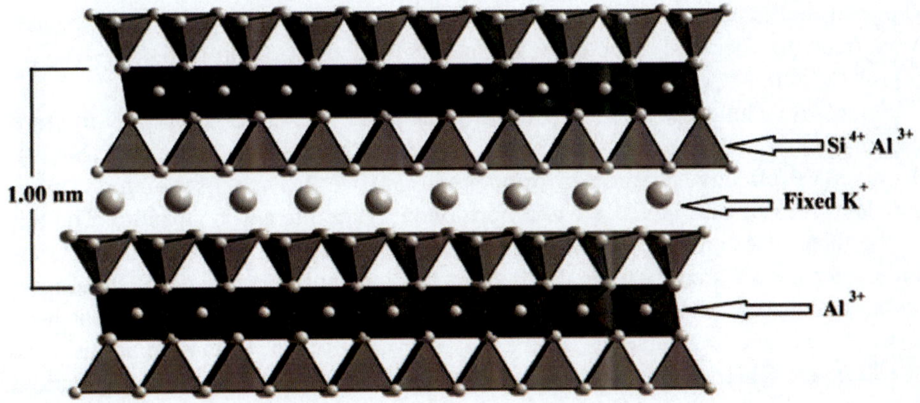

Fig. 7 Mica clay structure [7]

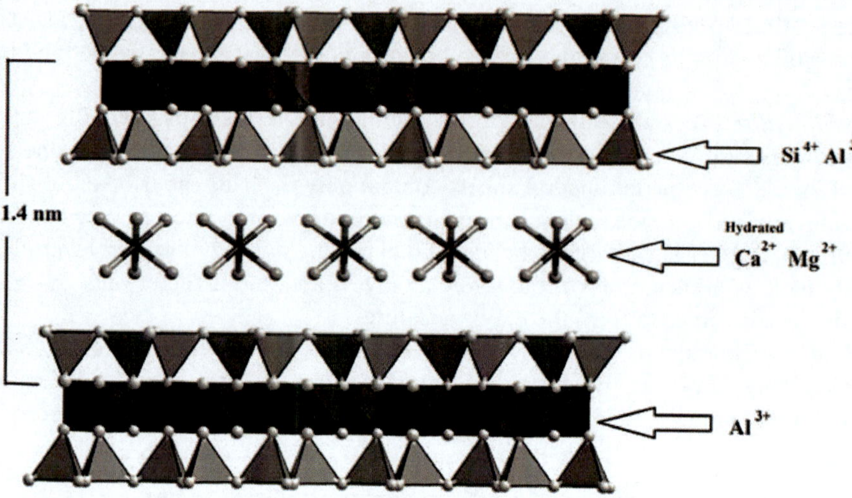

Fig. 8 Vermiculite [7]

Al^{+3} a negative charge is induced in the clay structure. This excess negative charge is balanced by monovalent cations such as Na^+, K^+ that occupy interlayer sites between two 2:1 layer. Taking all this into consideration the derived representative chemical formula for mica is $KAl_2(AlSi_3)O_{10}(OH)_2$. Al^{3+} in octahedral site resulted in dioctahedral clay. The octahedral site can also accommodate Mg^{2+} in place of Al^{3+} and resulted in tri-octahedral clay structure. In some of the mica clay, Fe^{2+} and Fe^{3+} can substitute for Mg^{2+} and Al^{3+} in the octahedral site and Na^+ and Ca^{2+} can substitute for K^+ in the interlayer.

Vermiculites: Vermiculites have 2:1 layer structure (T–O–T structure). They have an inter-layer charge of 0.9–0.6 per formula unit. These negative charges are neutralized by the interlayer cations like Na, K, Ca and Mg (Fig. 8) [7]. These interlayer cations loosely bound and are exchangable with other inorganic and organic cations. The high charge per formula unit gives vermiculite a high cation exchange capacity. A vermiculite chemical structure can be represented as: $M^+_{0.75}$ $Al_2(Si_{3.25}Al_{0.75})$ $O_{10}(OH)_2$, here M^+ represents ex-changeable Na, K, Ca and Mg cations.

Smectites: The smectite group clays consist of 2:1 (T–O–T) layer structure similar to mica and vermiculite and contains hydrated exchange-able cations (Fig. 9). However, the effective charge per formula unit is 0.6–0.2. The most common smectite clay minerals observed are Montmorillonite, beideilite and nontronite. Though all the three class are di-octahedral clay, the metal ions at tetrahedral and octahedral center are different. An idealized smectites molecular formula is Smectite ion exchange capacity is low compare to vermiculite as the net interlayer charge is less. However, the swelling capacity is very high for smectite type of clay minerals.

Chlorites: Chlorite is recognized by its green color. A generalized chemical formula of chlorite is $(Mg,Fe,Al)_3(Si,Al)_4O_{10}(OH)_6$. Here the octahedral Mg^{2+} is substituted by Fe^{3+}, and Al^{3+}. The elements Al^{3+} can occupy both octahedral sites

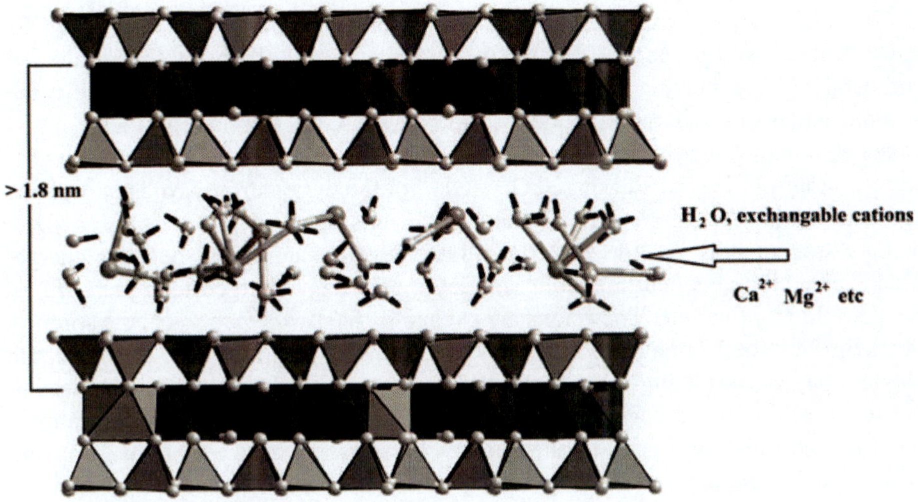

Fig. 9 Smectite clay structure [7]

as well as tetrahedral Si^{4+} sites. It too has a 2:1 (T–O–T) layer structure with excess negative charge (Fig. 10) [7]. This excess negative charge is neutralized by interlayer Mg^{2+} ions that form a brucite like octahedral structure. Hence, in chlorite clay the T–O–T layers sandwich the octahedral $Mg(OH)_2$ layer.

Allophanes: Till now we were discussing well-ordered x-ray crystalline clays. However, the allophanes are poorly crystalline or amorphous in nature. Its idealistic chemical formula is $Al_2O_3{\cdot}(SiO_2)1.3–2{\cdot}(2.5–3)H2O$.

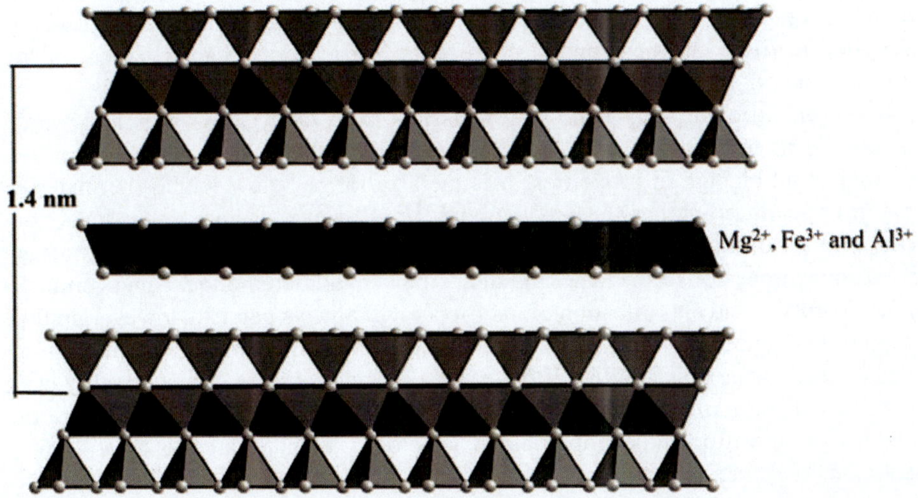

Fig. 10 Chlorite clay structure [7]

Two characteristic aspect of any clay structure is important to understand its microstructure and properties. First is the lattice mismatch between the tetrahedral and octahedral sheets and its effects on the clay microstructure. And second is the polymorphism due to deferent modes of laminate stacking. Tetrahedral silicate sheets have two planes of oxygen/hydroxyl ions. One plane is formed by apex oxygen O^{2-} and the other plane is by the shared O^{2-}, OH^{-} of the tetrahedrons. An isomorphous substitution of silicon atom by bivalent metal ions like Fe^{2+}, Mg^{2+} or a trivalent metal cation such as Fe^{3+} or Al^{3+}, the silicate develops an excess negative charge. This excess charge is neutralized by interlayer positively charged ions such as Li^{+}, Na^{+}, K^{+}, Mg^{2+}, Ca^{2+} etc. These ions are exchangeable by the cations in the solution some time can be from a organic cations. Sodium Na^{+} or lithium Li^{+} cations in the clay interlayer leads to high osmotic pressure between the platelets. This may lead to infinite swelling in presence of water (swelling clay). The clay platelet formations, stacking, agglomeration and exfoliation are characterized by the d_{001} basal spacing (along c-axis) and is determined by XRD and electron micrographs.

2 Nanoclay

Recently, science and technology in the field of nanotechnology has become one of the most interesting and fast-growing multidisciplinary research areas in both science and engineering. Nano-materials are classified as materials with at least 50% particles are between 1–100 nm scale [8]. Materials on the nanometer (atomic) scale possess different structural, physical, chemical or biological properties than that of bulk of the same materials [9]. At nanometric scale, atoms located at a surface is considerably high compared to total number of atoms in the material [9]. Hence, many chemical and physical properties that are controlled by surface phenomena such as catalytic, electrical, magnetic and mechanical properties, crystal growth mechanism and thermal conductivity properties are different than the bulk materials.

However, differentiating nano-clay materials from bulk is not straightforward. Nano-scale material can also be differentiating than the bulk materials based on the significant change in properties. It is the size range below which the material properties are different than the bulk materials [8]. However, the properties transition is abrupt and different for different materials. Another way to define nanomaterials is; at least one dimension of the materials should be in nanometer range. An uncertainty in the definition and classification of nano-clay will always persist. However, in this chapter, the Nano clays term is used for clay minerals with at least one dimension in nanometer range that is below 100 nm. So, depending on the dimensionality clay particles will be termed as nanoparticles, nano-plate or nano-fiber. Depending on chemical composition and nanoparticle morphology nanoclays are montmorillonite, bentonite, kaolinite, hectorite, and halloysite.

2.1 Clay Nanostructure

Nano-structured clay exists in nature by its own. If the clay microstructure is analyzed closely, one can find the bulk clay is agglomerations of nano-structured units like rods, tubes and plates. With little effort these agglomeration can be breaks into individual nano-structures to exploit the nano-properties to its fullest [10]. More obviously, making clay nano-plates is interesting from the clay layers which are nanometers thickness stacking together makes the clay structure. The clay micro-structural studies have attracted much attention in recent years due to their widespread applications.

Clay Nanorods: One-dimensional nanoscale clay minerals with nanorod-shape morphology have also received more and more research attention because of the low cost and safer chemicals. Nanorods have one-dimensional nano size effect and been widely used in polymer composites, catalyst, adsorbents and others. Palygorskite (also called as attapulgite) is a representative natural nanorod with the theoretical formula of $Si_8Mg_8O_{20}(OH)_2(H2O)_4{\bullet}4H_2O$. Palygorskite is from the family of sepiolite. It has a 2:1 ribbon-layer structure. However, unlike the ideal 2:1 phyllosilicates, in palygorskite–sepiolite group clay the apical oxygen atoms point away from the basal oxygen atom plane in opposing directions to form ribbons of joined pyroxene-like chains. The apical oxygen atoms pointing along either the [100] or the [−100] direction, that is, in opposing directions (Fig. 11) [11]. Four apical oxygen atoms of two opposing tetrahedral strip and two OH groups (or by OH_2 groups in some cases) link to metal cations (typically Mg or Al in palygorskite and sepiolite) to form octahedral coordination. Strips that are eight octahedra wide link to tetrahedra via apical oxygen atoms in sepiolite, and strips that are five octahedra in width occur in palygorskite.

The perfect palygorskite crystal is a tri-octahedral mineral in which all the octahedral sites are occupied by Mg^{2+} ions. However, isomorphous substitution of all Mg^{2+} by trivalent metal ions like Al^{3+} and Fe^{3+} leads to the formation of di-octahedral layer. If Mg^{2+} ions are partially replaced by Al^{3+} and Fe^{3+} can be resulted in mix

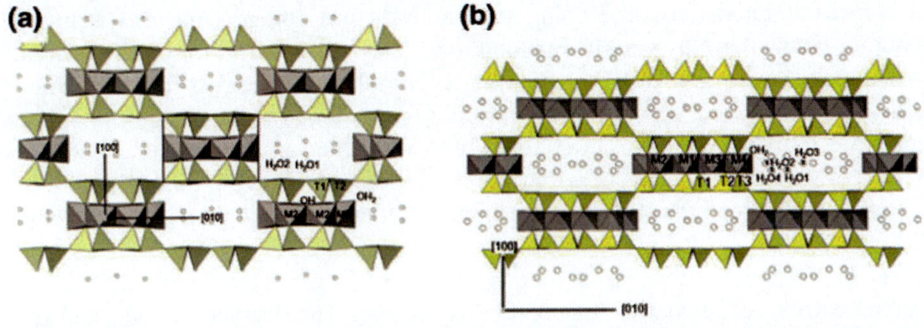

Fig. 11 Projection of the monoclinic palygorskite (**a**) and sepiolite (**b**) structures along the [001] direction [11]

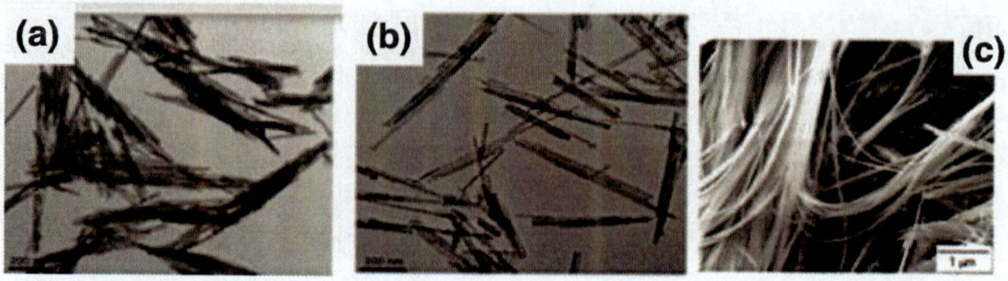

Fig. 12 Clay nano-rods and nanofibers [12, 14]

di and tri octahedral layers. As a result, crystallographic defects are observed in the octahedral sheets of natural palygorskite. The charge imbalance leads to high inter-layer negative charges that are compensated by exchangeable cations. The natural palygorskite microstructurally preferred nano-rods shape and the aggregated rod bundles (Fig. 12) [12]. However, sepiolite has a fibrous structure that has made it a potential candidate to be used for the preparation of eco-materials such as super-absorbent composite, environmental repair materials and flame-retardant materials. The palygorskite unique crystal structure, the layer stacking mode and nano-rod microstructure resulted in plentiful pores, higher aspect ratio, better ion exchange capacity (about 30–40 meq/100 g) in it. Palygorskite has excellent colloidal stabilization properties, adsorption, reinforcing properties, thermal and mechanical stability. And is explored as a potential candidate in these applications. In order to achieve the optimum properties and to extends its applications nano-structure processing is very important". Many physical methodology such as ball milling, extrusion, high speed shearing and ultra-sonification along with surface modifications are studied to obtain optimized nano-structure that can have better properties [12, 13]. However, such treatment will leads to reduce in aspect ratio of these nano-rods. However, sepiolite prefers nano-fibrous morphology (Fig. 12) [14].

Clay Nanotubes (halloysite and imogolite nanotubes): Naturally occurring nanotube-structured materials are predominantly imogolite and halloysite clay. "The outer and inner surfaces of clay-based nanotubes can be easily modified with organic groups, provides new chemical functionalities. These natural nanotubes have enormous research interest for developing nanocatalysts, gas storage, controlled delivery systems, nanowires, clay polymer nanocomposite agents etc. The structural formula of halloysite is $Al_2Si_2O_5(OH)_4{\bullet}nH_2O$; n is 2 for hydrated and 0 for dehydrated halloysite [15]. It is a 1:1 di-octahedral clay that belongs to kaolin group. It can retain the morphology even after losing the water, and only the lattice parameters slightly changed. The schematic representations of the crystalline structure of halloysite-(10 Å) and the structure of a single tubular halloysite particle is shown in (Fig. 13)".[16].

The halloysite clay consists of "one layer of corner-shared tetrahedral SiO_4 sheet stacked with an edge-shared octahedral $AlO_2(OH)_4$. The octahedral layer and the SiO_4 tetrahedral layer are linked by sharing of apical-oxygen of the SiO_4 tetrahedron. Hence, one surface of the octahedral unit layer consists entirely of the apical oxygen of the tetrahedral sheet and the other surface is composed of the OH groups of the

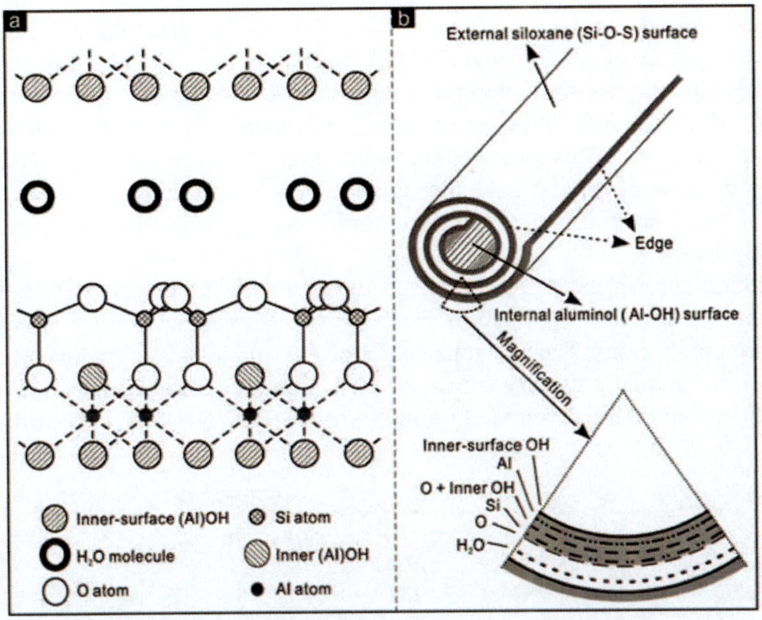

Fig. 13 Schematic of halloysite-(10 Å), $Al_2Si_2O_5(OH)_4{\bullet}2H_2O$ [16]

octahedral sheet. There is a lattice mismatch in between octahedral $AlO_2(OH)_4$ layer and SiO_4 tetrahedral layer in both a and b directions. For silica SiO_4 tetrahedra a is 5.02 Å, b is 9.16 Å, and for alumina octahedral $AlO_2(OH)_4$ layer a is 5.07 Å, b is 8.66 Å. This lattice mismatch between the octahedral and tetrahedral sheets and weak interlayer bonding resulted in clay layer rolling to nano-tube formation [17–19]. The halloysite layer rolling in the b direction is preferred over a direction. In halloysite nanotube (HNT), the outer tube surface is made up of siloxane (Si–O–Si) groups and the internal surface is made by a gibbsite-like array of Al-H groups".

When hydrated, n = 2 the mineral is called halloysite-10 Å due to its d_{001} spacing of 10 Å (Fig. 14) [20]. However, heating halloysite-10 Å to 120 °C leads to irreversible water loss (n = 0) [15, 20]. The interplanar d_{001} spacing reduced to 7 Å forming

Fig. 14 Halloysite 10 Å [20]

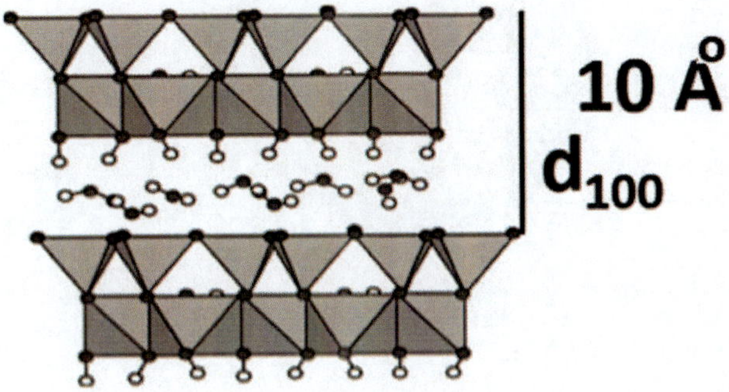

halloysite-7 Å. The structural differences can directly be observed by XRD study (the basal d spacing of 0.72 nm and 1.05 nm for halloysite-7 Å and 10 Å respectively [16]. The TEM images show that the halloysite particles have a cylindrical shape (Fig. 15). The cylindrical particles contain transparent central area longitudinally align to the tube axis. This indicates that the tubular particles are hollow and open-ended. In the case of dehydrated halloysite, the clay layers are rolled over and over to form a multi-layered nanotubed very similar to a multiwall carbon nanotubes structure.

Imogolite is also a naturally occurring aluminum silicate that is originated from volcanic materials such as pumice and volcanic ash. Microstructurally it is too prefer a nanotube morphology. The tube outside diameter is about 2–2.5 nm, inner diameter is about 1 nm and lengths vary from tens of nanometers to micrometers (Fig. 16). The estimated structural formula of imogolite is $Al_2SiO_3(OH)_4$ [21]. The tube walls

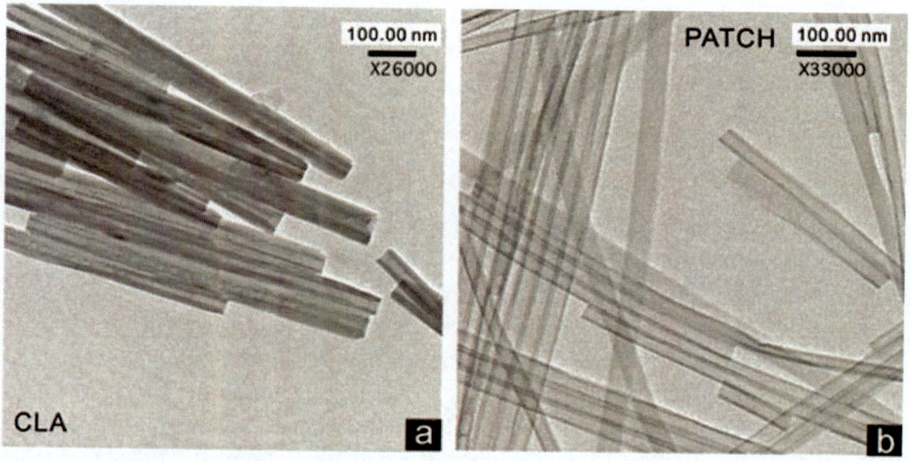

Fig. 15 TEM images of halloysite [16]

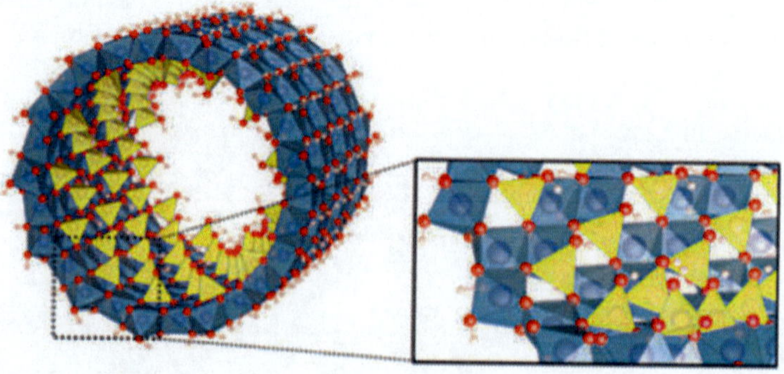

Fig. 16 Single-walled imogolite nanotube: aluminum (blue); silicon (yellow); oxygen (red); hydrogen (white).[22]

consist of a single continuous $Al(OH)_3$ (gibbsite) sheet with the inner hydroxyl group is replaced by oxygen. Three of these oxygen is commonly shared by Si–OH group through covalent bond forming the O_3SiOH tetrahedron units [21, 22]. Hence, both Al-octahedra and Si-tetrahedra are interconnected by sharing three oxygen atoms forming the 1:1 clay layer with free OH group at both outer and inner surface of the tube (Fig. 16) [22]. The imogolite structure is very different from halloysite structure. In halloysite, the curved octahedral $[Al(OH)_3]$ layer forms the internal surface of the nanotubes whereas the tube outer surface is composed of Si–O–Si groups. HRTEM shows tubes bundle forming characteristic patterns containing holes (Fig. 17) [21]. The imogolite tubes may now extend for microns.

Clay Nano sheets: Compared to one-dimensional nano-scale tubes/fibers, the two-dimensional layered structure is more predominant. The typical 1:1-type sheet-shaped clay minerals are kaoline and halloysites. Though halloysites tube/fiber microstructure is dominant, the kaolinites usually show planar morphologies (Fig. 18) [19]. In 1:1 clay, one octahedral and one tetrahedral layer make the clay laminates (T–O). In a 2:1 type clay, one octahedral and two tetrahedral layers make the clay laminates/platelets (T–O–T). Examples includes montmorillonite, bentonite, saponite, and hectorite clays [23]. "The thickness of single-layer platelets/laminates in the dry

Fig. 17 HRTEM of imogolite [21]

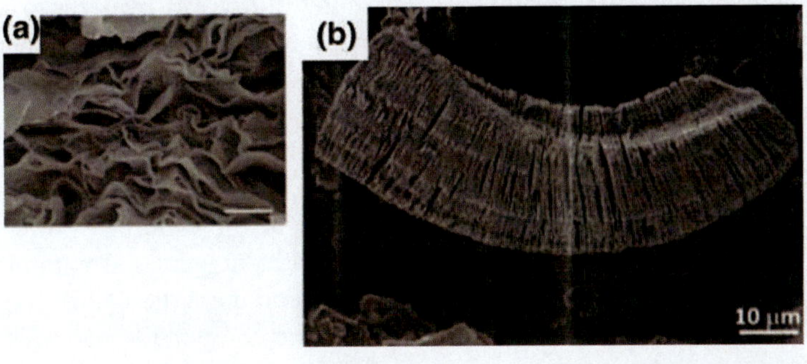

Fig. 18 **a** MMT Film, **b** Nanosilicate sheet [19, 25]

state is about 1 nm. Their lateral size can exceed several tens of nanometers. The primary unit of clay structures are of multiple alumino-silicate sheets with irregular polygonal shapes at average dimension of 100 nm × 100 nm × 1 nm for individual sheets [24]. This high aspect ratio drives the platelets to self assemble to a regular film like clay structure. SEM micrographs showed that a cross section of an MMT film has a wavy texture" (Fig. 18) [25].

3 Nanoclay–polymer Composites

Composite materials can be defined as a mixture of physically distinct two or more phases (micro-structurally) in such a way that the bulk properties are different and superior to those of the individual components [26]. In composite materials the reinforcement fibre/particles are embedded in a continuous phase called matrix. The reinforcement are hard materials and they can withstand maximum load and serve the desirable properties. The matrix binding the reinforcement phases in place and distributes the stresses among the constituent reinforcement materials. Another kind of layer composite is termed as laminates. In this laminates, reinforcement layers are bound together by alternate matrix layers.

In composites, Polymers make ideal matrix materials as they are easily available, lightweight and can easily be mould into desirable shape. The two different important class of polymers are; thermoset-plastics and thermo-plastics. Thermoset-plastics irreversibly become rigid upon heating. Thermoset plastics are in general soft-solid or liquid. Curing/heating increase the cross-linkage and develops a well-bonded three-dimensional molecular structure. The curing process is irreversible; it decomposes instead of melting. Thermoset plastics are; vulcanized rubber, fibre glass, polyster resin, polyurethane, melamine, Bakelite, silicon and epoxy resin. Thermoplastics are molten at elevated temperature but harden upon cooling. The materials can be reheated and reshaped repeatedly. Thermoplastic includes; polyethylene (PE), polypropylene (PP), polyvinyl chloride (PVC), polystyrene (PS) and polyester (PET).

One disadvantage of both kind of plastic is the low load bearing capacity (low elastic modulus G). Thermoplastics soften appreciably as they are heated. Their modulus decreases and they begin to creep (slowly deform over time) at higher temperatures. The effort is to overcome this limitation. Filling polymer with inorganic filler increases the modulus and increases their heat distortion temperature. It is commonly thought that adding fillers is to lower the overall materials cost of the composite. This is, however, rarely the case. Polyethylene (PE) and PP are the least expensive per unit volume compared to fillers [27]. Addition of any common filler increases the material cost of these polymers. However, the addition of filler changes nearly every property of the polymer with many advantages, and naturally, some disadvantages. The properties of the composite depend upon the volume percentage of filler added and are usually in between those of the component materials. Several of the properties such as density, modulus and yield strength can be predicted by the rule of mixtures (volume fraction). To make good composites one needs to know about polymers, engineering, fillers and surface science. That is why the study and application of composites so challenging, fascinating and rewarding.

3.1 *Bulk Clay Materials to Nano-clay Transition*

If we go by definition, even natural clays can be termed as nano-structured clay as the layer thickness is in nano scale range. In clay the different way of assembly of silica-oxygen tetrahedrons and aluminum or magnesium-oxygen octahedrons lead to the formation of a variety of nano layered structures. These nanosheets are stacked on top of each other like sheets of paper to form sandwiched structure with the interlayer cations [28]. However, the advantages of nano-structured materials have not been fully realized due to the larger size in their natural existence. This drives the research and innovation pertaining to nano-clay synthesis. If these layers can be separated in a system, the nanomaterial properties can be fully explored. Recently only a large number of research work has been focusing on techniques to separate out individual two dimensional layers of clay material called exfoliation. The individual layers in bulk clay materials are separated out by either inserting polymer (polymer–clay composite) or water molecule in the interlayer (swelling clay) [2].

The polymer/clay composites are broadly divided into three categories. First one is the conventional composites; clay materials are just mixed with polymer matrix. The clay aggregated particles lose their original aggregated states in polymer systems. The individual clay particles are called tactoids. Second one is intercalated nanocomposites; polymer molecules get into clay inter-layer and are sandwich in between clay platelets. The third one is exfoliated nanocomposites; individual 1 nm thick clay layers are separated in the continuous polymer/matrix [9]. These three different state clay particles state can be studied by XRD. As the basal spacing along c-axis increase the XRD peaks related to it will move to a lower theta value and completely disappear at exfoliation. Based on the above discussion, polymer-nano-clay system can be classified as tactoid, intercalated clay and exfoliated clay-composites (Fig. 19) [29, 30].

Disaggregation of Crystal Bundles or Aggregates: So, to start with we can visualize the clay nano-system as aggregated clay nano-particles. In aggregated form the nanoparticles (nanorods, nanofibers, nanotubes, nanosheets) are closely associated with may be through week interactions. The other forms of clay nano systems are; (1) tactoid; individual clay particles, (2) intercalated clay; polymer in between clay platelet and (3) exfoliation; platelet loose the regular order stacking. To achieve better mechanical properties of nanoclays system the aggregates need to be separated by an efficient method. Naturally occurred clay nano-rods or nano-fibers usually exist as bulk crystal bundles or aggregates. It is very difficult to disperse in either water or common organic solvents [28].

Tactoids: Some of the cases clay gallery does not expand primarily due to strong interaction between clay layers and its poor affinity to the polymers or solvents [31]. However, the particles aggregates are weakly bonded and can be separated with little

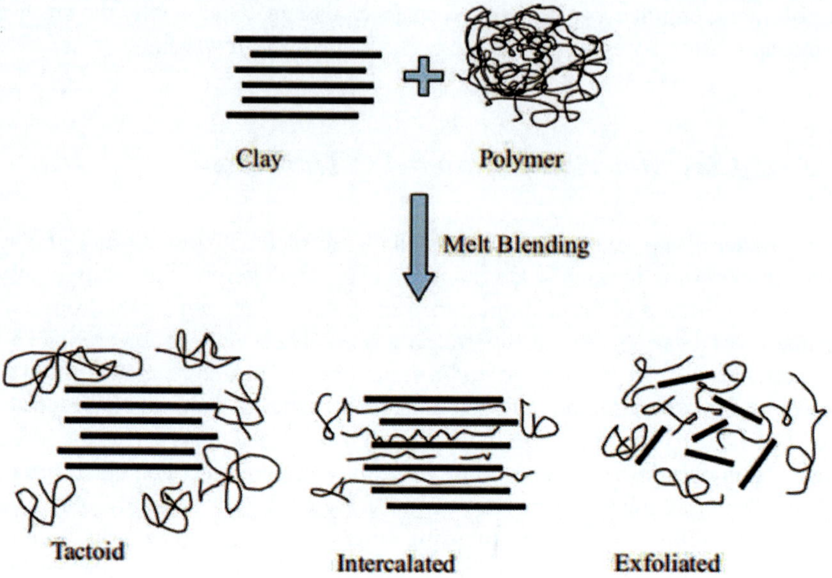

Fig. 19 Form of nano-clay system in a polymeric matrix [30]

force in the form of sear stress and sonification. Thus can be dispersed into well separated individual particles with less or no swelling. This leads to formation of tactoid structures; individual clay particles are dispersed in a matrix system. Nano-clay particulates are dispersed uniformly. The interplanar distance along c-axis (d_{100}) is defined as the basal plane spacing. If the polymer enter in between these inter-plane termed as intercalations. If it does not enter, d_{100} spacing of clay remains unchanged and the composite is tactoids. In this composite, 2–20 layers of clay (small tactoids) are well dispersed in the polymer matrix.

Intercalated Nano Composites: If the clay platelets are weakly bound and have high affinity to polymer, it is highly likely that the polymer chains penetrate into the interlayer region of the clay/tactoids. The interlayer spaces expand slightly, it allows polymer chains or solvent to penetrate in between the clay interlayers. The interlayer spacing increases but layer stacking is still intact. The formed structure is called intercalated structure (Fig. 20) [32, 33]. This is predominant if affinity between polymer and clay is moderate. This results in an ordered multiple layer structure with alternating polymer/inorganic layers at a repeated distance of a few nanometers and is termed as intercalated composites [34]. Due to intercalation, though the interplanar spacing d_{001} increases the clay laminate remains stacked. The change in inter layer d_{001} spacing in z-direction of clay can be monitored by x-ray diffraction (XRD) measurements. But XRD cannot quantify how much is the intercalated matter. Even XRD is not efficient on the identity of the guest species in the interlayers. If the intercalated clay layers are attached by edge hydroxyl group resulted in another class of nano-composite termed as flocculated nanocomposites. The 'flocculation' could also be due to long polymer chains that can be intercalate into two or more

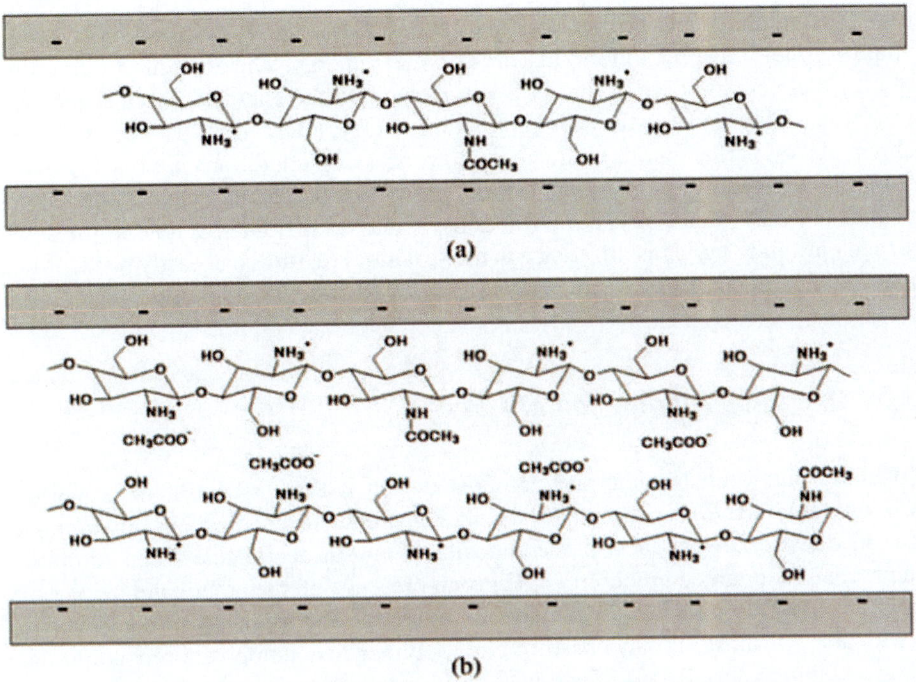

Fig. 20 Intercalated montmorillonite [33]

clay galleries. Here, the log chain polymer acts as a bridging agent to attach these clay gallery through their edges. In fact, most of the intercalated nanocomposites are combination of single stacks and flocculated nanostructure. The micro-structural difference directly can be observed using electron microscopic (TEM) analysis of composites.

Exfoliated Nano composites: In the case of exfoliated structures, clay layers are well separated and clay loss it layered stacking identity. Clay sheets are well separated and homogeneously dispersed within the continuous polymer phase (nano-composite) or may be in a solvent (swelling clay) [29, 31]. X-ray diffraction is a very useful technique to differentiate the exfoliation from the intercalation structure. In the case of exfoliation, the x-ray peak related to the inter-layer spacing that appears at the lowest 2θ value will disappear. Clay-polymer nano-composites when exfoliated is known to have the best properties. The optimal properties are due to the optimal interaction between clay platelets and polymers [34]. A composite for which basal spacing d_{001} is more than 10 nm is designated as exfoliation. Conventional XRD is not efficient to monitor this change and electron microscopy (TEM) is a useful tool to study this change.

There are two types of exfoliation namely ordered and disordered. In ordered exfoliation, the inter layer spacing is increased but they are still stacked together (intercalated, but interplanar spacing is 10 nm). In disordered structure the clay laminates are randomly oriented. Most often clay-polymer nano-composites are mixture of both intercalated and exfoliated clay structures. This type of mixed microstructure could be due to insufficient mixing time or shear energy. The amount of dispersion of clay platelets in polymer matrix is determined by the clay and polymer intrinsic properties. Clay aspect ratio, volume fraction of clay, interactions between polymer, clay and clay surface modifying agent and processing conditions are key factors to achieve the complete exfoliation. Generally fully exfoliated polymer–clay nanocomposites are only found in volume fractions of clay lower than 3% due to the small size of clay platelets. This difference in the structural features can easily be identified using TEM but cannot be distinguished by X-ray diffraction studies.

3.2 *Organic Modification of Clay*

To obtain the nanoclay-polymer composites and realize its optimum properties, mixing and interactions of polymers and clays at atomic level is needed. However, just mixing does not always form nanocomposites. Clay surface/platelets are hydrophilic in nature due to large number of −OH group presence on the other hand in majority polymers are hydrophobic. Hence, modification of either the clay or the polymer is necessary. All most all clays used in clay-polymer nanocomposite preparation are pre-modified materials. Modification starts with a purification and surface treatment steps.

Hydroxylation is conducted to develop more −OH group at clay surface at basic medium [35]. Alkali activation can moderately break the Si–O–M and Si–O–Si bonds and remove the metal cations (i.e., Al^{3+}, Mg^{2+} and Si^{4+} from the phyllosilicate backbone. Hence, new adsorption sites are created and the surface negative charges are increased. In contrast, acid treatment leads to an increase in the surface area of clay [36]. However, dissolution of the inner M-OH catalyzed by acid or basic treatments may alter the structure of clay.

To enhance clay nanostructure organo-philicity, organic modification is very important. Cation exchange using quaternary ammonium salt of organic compound is the most commonly implemented method for organo-philic modification [37]. As organic cation size is greater than that of interlayer cations present, exchange increases the basal distance. This facilitates the intercalation of inorganic solvents/monomer/polymer melt in the interlayer space. Intercalated montmorillonite (MMT) with organo-quaternary ammonium salts with terminal carboxylic acid groups increases the clay nylon-6 polymer interaction and provide intercalated clay-polymer nanocomposites [38]. However, it is important to control the amount of cation exchange to have optimum organo-philicity to avoid steric hindrance that would prevent the intercalation of organic molecules. There are three different way for organic modifications [37]; (a) solid–solid process, involves grinding of clay powders and surfactant followed by heat treatment to ensure the diffusion of surfactant molecules in the interlayer space, (b) liquid–solid process, the clay solid is mixed with surfactant in a solvent with constant stirring and (c) liquid–liquid process, the clay and surfactant both are dispersed in a common solvent. The liquid- liquid

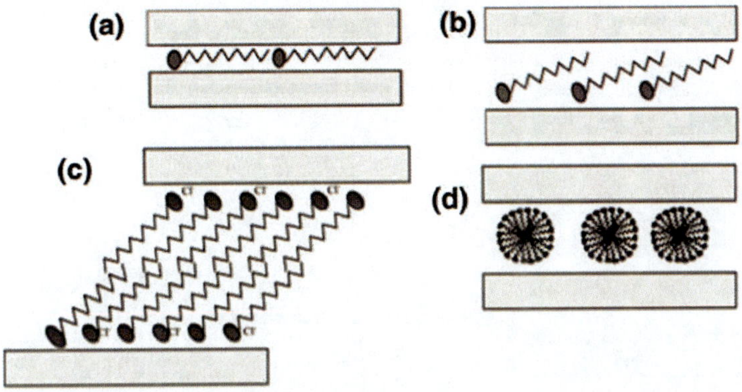

Fig. 21 Conformation of alkyl-ammonium salts in clay inter-layers; **a** mono-layer parallel to clay sheet, **b** bi-layer parallel to clay sheet, **c** paraffin structures, and **d** micelles [42]

process is much more effective than the solid–solid mixing in organo-clay modifications. Other interesting modifications methods of clay includes; organo-silane [39], organo-phenyl iodonium [40], and diazonium salt grafting [41].

The arrangement of the alkyl ammonium salts in interlayer spaces depends on the alkyl ammonium initial concentration, alkyl chain length and cation exchange capacity (CEC). The arrangement can be; (a) mono-layer parallel to clay sheet, (b) bi-layer parallel to clay sheet, (c) paraffin structures, and (d) micelles as shown in (Fig. 21) [42]. Organo-silane modification is the creation of covalent bonds between organo-silane and the clay matrix. Most often, after hydrolysis the silanes become silanols and react with clay hydroxyl groups to form siloxane bonds [43]. The most accessible hydroxyl groups of clay are located on sheet edges resulted from broken links of alumino-silicate crystal. Organo-silane attached to clay surface through ether linkage by binding to these –OH groups. The three different possible grafting/intercalation of organo-silane is illustrated in (Fig. 22) [43]. It is possible to achieve new interfaces between the filler and the matrix by selecting the proper chemical structure of organo-silane. The organo-silane with a functional group will react with polymer or with its monomers. A covalent coupling between clay and polymer will take place. Organo-silane with a carbon chain without a reactive functional group will form van der Waals bonds with the polymer matrix in the nanocomposites. Diazonium salt is initiator for many polymerization reactions". It is anticipated that clay-polymer nanocomposites can be prepared by in-situ radical and photo polymerization methods using the anchored diazonium salt [44, 45]. The incorporation of diazonium salt in clay layers is through grafting and/or intercalations (Fig. 22) [44].

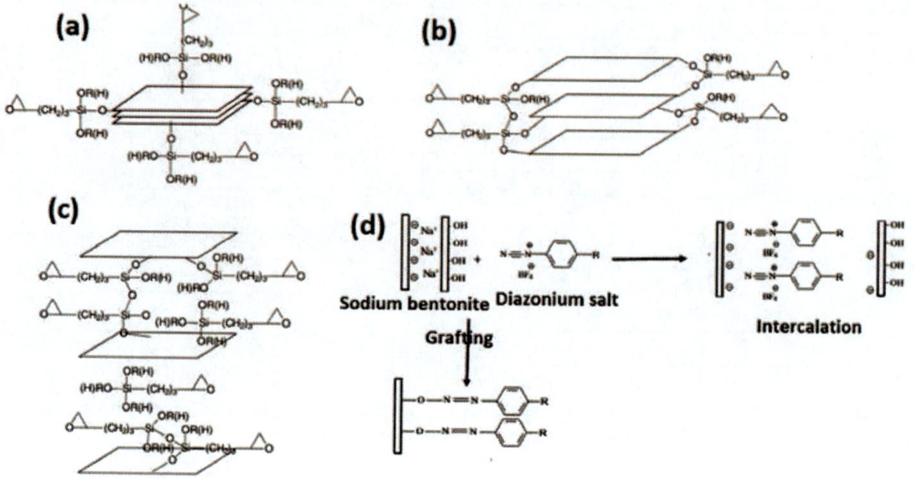

Fig. 22 Intercalated/grafted organosilane in clay layers (**a**–**c**) and diazonium salts (**d**) [43, 44]

3.3 *Nanoclay-Polymer Composites Synthesis*

Broadly fabrication of clay/polymer nanocomposite can be visualized as dispersion of clay nano-particles in a polymer. It will leads to either just dispersion of clay in polymer matrix or intercalation of polymer in between clay layers or exfoliation of clay layer. But in real situation it can be a combination effects depending on the clay-polymer properties and processing conditions. However, the selection of synthesis strategies is decided by the nature and compatibility of the clay and polymer to be used. If the clay layer interfaces are compatible with the polymer intercalated clay-polymer composites can be synthesized without the surface pretreatment. In real situation, clay surfaces and interfaces are hydrophilic because of Si–OH group. However, the polymers are predominantly hydrophobic in nature due to long chain hydrocarbons. Hence, both are not compatible as such. Pretreatment of either the clays or the polymers or both is necessary to achieve a uniform dispersion, strong bonding and optimum composites with high mechanical properties. The most common clay-polymer composites synthesis techniques are; (a) melt blending, (b) solution blending, (c) in situ polymerization, (d) electro-spinning; and (e) layer by layer (LBL) assembly/lamination. Depending on the clay/polymer properties and to develop the desired microstructures with desired properties either one or combination of multiple techniques are used in synthesis.

In situ polymerization: In insitu polymerization, the monomer is first introduced between clay platelets by homogeneous mixing. Mixing leads to monomer intercalation or swelling of the clay in the monomer solution. Followed by polymerizing the monomers in between the particles. The polymerization is carried out by the heat, electromagnetic radiation or a chemical activation [46]. The other polymerization process such as radical polymerization, photo-polymerization, click coupling chemistry, ring opening polymerization and emulsion polymerization methods are also used for clay-polymer composites synthesis [46]. Controlled radical polymerization has an advantage over other process in minimizing the side reactions and resulted in uniform molecular weight polymers.

This technique was first used by Toyota research group for the preparation of Nylon-6 polymer/clay nanocomposite from caprolactam monomer [47]. Exfoliated nanocomposite of nylon-6 and montmorillonite was obtained from the polymerization of caprolactam in the interlayer space of montmorillonite. This method is suitable for the preparation of clay-composites based on mainly thermoset polymer such as epoxies and styrene. The degree of dispersion and exfoliation is controlled by the clay and monomer chemistry. The interlayer connectivity and affinity between monomer and clay laminates polarities determine the diffusion rate and equilibrium concentration of monomer within the clay galleries [48]. Advantages attributed to in situ polymerization are; (1) clay-polymer composites based on both thermoplastic and thermoset plastic can be synthesized, (2) surface grafting of polymers on clay laminates; surface grafting improves the clay polymer interactions and hence the composites properties, (3) enhance clay dispersibility and (4) less stringent thermal process requirements.

In situ polymerization is "the most suitable preparation method for polyolefin-clay nanocomposites because of the freedom in selecting broad processing temperature and solvent [49]. An improved method for producing nylon 6/clay nanocomposites using an in situ polymerization process that effectively exfoliates the clay layers is illustrated in (Fig. 23) [50]. Amino-lauric acid (ω-amino acid) is first added to aqueous hydrochloric acid. The amine groups are attached to H^+ ions. This protonated amino-lauric acid is positively charged and can be exchanged with interlayer sodium that facilitates the polymer intercalation in sodium montmorillonite. These intercalated amino acids have terminal free carboxyl groups. Under appropriate reaction conditions, these carboxyl groups will initiate ring-opening polymerization of caprolactam. Thus, resulted in nylon 6 polymer chains that are ionically bonded to the clay platelets. The length of these polymers chain is driven by the free energy of polymerization". The polymer growth forces the platelets apart until complete exfoliation is attained.

Solution intercalation methods: This method is suitable if clay is expandable or swell-able in the solvent. In this method a solvent is selected in such a way that clay can be dispersed in it. The polymer or pre-polymer is dissolved in the solvent. Solvent penetrates the interlayer making the clay interlayer expand and makes passage for the polymer chains. The polymer chains are absorbed onto the delaminated sheets. The solvent can be removed by vaporization or precipitation leaving the polymer chain in between the clay laminates resulted in intercalated nanocomposite formation. Solvents such as water, acetone, chloroform, acetonitrile and toluene can be used for such process. The solvent evaporation allows the polymer chains to diffuse between the layers and trapped in between as a sandwich structure. It may forms intercalated, exfoliated or combination of both depending on the nature and extend of the interactions between polymer and clay sheets.

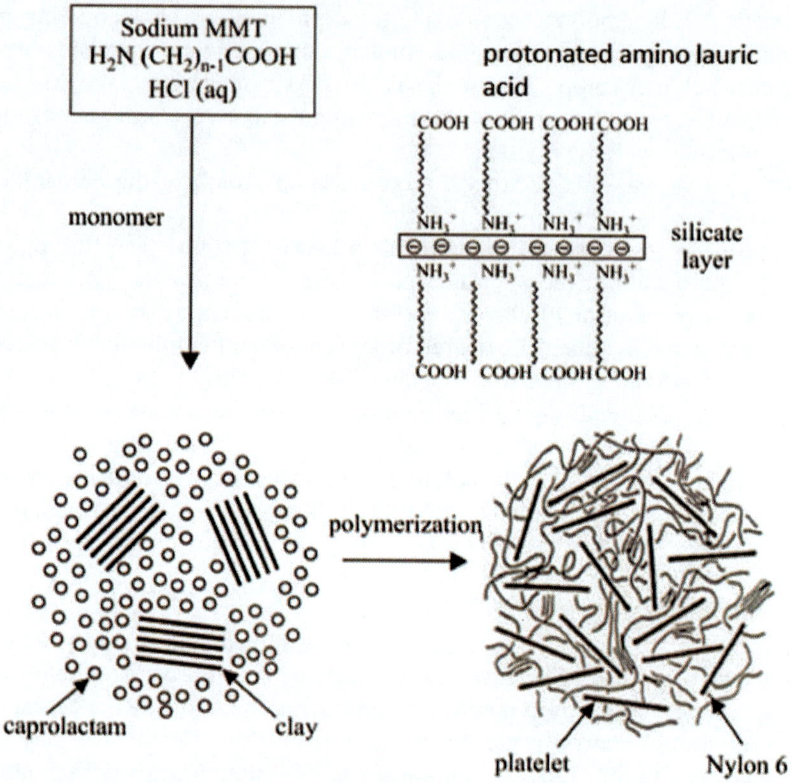

Fig. 23 In situ formation of nylon-6/MMT polymer [50]

The process consisted of three stages; (1) clay dispersion in polymer solution, (2) controlled removal of the solvent and (3) composite film casting [51]. The clay and polymer dispersion is accelerated with energetic agitation such as stirring, reflux, shear mixing and ultra-sonification. Ultra-sonification is an efficient process for the dispersion of nano clays in a polymer solution. Ultrasound breaks up nano clay clusters through cavitation in the solution and/or exciting resonance vibrations of the clusters. However, this process is less applicable in industry as it requires large volume of solvent.

Melt blending methods: The process is also termed as melt-intercalation. This method is highly effective for thermoplastic (polypropylene, polycarbonate, Nylon-6) nanocomposites preparation. Clays are organically modified and polymer chains are surface modified to enhance compatibility that promotes the exfoliation. In this process in general the polymer is melted and combined with the desired amount of the intercalated clay. The melt-blending process is two types static and dynamic. In static melt blending; the clay-polymer mixture is heated in vacuum at a temperature approximately 50 °C above polymer glass transition temperatures [51]. In dynamic process; the composite is prepared using a melt mixer (extruder) in the presence of an inert gas such as argon or nitrogen gas. In this method polymer mixed with clay filler is heated with constant shear stress sufficient to pull apart clay-platelets". The process is termed as extrusion process. In this extrusion process, by changing the extruder screw configuration better control over shear force is achieved and hence resulted in better clay-polymer mixing. Higher shear rates tends to provide better dispersion. Processing conditions, surface modification and compatibility of clay and polymer matrix all are very important in attaining the optimum dispersion with optimum properties. Extrusion is preferable and popular methods among industries to prepare clay-polymer composites due to flexibility in process parameters and versatility.

Melt intercalation is considered environmentally friendly as no solvent is involved in this process. It is preferred for preparing clay/polymer nanocomposites of thermoplastics matrix. However, to attain a homogeneous dispersion of nanoparticles in the polymer matrix is still challenging. Melt-blending method often results in incomplete filler dispersion. This leads to clay particle aggregations at higher clay loading. In melt blending, mixing process needs to be optimized for optimal microstructure and properties. The process optimization includes varying temperature, screw-speed, residence time and shear stress. Again, the degree of dispersion, intercalation and exfoliation is controlled by; (1) clay agglomerate structure, packing density, aspect ratio, purity, (2) polymer matrix melt viscosity and (3) shear force. However, the shear forces generated in most extruders are not sufficient enough to delaminate and disperse the clay in the polymer matrix uniformly.

An extruder is consisted of a materials feed system (hopper), melting or plasticizing system, a screw that simultaneously mix the molten materials and push through a die opening (nozzle head). Extrusion is the process in which a molten material is pressurize through a nozzle to produce an long product of constant cross section [52]. Single screw extruders are the most commonly used in the polymer industry due to the straight forward design, low cost and it reliability. Figure 24, shows schematic of a single screw extruder [52]. Polymer -clay mixture can be fed through the hopper. The materials than conveyed forward by the screw. Heating elements over the barrel soften or melt the polymer. In this case, the screw has dual role; mixing the polymer

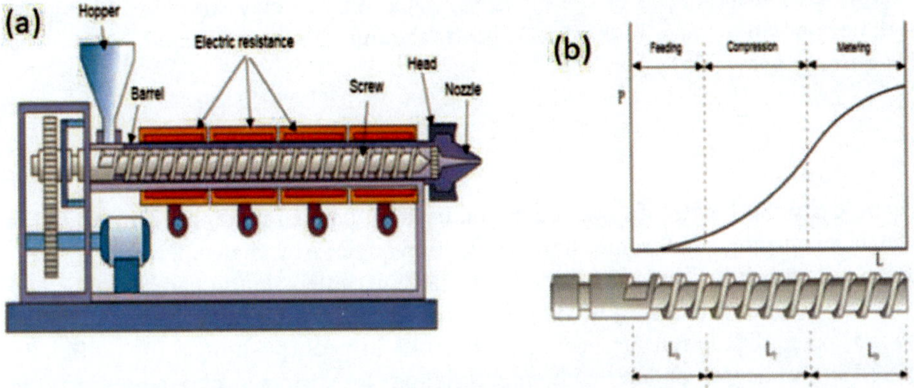

Fig. 24 **a** Schematic of an extruder and **b** Pressure zone in extruder [52]

mixture and push forward the molten polymer or polymer mixture through the nozzle. Depending on the pressure exerted by the screw and barrel system three zones are identified as indicated in Fig. 24b [51]. (1) The feeding zone (close to the hopper opening); polymer is mixed with the reinforcement and gradually compacted at a certain speed. (2) The transition zone; the polymer mix will be in molten state and the fusion takes place. The trapped air escapes by means of the feed hopper. (3) Compression section; the molten material is homogenized and pressurized to exit the extruder through the nozzle.

Ultrasound assisted extrusion is a new development. Use of ultrasound energy during extrusion process improves the additives dispersion in polymer melt. The use of ultrasound during extrusion process disturbs the flow pattern of molten polymer mixture. This agitation leads to increase in the movement of the molecular chains, so the dispersion of additives in polymer matrix is improved. Intercalation under ultrasound is higher for low feed rate [53]. The application of ultrasound leads to 33% increase in flexural modulus of poly-lactic acid-sepiolite composites [54]. Ultrasound disaggregated the sepiolite agglomerates, resulting in a better dispersion of the filler for a determined concentration of clays.

4 Structural Characterization of Polymer/clay Nanocomposites

In polymer/clay nanocomposites properties are direct results of the microstructures. It is important to know "the degree of intercalation/exfoliation and its effect on the nanocomposite properties. Hence, the micro-structural analysis of the clay, modified clay and the prepared nanocomposite is important. Common techniques such as X-ray diffraction (XRD) analysis, scanning electron microscopy (SEM) and transmission electron microscopy (TEM) are widely used to characterize the micro structure. Clays due to the layered structure shows a characteristic peak in XRD analysis. The peak position (2θ) is indicative of the interlayer separation in clay structure. Using the peak peak position value in the XRD spectra the inter layer space can be calculated using Bragg's law (Eq. 1).

$$n\lambda = 2d \sin\theta \tag{1}$$

where λ is wave length of X-ray radiation used in the diffraction experiments, d is inter-layer spacing and θ is measured diffraction angle. Any change in the d-spacing of a clay layers by organic modification or polymer intercalation can be confirmed by XRD spectra. Increasing of d-spacing results in shifting of related XRD peak toward lower diffraction angles (2θ). The shift in peak position (2θ), is directly related to the degree of intercalation/exfoliation. In exfoliated clay composite the XRD peak related to the interlayer spacing will disappear completely. For example Fig. 25 shows the XRD patterns of polystyrene-hectorite nanocomposites [30]. It

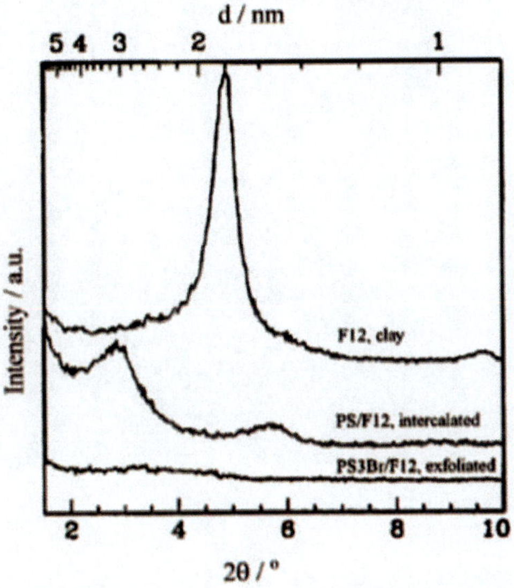

Fig. 25 XRD patterns of polystyrene/fluoro-hectorite nano-composites [30]

indicates the increasing of layers spacing due to the intercalation or exfoliation. The (001) peaks for hectorite clay is at $2\theta = 5.5°$, corresponding $d_{(001)}$ is 1.6 nm". As the polymer intercalated in clay layer $d_{(001)}$ is increased to 2.9 nm due to increase in layer gap. In exfoliation the $d_{(001)}$ XRD peak is completely lost [30].

Transmission electron microscopy (TEM) technique is also an essential tools for evaluation of the nanocomposite structure and complementary to XRD observation. It shows the structural features of polymer/clay nanocomposite. One can visually observe the nanostructure of nanocomposites and clay d-spacing. In the TEM micrographs of clay-polymer composites the darker lines represent the clay laminates. The clay laminates are made of heavier elements such as Al, Si and O generates darker contrast in TEM bright field imaging. The polymer matrix or the polymer present in inter layer spacing of clay is composed of lighter atoms such as C, H, N and Na. Hence these regions are identified by the bright contrast. Therefore, in a TEM bright field image of a clay-polymer composite; the spacing between darker line shows the d-spacing and dispersion status. The overall structure of the nanocomposite including intercalation, exfoliation, dispersion and defects of clay layer can be conclusively obtained using TEM technique. For example Fig. 25 shows the TEM micrograph of montmorilonite (MMT)-epoxy nanocomposites [55]. The montmorilonite (MMT) interlayer increased to 5.8 nm an indicative of partial exfoliation. The interlayer distance increased to 10–25 nm in fully exfoliated structure (Fig. 26).

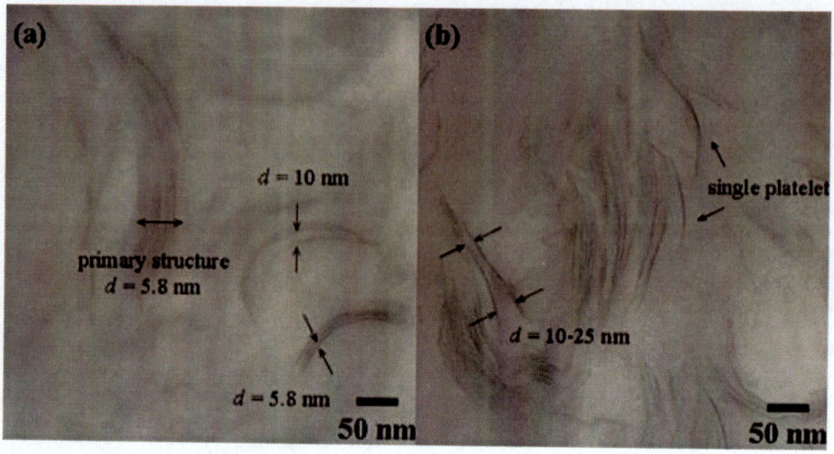

Fig. 26 TEM micrograph of organically modified montmorilonite (MMT)-epoxy nanocomposites [55, 56]

5 Properties and Applications of Nanocomposites

The polymer–clay nanocomposites research is driven by the properties improvements in: mechanical, barrier, flame retardant, electrical, and biodegradable. Because of the low price, availability, high aspect ratio as well as desirable nanostructure and interfacial interactions, clays is highly desirable as filler in polymers. The addition of clay-nanostructures to the polymers aims to produce the polymer/clay nanocomposites with desired properties for desired applications. It can provide improved tunable properties at very lower percentage of loadings.

5.1 *Mechanical Properties*

The mechanical properties enhancements are summarized in Table 1. Additions of clay (1–5 wt%) increased the tensile strength, modulus of many polymers. The important aspect in the application of fillers to the polymer is improvement in the mechanical properties and hence fillers are called as reinforcement agents. The reinforcement is due to the higher resistance of rigid filler against applied load. In general, addition of clay to polymers resulted in improved and new unexpected properties. As a result, the clay-nanocomposites have wide applications in automobile components, packaging materials, construction materials, flame retardants, protective films etc. [30].

Nanoclay-polymer composites achieved high tensile strength, tensile modulus and shear strength compared to polymer only with little or no loss in impact resistance [30]. Improvements in mechanical properties are achieved without loss of optical clarity by 1–5% clay addition to the polymers. The reinforcing ability of clay fillers

Table 1 Mechanical properties of polymer–clay nano-composites

Polymer matrix	Clay-percentage loading (wt%)	Mechanical properties comparison		References
		Virgin polymer	Nano-composites	
PP	2.1 (montmorillonite-MMT)	Modulus-429 MPa Tensile strength-21.1 MPa	Modulus-578 MPa Tensile strength-23 MPa	[57]
	5.3 (montmorillonite-MMT)		Modulus-797 MPa Yield strength-25 MPa	
PE	3.5 (montmorillonite-MMT)	Modulus-102 MPa Yield strength-7.3 MPa Yield strain-7.1%	Modulus-140 MPa Yield strength-9.4 MPa Yield strain-6.8%	[57]
	5.4 (montmorillonite-MMT)		Modulus-180 MPa Yield strength-10.3 MPa Yield strain-5.8%	
PET	1 (Cloisite-15A) (modified-MMT)	Young's Modulus-19 MPa Tensile strength-40 MPa	Young's Modulus-24 MPa Tensile strength-45 MPa	[58]
	5 (Cloisite-15A) (modified-MMT)		Young's Modulus-26 MPa Tensile strength-41 MPa	
Epoxy resin	3 (montmorillonite-MMT)	Tensile strength-15 MPa Rockwell hardness-32	Tensile strength-20 MPa Rockwell hardness-40	[59]
	5 (montmorillonite-MMT)		Tensile strength-17 MPa Rockwell hardness-32	
Polyvinyl alcohol (PVA)	2 (Halloysite-nanotubes)	Young's Modulus-5.5 MPa Tensile strength-115 MPa	Young's Modulus-5.8 MPa Tensile strength-112 MPa	[60]
	10 (Halloysite-nanotubes)		Young's Modulus-6 MPa Tensile strength-108 MPa	
Chitosan	2 (Halloysite-nanotubes)	Young's Modulus-750 MPa Tensile stress-20 MPa	Young's Modulus-900 MPa Tensile stress-30 MPa	[61]

(continued)

Table 1 (continued)

Polymer matrix	Clay-percentage loading (wt%)	Mechanical properties comparison		References
		Virgin polymer	Nano-composites	
	7.5 (Halloysite-nanotubes)		Young's Modulus-1240 MPa Tensile stress-54 MPa	
Polyimide	1 (modified-smectite)	Storage modulus-2006 MPa	Storage modulus-2216 MPa	[62]
	5 (modified-smectite)		Storage modulus-2852 MPa	

is controlled by particle size, structure and surface properties. Smaller clay particles impart greater reinforcement, increase elastic modulus and tensile strength. Smaller in size larger is the surface area. Hence, larger is the contact area with polymer matrix that leads to increase in reinforcement [5]. Specific surface area of clay particle is proportional to the shape/geometry of the clay. Spherical particles of 1 μm have surface area of 6 m^2/g. However, nano-clay platelets have the highest surface area in order of 300–450 m^2/g which makes it more effective for reinforcement [9]. At exfoliation the clay platelet surface area can be up to 1200 m^2/g. In polymer–clay composites the clay filler will carry the maximum portion of applied load, if the interfacial interactions between filler and matrix are adequate. However, it is widely recognized that the excellent properties of nano-composites have yet to be achieved. This is particularly true at higher volume fraction.

5.2 *Thermal Properties and Flame Retardancy*

Thermo-gravimetric analysis (TGA) is frequently used to analyze thermal stability of a polymeric material. The weight loss is monitored as a function of "temperature after formation of volatile materials at high temperatures. Clay minerals are almost stable in the temperature ranges that organic polymers degrade in to volatile compounds. Therefore, in TGA experiments the clay content of nanocomposites is remain as residue. Generally, presence of the clay in the polymer matrix is considered to improve the thermal stability. Clay layer is better insulator and mass transport barrier to the volatile compounds released during degradation. TGA result showed that PMMA clay nanocomposite had 40–50 °C higher decomposition temperature than the virgin PMMA". The increment in decomposition temperature is attributed to restricted thermal motion of PMMA in the gallery [63]. If heated in nitrogen or even in air, polystyrene (PS) -clay nanocomposites degrades 30–40 °C higher than the pure PS degradation temperature [64].

The flame/fire retardancy of polymers is substantially important in many applications. In past years halogen based/phosphorus based flame retardants additives were frequently used in polymers. These compounds improve the flame retardancy of polymers without changing the polymer properties. However, the halogenated compounds release toxic gases and have pushed the market trends to halogen-free flame retardants. Studies showed that clay nano-structures show some degree of flame retardancy along with the mechanical properties improvements. In the case clays appear to help nanocomposite to form layered carbonaceous char and this char acts as a barrier for O_2 as well as combustion products generated during decomposition to prevent further burning. However, it is observed that clay materials are beneficial for retarding the flame spread but no improvement in the ignition retardancy [65, 66]. Therefore, the clay minerals are used together with a low fraction of conventional flame retardants. The clay minerals and halogen/phosphorous based flame retardants have considerable synergistic effect in the reduction of ignitability of polymers.

5.3 Gas Barrier Properties

Polymer based packaging materials have low barrier properties compare to glass or metal packaging materials. Polymer packages are permeable to small molecules like gases, water vapor and low molecular weight compounds like aromas, flavors and additives present in food. The degree of permeability is different for different polymers and different molecules. Clays are considered to enhance the barrier properties. Clay sheets are naturally impermeable to all these molecules. When present in polymer (polymer–clay composites), they form a maze like structure or create a tortuous path. Thus, presence of clay laminates creates a longer pathway in the polymer matrix for diffusing gases thus reduces the permeability. The barrier properties enhancement depends on the degree of exfoliation of clay into clay-laminates in polymer matrix. This creates a impermeable laminates network in the gas molecules diffusion path. Increasing lateral length of clay sheet, degree of exfoliation and degree of dispersion increases barrier properties of the polymer matrix.

6 Conclusions

Polymer–nanoclay composites are attractive due to novel materials properties than the only polymeric materials. Clay is preferred as nano-fillers due to easy availability, low cost and less toxicity. Adding clay nano materials to polymer matrix resulted in tactoids, intercalations and exfoliations depending on the processing conditions. Polymer–clay composites obtain better mechanical properties, fire retardant and gas barrier properties. They are mainly dependent on the type of nanoclay, modification and synthesis approach. In general, melt blending is considered to be an industrially viable, efficient and environmental friendly process. The in-situ polymerization technique has more control on the clay intercalation, dispersion of nanoclays in the polymer matrix.

References

1. Barton, C.D.: Clay minerals. In: Rattan Lal (ed.) Encyclopedia of Soil Science, pp 187–192. Marcel Dekker Publication, New York (2002)
2. Aboudi Mana, S.C., Hanafiah, M.M., Chowdhury, A.J.K.: Environmental characteristics of clay and clay-based minerals. Geol. Ecol. Landscapes **1**(3), 155–161 (2017)
3. Guggenheim, S., Martin, R.T.: Definition of clay and clay mineral: joint report of the AIPEA nomenclature and CMS nomenclature committees. Clays Clay Miner. **43**(2), 255–256 (1995)
4. William, F.B. (ed.): Clay mineralogy and clay chemistry. In: Soil and Environmental Chemistry, pp. 85–116. Elsevier (2012)
5. Martin, R.T., Bailey, S.W., Eberl, D.D., Fanning, D.S., Guggenheim, S., Kodama, H., Pevear, D.R., Środoń, J., Wicks, F.J.: Report of the clay minerals society nomenclature committee: revised classification of clay materials. Clays Clay Miner. **39**, 333–335 (1991)
6. Tournassat, C., Steefel, C.I., Bourg, I.C., Bergaya, F.: Surface properties of clay minerals. In: Tournassat, C., Steefel, C.I., Bourg, I.C., Bergaya F (eds.) Developments in Clay Science, pp. 5–31. Elsevier (2015)
7. Schulze, D.G.: Clay minerals. In: Hillel, D (ed.) Encyclopedia of Soils in the Environment, vol. 1, pp. 246–254. Elsevier/Academic Press, Boston
8. Nanomaterials definition matters. Nat. Nanotechnol. **14**(3), 193–193 (2019)
9. Yaya, A., Agyei-Tuffour, B., Dodoo-Arhin, D., Nyankson, E., Annan, E., Konadu, D.S., Sinayobye, E., Baryeh, E.A., Ewels, C.P.: Layered nanomaterials-a review. Global J. Eng. Des. Tech. **1**, 32–41 (2012)
10. Mousavi, S.M., Hashemi, S.A., Salahi, S., Hosseini, M., Ali, A.M., Babapoor, A.: Development of clay nanoparticles toward bio and medical applications. In: Zoveidavianpoor, M. (ed.) Current topics in the utilization of clay in industrial and medical applications, p 807. Intech Open (2018)
11. Guggenheim, S., Krekeler, M.P.S.: The structures and microtextures of the palygorskite–sepiolite group minerals. In: Galàn, E., and Singer, A. (eds.) Developments in Clay Science, pp. 3–32. Elsevier (2011)
12. Wang, W., Wang, A.: Recent progress in dispersion of palygorskite crystal bundles for nanocomposites. Appl. Clay Sci. **119**, 18–30 (2016)
13. Xu, J., Wang, W., Wang, A.: A novel approach for dispersion palygorskite aggregates into nanorods via adding freezing process into extrusion and homogenization treatment. Powder Technol. **249**, 157–162 (2013)
14. De Lima, J. A., Camilo, F. F., Faez, R. and Cruz S. A., A new approch to sepiolite dispersion by treatment with ionic liquids, Applied Clay Science, Elsevier, 2017, 234–240.
15. Massaro, M., Lazzara, G., Milioto, S., Noto, R., Riela, S.: Covalently modified halloysite clay nanotubes: synthesis, properties, biological and medical applications. J. Mater. Chem. B **5**(16), 2867–2882 (2017)
16. Yuan, P., Southon, P.D., Liu, Z., Green, M.E.R., Hook, J.M., Antill, S.J., Kepert, C.J.: Functionalization of halloysite clay nanotubes by grafting with γ-Aminopropyltriethoxysilane. J. Phys. Chem. C **112**(40), 15742–15751 (2008)
17. Lvov, Y., Abdullayev, E.: Functional polymer–clay nanotube composites with sustained release of chemical agents. Prog. Polym. Sci. **38**(10), 1690–1719 (2013)
18. Pasbakhsh, P., Churchman, G.J., Keeling, J.L.: Characterisation of properties of various halloysites relevant to their use as nanotubes and microfibre fillers. Appl. Clay Sci. **74**, 47–57 (2013)

19. Bauluz, B.: Halloysite and kaolinite two clay minerals with geological and technological importance. Rev. Real Academia De Ciencias Zaragoza **70**, 1–33 (2015)
20. Joussein, E., Petit, S., Churchman, J., Theng, B., Righi, D., Delvaux, B.: Halloysite clay minerals—a review. Clay Miner. **40**(4), 383–426 (2005)
21. Bursill, L.A., Peng, J.L., Bourgeois, L.N.: Imogolite: An aluminosilicate nanotube material. Philos. Mag. A **80**(1), 105–117 (2000)
22. Paineau, E.: Imogolite nanotubes: a flexible nanoplatform with multipurpose applications. Appl. Sci. **8**(10), 1921 (2018)
23. Starodoubtsev, S.G., Lavrentyeva, E.K., Khokhlov, A.R., Allegra, G., Famulari, A., Meille, S.V.: Mechanism of smectic arrangement of montmorillonite and bentonite clay platelets incorporated in gels of Poly(Acrylamide) induced by the interaction with cationic surfactants. Langmuir **22**(1), 369–374 (2006)
24. Usuki, A., Hasegawa, N., Kadoura, H., Okamoto, T.: Three-dimensional observation of structure and morphology in Nylon-6/Clay nanocomposite. Nano Lett. **1**(5), 271–272 (2001)
25. Wang, Y.C., Huang, T.K., Tung, S.H., Wu, T.M., Lin, J.J.: Self-assembled clay films with a platelet–void multilayered nanostructure and flame-blocking properties. Sci. Rep. **3**, 2621 (2013)
26. Hull, D., Clyne, T.W.: An introduction to composite materials. Cambridge Solid State Science Series. Cambridge University Press (1996)
27. DeArmitt, C., Hancock, M.: Particulate-filled polymer composites, Rothon, R.N. (ed.) pp. 357–424 (2003)
28. Wang, W., Wang, A.: Nanoscale clay minerals for functional ecomaterials: fabrication, applications, and future trends. (2018)
29. Arora, A., Padua, G.W.: Nanocomposites in food packaging. J. Food Sci. **75**(1), R43–R49 (2010)
30. Chen, B., Evans, J.R.G., Greenwell, H.C., Boulet, P., Coveney, P.V., Bowden, A.A., Whiting, A.: A critical appraisal of polymer–clay nanocomposites. Chem. Soc. Rev. **37**(3), 568–594 (2008)
31. Ray, S.S., Okamoto, M.: Polymer/layered silicate nanocomposites: a review from preparation to processing. Prog. Polym. Sci. **28**(11), 1539–1641 (2003)
32. Cui, Y., Kumar, S., Konac, B.R., Houckec, D.V.: Gas barrier properties of polymer/clay nanocomposites. RSC Adv. **5**(78), 63669–63690 (2015)
33. Monvisade, P., Siriphannon, P.: Chitosan intercalated montmorillonite: Preparation, characterization and cationic dye adsorption. Appl. Clay Sci. **42**(3), 427–431 (2009)
34. Sunil, B.H., Pushpalatha, M., Basavaprasad, V.M., Huvanna, T.P.: Modified nano-clay formulation and their application. Int. J. Chem. Stud. **6**(4), 705–710 (2018)
35. Wang, W., Wang, F., Kang, Y., Wang, A.: Enhanced adsorptive removal of Methylene Blue from aqueous solution by alkali-activated palygorskite. Water Air Soil Pollut. **226**(3), 83 (2015)
36. Irani, M., Fan, M., Ismail, H., Tuwati, A., Dutcher, B., Russel, A.G.: Modified nanosepiolite as an inexpensive support of tetraethylenepentamine for CO2 sorption. Nano Energy **11**, 235–246 (2015)
37. Jlassi, K., Krupa, I., Chehimi, M.M.: Overview: clay preparation, properties, modification. In: Jlassi, K., Chehimi, M.M., Thomas, S., (eds.) Clay-Polymer Nanocomposites, pp. 1–28. Elsevier (2017)
38. Fukushima, Y., Inagaki, S.: Synthesis of an intercalated compound of montmorillonite and 6-polyamide. J. Incl. Phenom. **5**(4), 473–482 (1987)
39. Bruce, A.N., Lieber, D., Hua, I., Howarter, J.A.: Rational interface design of epoxy–organoclay nanocomposites: role of structure-property relationship for silane modifiers. J. Colloid Interface Sci. **419**, 73–78 (2014)
40. Bayram, I., Oral, A., Sirin, K.: Synthesis of poly(cyclohexene oxide)-montmorillonite nanocomposite via in situ photoinitiated cationic polymerization with bifunctional clay. J. Chem. **2013**, 6 (2013)
41. Salmi, Z., Benzarti, K., Chehimi, M.M.: Diazonium cation-exchanged clay: an efficient, unfrequented route for making clay/polymer nanocomposites. Langmuir **29**(44), 13323–13328 (2013)

42. Othmani-Assmann, H., Benna-Zayani, N., Geiger, S., Fraisse, B., Kbir-Ariguib, N., Trabelsi-Ayadi, M., Ghermani, N.E., Grossiord, J.L.: Physico-chemical characterizations of Tunisian organophilic bentonites. J. Phys. Chem. C **111**(29), 10869–10877 (2007)
43. Di Gianni, A., Amerio, E., Monticelli, O., Bongiovanni, R.: Preparation of polymer/clay mineral nanocomposites via dispersion of silylated montmorillonite in a UV curable epoxy matrix. Appl. Clay Sci. **42**(1), 116–124 (2008)
44. Jlassi, K., Mekki, I., Benna-Zayani, M., Sigh, A., Aswal, D.K., Chechimi, M.M.: Exfoliated clay/polyaniline nanocomposites through tandem diazonium cation exchange reactions and in situ oxidative polymerization of aniline. **4**, 65213–65222 (2014)
45. Msaadi, R., Yilmaz, G., Allushi, A., Hamadi, S., Ammar, S., Chechimi, M.M., Yagsi, Y.: Highly selective copper ion imprinted clay/polymer nanocomposites prepared by visible light initiated radical photopolymerization. Polymers **11**, 286 (2019)
46. Atilla Tasdelen, M., Kreutzer, J., Yagci, Y.: In situ synthesis of polymer/clay nanocomposites by living and controlled/living polymerization. Macromol. Chem. Phys. **211**, 279–285 (2010)
47. Usuki, A., Kawasumi, M., Kojima, Y., Okada, A.: Swelling behavior of montmorillonite cation exchanged for ω-amino acids by e-caprolactam. J. Mater. Res. **8**(5), 1174–1178 (1993)
48. Pavlidou, S., Papaspyrides, C.D.: A review on polymer–layered silicate nanocomposites. Prog. Polym. Sci. **33**(12), 1119–1198 (2008)
49. Abedi, S., Abdouss, M.: A review of clay-supported Ziegler-Natta catalysts for production of polyolefin/clay nanocomposites through in situ polymerization. Appl. Catal. A **475**, 386–409 (2014)
50. Fornes, T.D., Paul, D.R.: Formation and properties of nylon 6 nanocomposites. Polímeros **13**, 212–217 (2003)
51. Guo, F., Aryana, S., Han, Y., Jiao, Y.: A review of the synthesis and applications of polymer-nanoclay composites. Appl. Sci. **8**(9), 1696 (2018)
52. Avila-Orta, C.A., González-Morones, P., Agüero-Valdez, D., González-Sánchez, A., Martínez-Colunga, J.G., Mata-Padilla, J.M., Cruz-Delgado, V.J.: Ultrasound-assisted melt extrusion of polymer nanocomposites. Intech Open (2019)
53. Lapshin, S., Swain, S., Isayev, A.I.: Ultrasound aided extrusion process for preparation of polyolefin–clay nanocomposites. Polym. Eng. Sci. **48**, 1584–1591 (2008)
54. García, L., Castell, P., Peinado, V., Muniesa, M., Fernandez, A.: Improvement of mechanical properties of poly(lactic acid) by integration of sepiolite nanoclays: Effect of ultrasonication on clay dispersion. Mater. Res. Innovations **18**, S2-85–S2-89 (2014)
55. Lin, J.-J., Chan, Y.-N., Lan, Y.-F.: Hydrophobic modification of layered clays and compatibility for epoxy nanocomposites. Materials **3**(4), 2588–2605 (2010)
56. Jan, I.-N., Lee, T.-M., Chiou, K.-C., Lin, J.-J.: Comparisons of physical properties of intercalated and exfoliated clay/epoxy nanocomposites. Ind. Eng. Chem. Res. **44**, 2086–2090 (2005)
57. Kato, M., Usuki, A., Hasegawa, N., Okamoto, H., Kawasumi, M.: Development and applications of polyolefin–and rubber–clay nanocomposites. Polym. J. **43**, 583–593 (2011)
58. Dardmeh, N., Khosrowshahi, A., Almasi, H., Zandi, M.: Study on effect of the polyethylene terephthalate/nanoclay nanocomposite film on the migration of terephthalic acid into the yoghurt drinks simulant. J. Food Process Eng. **40**(1), e12324 (2017)
59. Puggal, S., Mahajan, S.: Evaluation of mechanical properties of polymer-clay nanocomposites subjected to different environmental conditions. Int. J. Appl. Eng. Res. **10**, 22069–22076 (2015)
60. Zhou, W.Y., Guo, B., Liu, M., Liao, R., Rabie, A.B., Jia, D.: Poly(vinyl alcohol)/Halloysite nanotubes bionanocomposite films: properties and in vitro osteoblasts and fibroblasts response. J. Biomed. Mater. Res. Part a **93**, 1574–1587 (2009)
61. Liu, M., Zhang, Y., Wu, C., Xiong, S., Zhou, C.: Chitosan/halloysite nanotubes bionanocomposites: structure, mechanical properties and biocompatibility. Int. J. Biol. Macromol. **51**(4), 566–575 (2012)
62. Wang, H.W., Dong, R.X., Liu, C.L., Chang, H.Y.: Effect of clay on properties of polyimide-clay nanocomposites. J. Appl. Polym. Sci. **104**(1), 318–324 (2007)

63. Blumstein, A.: Polymerization of adsorbed monolayers. II. Thermal degradation of the inserted polymer. J. Polym. Sci. A Gen. Pap. **3**(7), 2665–2672 (1965)
64. Vyazovkin, S., Dranca, I., Fan, X., Advincula, R.: Kinetics of the thermal and thermo-oxidative degradation of a polystyrene-clay nanocomposite. Macromol. Rapid Commun. **25**, 498–503 (2004)
65. Levchik, S.V., Weil, E.D.: Combustion and fire retardancy of aliphatic nylons. Polym. Int. **49**(10), 1033–1073 (2000)
66. Gilman, J. W.: Flammability and thermal stability studies of polymer layered-silicate (clay) nanocomposites1 this work was carried out by the National Institute of Standards and Technology (NIST), an agency of the U. S. government, and by statute is not subject to copyright in the United States.1. Appl. Clay Sci. **15**(1), 31–49 (1999)

Metal-Organic Frameworks (MOFs)

Nurul N. M. Ishak, N. N. M. Khiruddin, N. Nasri, and T. B. S. A. Ravoof

Abstract This chapter outlines the design and synthesis of a class of highly functional porous materials known as MOFs (Metal-Organic Frameworks) by various techniques including conventional, microwave-assisted, electrochemical, mechanochemical and sonochemical methods. The physical attributes of various MOFs like the strength of the frameworks and surface area, thermal stabilities, chirality, luminescence and magnetic properties are also elaborated upon, citing recent literature. MOFs as composites, specifically with nanoparticles, metal oxides, organic polymers and polyoxometalates are outlined. The use of MOFS in engineering applications, especially in CO_2 capture, storage of gases, catalysis, sensing, drug delivery and as semiconductors are discussed, using specific and recent examples.

Keywords Porous materials · Metal-organic frameworks · MOF-composites · Gas storage · Catalysis · Drug delivery

1 Introduction

Porous materials are a class of materials with low density, large specific surface area, and interesting physical, mechanical, thermal and electrical properties [67]. According to International Union of Pure and Applied Chemistry (IUPAC), these materials can be classified based on their pore diameters, which is microporous (<2 nm), mesoporous (2–0 nm) and macroporous (>50 nm) [99]. Porous materials such as zeolites, mesoporous silica, activated carbon and microporous polymer have

N. N. M. Ishak · N. N. M. Khiruddin · N. Nasri · T. B. S. A. Ravoof (✉)
Department of Chemistry, Faculty of Science, Universiti Putra Malaysia (UPM), 43400 Serdang, Selangor, Malaysia
e-mail: thahira301@yahoo.com

T. B. S. A. Ravoof
Foundry of Reticular Materials for Sustainability (FORMS), Materials Synthesis and Characterisation Laboratory, Institute of Advanced Technology, Universiti Putra Malaysia (UPM), 43400 Serdang, Selangor Darul Ehsan, Malaysia

N. M. Mubarak et al. (eds.), *Contemporary Nanomaterials in Material Engineering Applications*, Engineering Materials,
https://doi.org/10.1007/978-3-030-62761-4_5

been intensively investigated as ideal platforms for various applications especially in gas storage, chemical sensing, energy conversion and catalysis.

Since the early 1990s, research into materials with polymeric and porous structures based on organic/inorganic hybrid materials have been on the rise. In relation to this, Metal organic frameworks (MOFs) are a rapidly emerging unique type of crystalline porous materials. MOFs are self-assembled from inorganic metal clusters and organic bridging ligands via strong covalent bonds. Via a self-assembly mechanism, the extended one dimensional (1D), two dimensional (2D) and three dimensional (3D) structures are built where the metal centres (known as the connector) are linked by ditopic or multitopic organic ligands, known as linkers (Fig. 1). An important feature of MOFs is that their framework structures, pore environment, functionality can be fine controlled by the choice of metal and organic building unit. Therefore, the topology of the network is determined by the intrinsic structural features of the choice of metal ions with different oxidation states and organic linkers.

In the synthesis of such inorganic-organic hybrid materials, transition metal or lanthanide salts are reacted with rigid, often aromatic, organic donor ligands, which feature two or more Lewis-basic functional groups, usually neutral nitrogen donors (e.g., pyridyl, cyanide groups) or neutral/anionic oxygen donors (e.g., carbonyl, alkoxy, carboxylate groups) [78]. The synthesis of MOFs are normally conducted under mild conditions. A few methods to synthesise MOFs have been reported including conventional synthesis which used electrical heating and non-conventional synthesis at room temperature. Other alternative synthetic routes including microwave-assisted, electrochemical, mechanochemical and sonochemical methods have also been used in the synthesis of MOFs. These methods can produce MOFs with varying particle sizes and size distribution as well as morphologies

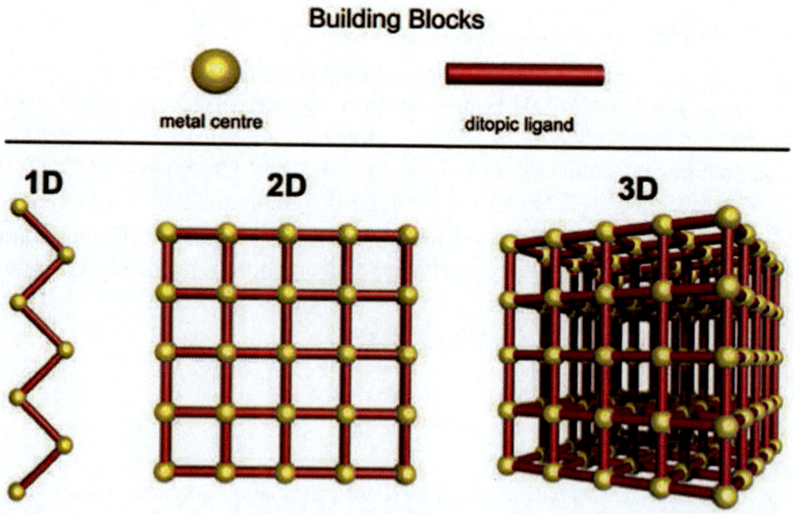

Fig. 1 Extended structures of Metal Organic Frameworks (MOFs) based on metal centre and linear ditopic ligands featuring different structural dimensionalities. Sourced from Kitagawa et al. [44]

that can influence the material properties. In fact, the pore size, pore shape, network topology and surface functionality are more important to develop the unique structure of MOFs [26] The structure of MOFs can also have influenced by other factors, such as the coordination environment of metal ions, the metal-to-ligand molar ratio, ligand structure, the presence of solvent molecules, counter ions, reaction temperature, pH value of the solution and many other factors [95].

MOFs have numerous advantages compared to conventional porous materials like zeolites or activated carbon, in terms of their ability to fine-tune the structures by rational design and incorporate functionalities into the molecular material. MOFs possess a highly crystalline nature, extraordinarily low densities (1.0–0.2 g/cm^3), large pore sizes (up to 29 Å), large free volume, high surface area (500–4500 m^2/g), and fascinating topologies [57]. The choice of the initial metal cluster and organic linkers makes it possible to vary some parameters will leads to new ways of designing materials with tailored physicochemical properties.

Composites are multi-component materials with multiple phases with at least one continuous phase [96]. MOF-composites have received astounding interest due to their various applications especially in separation and catalysis. By compositing MOFs with active components, the properties of MOF, stability, morphologies as well as their potential applications could be improved. In general, there are two basic types of MOF composites, which are discontinuous phase and continuous phase [2]. Discontinuous phase occurs when a component (such as a polymer) is unable to form an inter-grown and continuous layer on MOFs surface. This type of MOF-composite tends to crack and is not stable for applications especially in gas separation. Continuous phase MOF-composites are MOFs that contain special and specific chemical functionalities and are selective, thus they have good potential in gas adsorption studies. The design of MOF-composites by combining the MOF precursor with nanoparticles, metal-oxides, polymers and polyoxometalates creates materials that can be used in various applications such as hydrogen storage, catalytic processes, energy storage, acidic gas adsorption and ammonia adsorption [103].

Owing to their porous structure, large surface area, good stability and ability to be functionalized, MOFs are good candidates in catalysis, gas separation, and ion exchange reactions where zeolites are commonly used. Since MOFs have continuous, uniform and permeable channels, they can provide active sites on the pore surface and transport reactants/products to or from inner reactive vessels in catalytic applications [60]. In addition, the high porosities, tunable framework structures, and immobilized functional sites of MOFs can allow gas storage in pore spaces and interact with ions gas molecules [53]. Therefore, the high surface areas of MOFs also could concentrate analytes to high levels and enhance detection sensitivity making them useful as chemical sensors. Moreover, the specific functionality of MOFs allowed specific recognition with exceptional selectivity by host–guest interactions or size exclusion. Thus, the increased regeneration and recycling uptake and release of substrates are due to the flexible porosity of the frameworks [101].

Recently, MOFs have been given much attention in the improvement of drug delivery and cancer theranostic carriers [97]. MOFs possess the following obvious advantages compared to traditional drug carriers: the versatile structures of MOFs with various morphologies, sizes, compositions, chemical properties, and multifunctionalities allow stimuli-responsive drug controlled release [33, 59] Notably, the modification of MOFs has no significant alteration in their desirable physicochemical properties, where the controllable of sizes, shapes, and high uniformity are still maintained as in the parent material [69].

In this chapter, the synthesis and design of metal organic frameworks via several synthetic methods and their physical and chemical properties will be discussed. The improvement of the properties of these materials as MOF-composites also will be detailed. MOFs as promising porous materials, their engineering applications in gas storage, catalysis, chemical sensing and drug delivery will also be outlined.

2 Design and Synthesis of Metal Organic Frameworks (MOFs)

Synthetic method development plays an important role in this area of research where many types of functional groups can be incorporated directly into the structure which may be tailored to suit the need for specific applications. To understand the direct synthesis effort, the knowledge of possible topologies, functionality of organic linkers, metal coordination environments and the formation of desired inorganic building blocks are vital. This section details the variety of synthetic methods employed in MOFs synthesis including the chemistry concepts behind solvothermal reaction conditions, microwave assisted synthesis, electrochemical and mechanochemical synthesis as well as sonochemical synthesis (Fig. 2).

2.1 *Metals with Vacant Sites*

In the early stages of MOF synthesis, direct assembly of particular metal nodes and organic linkers was the usual main approach. Transition metal ions are often used as the inorganic components of MOFs. Generally, different metal ions are known to have different coordination numbers and geometries, including linear, square planar, tetrahedral, square pyramidal, trigonal bipyramidal, octahedral, trigonal prismatic and pentagonal bipyramidal [72]. In principle, the metal ion which has more than one vacant site or labile site has great potential to react with bridging ligands in order to make MOFs [37]. It is also known that the more labile metal ions are Cu+, Cu^{2+}, Ag+, Cd^{2+}, Zn^{2+}, Co^{2+} and Ni^{2+} and these feature prominently in the field of coordination polymers. Among these geometries, metals with tetrahedral structures are usually used to generate network topologies of MOFs.

Therefore, the Secondary Building Units (SBUs) approach in MOF chemistry is the most significant contributor to its rapid development. Basically, SBUs are the formation of polynuclear cluster nodes which are often based on metal-carboxylate

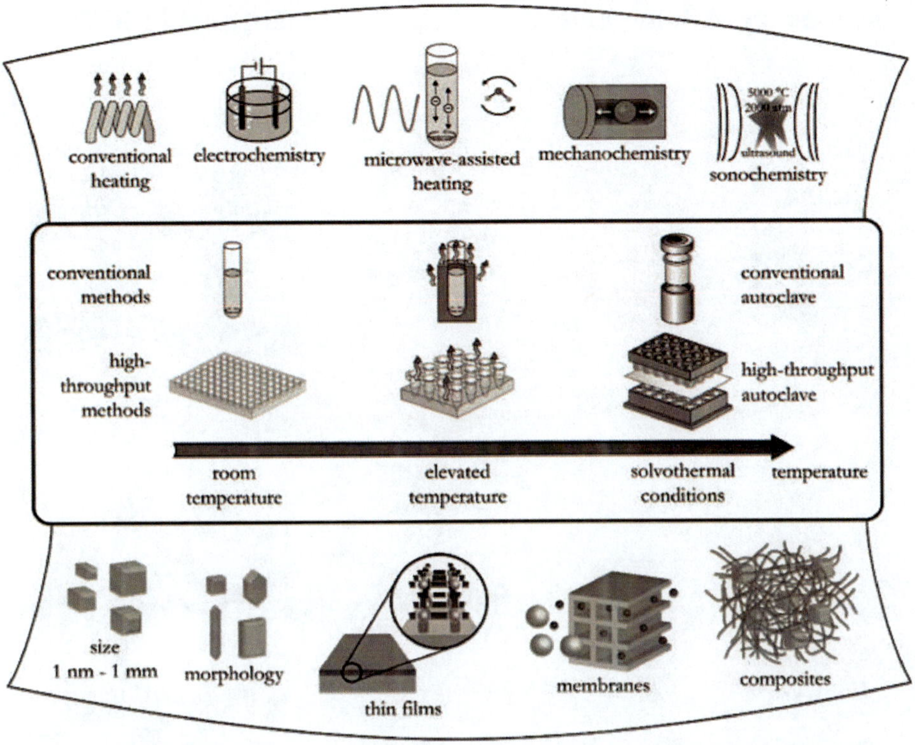

Fig. 2 Overview of the synthesis methods and final products in MOF synthesis. Sourced from Stock and Biswas [86]

bonding. The structures will be thermodynamically stable via strong covalent bonds and mechanically stable because of strong directional bonds that can lock down the position of metal centres in MOFs [40]. The use of multidentate linkers such as carboxylates allow the formation of more rigid frameworks due to their ability to aggregate metal ions into M-O-C clusters. The SBUs are sufficiently rigid because the metal ions are locked into their positions by the carboxylates.

The SBUs are represented as large rigid vertices resulting from the joining of rigid organic links to produce extended frameworks with high structural stability. Metal clusters such as zinc (Zn), chromium (Cr) and zirconium (Zr) with different coordination numbers (CN) produce different morphologies when reacted, for example, with benzene dicarboxylate (BDC) linkers (Fig. 3). Another example is bridging 1,3,5-benzenetricarboxylates (trimesic acid) reacted with Cu^{2+} ions that formed structural $Cu_2(OCR)_4$ units assembled from a number of metal ions and ligands. The ligand and metal ions are considered the primary building units. This is analogous to aluminosilicate zeolite chemistry, where nine SBUs based on the tetrahedral AlO_4 and SiO_4 primary building units have been classified.

In the late eighties and nineties of the last century, pioneering work on the targeted synthesis of porous 3D structures was done by Robson and co-workers (1990). The

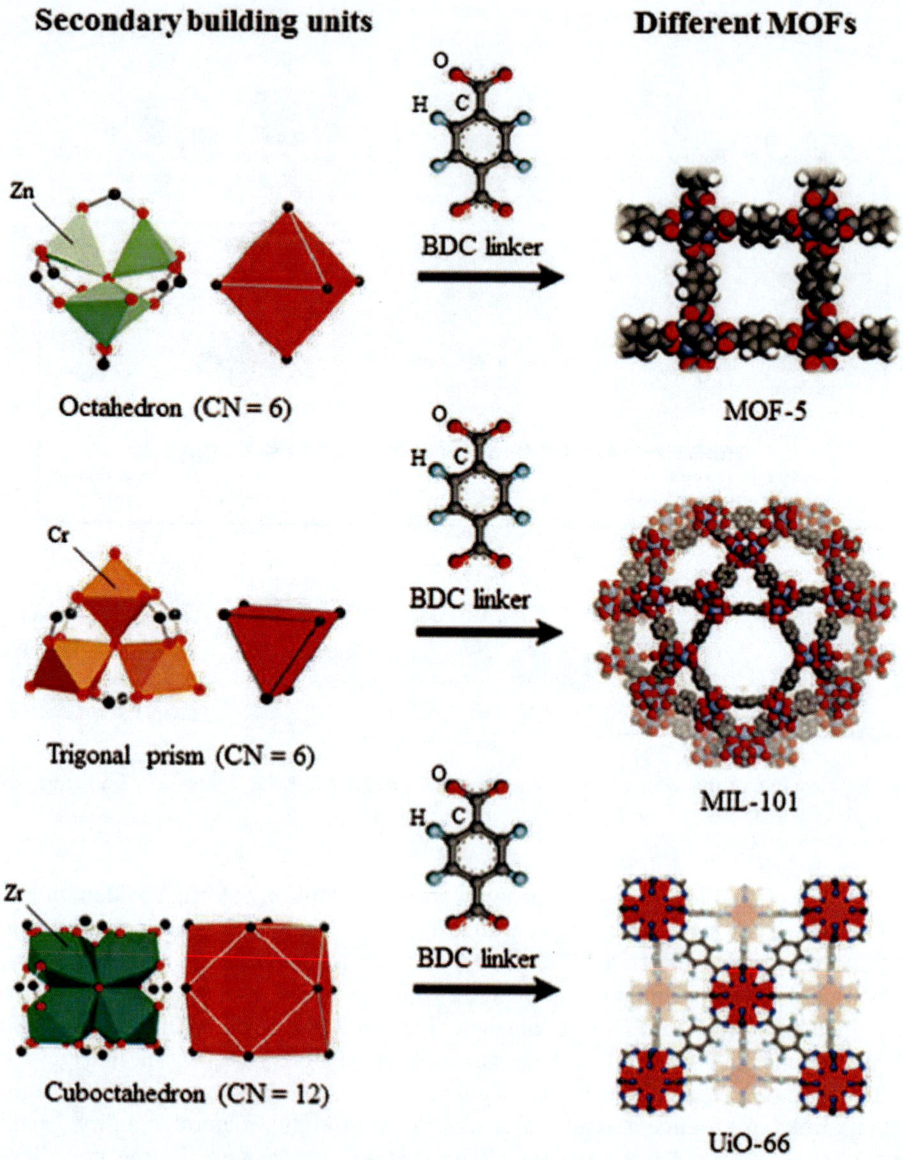

Fig. 3 Example of other SBUs with different MOFs structures. Sourced from Butova et al. [13]

utilisation of organic linkers of distinct geometries together with metal ions yielded supramolecular structures exhibiting a specific structure type, for example, diamond [35], rutile [5] or PtS-like [1] topology, which featured the same topology as the respective inorganic minerals, but with largely extended cell parameters.

2.2 Organic Linkers as Building Blocks of MOFs

The basic requirements for organic linkers in MOFs are that they can form coordination bonds with a central metal or secondary building units (SBUs). Linkers or ligands with rigid backbones are often preferred because the rigidity helps to sustain the pore structure after the removal of the guest molecules. As illustrated in Fig. 4, the bond between metal and linkers can be both nitrogen–metal coordination bonds and metal-carboxylates covalent bonds [29].

Organic linkers can be either electrically neutral, cationic or anionic. Figure 5 shows some examples of ligands that consist of rigid 3- and 4-connecting N-donor ligands [37]. These linkers are especially useful as pillars in the construction of pillared-layer 3D networks. Besides, carboxylate ligands which are known as anionic linkers are also widely used nowadays to construct 3D MOFs. This is due to their ability to aggregate metal ions into clusters and therefore forming more stable frameworks. However, cationic organic ligands have remained mostly unused, owing to their low affinities for cationic metal ions. A rigid 3-connecting ligand, 4,4′-azo-1,2,4-triazole has six potential atoms that can coordinate with metal ions to form a 3D structure as was reported by Zhang and Shreeve [104]. Similarly, a 4-connecting flexible tetrazole ligand, 1,5-bis(tetrazolo)-3-oxapentane was synthesised by Cui et al. [99] as an organic linker to construct porous MOFs to enhance CO_2 uptake. In 2008, Lin et al. successfully synthesised 3D MOFs constructed from polycarboxylate acid and flexible imidazole-based ligands.

In the synthesis of MOFs, the use of long organic linkers will provide large storage space and a greater number of adsorption sites inside a given material. However, two or more frameworks can grow and mutually intertwine together because the large space within the crystal framework makes it likely to form interpenetrating structures. Therefore, several methods have been developed to prevent the interpenetration and engineer more stable frameworks.

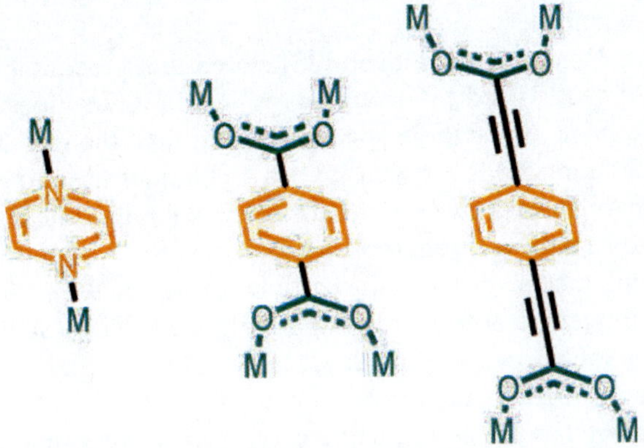

Fig. 4 The covalent bond formation between metal cluster and organic linkers. Sourced from Gonzalez-Nelson et al. [29]

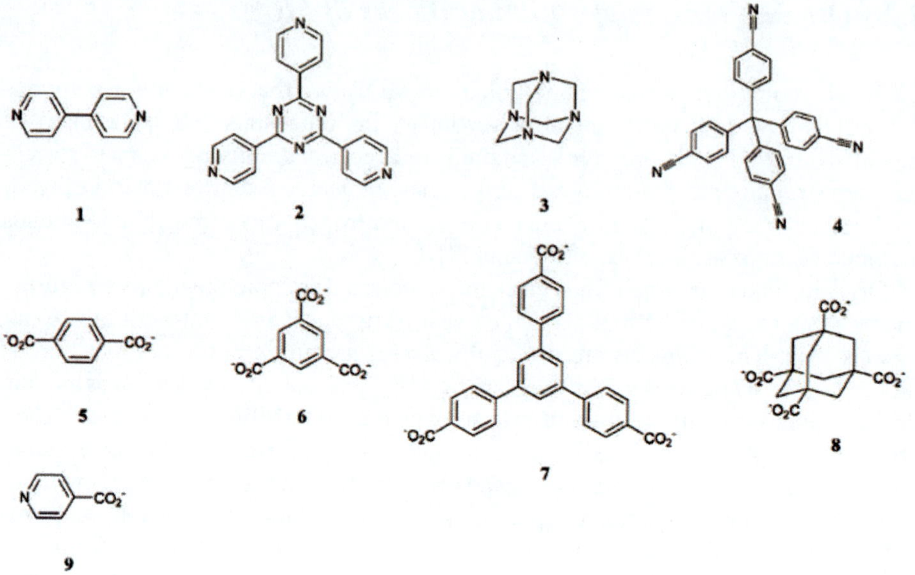

Fig. 5 Examples of some organic linkers used in the synthesis of MOFs. Sourced from James [37]

2.3 Conventional Synthesis

Conventional synthesis is commonly applied to reactions that are executed by conventional electric heating where the main parameter in the synthesis is the reaction temperature. MOFs are usually synthesised using conventional solvothermal and non-solvothermal methods. Solvothermal reactions basically occur in a closed vessel system where autogenous pressure is created above the boiling point of the solvent. Non-solvothermal reactions occur at the boiling point of solvent under ambient pressure which can also be classified as room-temperature or elevated temperature reactions.

Hoskins and Robson [35] stated that precipitation followed by recrystallization of synthesised MOFs mainly relied on low temperatures or slow evaporation of the solvent. Since there is possibility to tune the rate of nucleation and crystal growth, these methods are well-known techniques used to grow molecular or ionic crystals. In order to grow crystals from clear solutions, the concentration of the reactants need to be adjusted until the critical nucleation concentration is exceeded. Crystal growth will take place when the particles exceeded the critical radius by increasing the reaction temperature or by evaporating the solvent. Several crystallization methods such as solvent evaporation, layering of solutions, and slow diffusion could lead to concentration gradients which allow the formation of MOFs crystals [86]. Therefore, large crystals suitable for single crystal X-ray structure determination are often obtained by the concentration gradient method where temperature gradient or slow cooling process was applied to a reaction mixture. Some prominent MOFs such as MOF-5,

MOF-74, HKUST-1 or ZIF-8 were obtained at room temperature by just mixing solutions of the starting materials. This method is also known as direct precipitation where crystallization of MOFs occurs in a short period. The formation of product can be influenced by different reaction temperatures and therefore more condensed/dense MOF structures are obtained at high temperature reactions. In some ways, increasing reaction temperature is needed to obtain suitable crystallinity and reaction rates, especially when using more inert ions. Yet, the morphology of the crystals can also be affected by reaction temperature and prolonged the reaction times can result in the degradation of the MOF crystals.

2.4 *Microwave-Assisted Synthesis*

In the previous section, the synthesis of MOFs occurs while in a solution and temperatures ranging from 27 to 250 °C by conventional heating. Generally, energy is introduced when heat is transferred in the oven through convection. Alternatively, energy can also be introduced by electric potential, electromagnetic radiation, mechanical waves (ultrasound) and mechanochemical methods [83]. The energy source is related to the pressure, energy per molecule and also duration introduced in the system. Thus, the type of product formed and the morphology of the crystals depend on these parameters.

In synthetic chemistry, microwave irradiation is a well-known method used in the synthesis of MOFs. Microwaves (MW) can be described as a form of electromagnetic radiation with a frequency range between 300 and 3000 000 MHz. Generally, this method depends on electromagnetic interaction waves and the mobile charges generated. The mobile charges can be polar solvent molecules, ions in solution or electrons, ions in solid. During heating, an electric current is formed due to electric resistance of the solid. While in solution, an oscillating field causes the change of molecule orientations permanently when the polar molecules align themselves in an electromagnetic field. Thus, by applying appropriate frequencies and temperatures in the system, the kinetic energy will increase due to the collisions between molecules.

Generally, the synthesis of MOFs using MW has often been carried out at a temperature above 100 °C with reaction times less than 1 h. There are only several reports that describe the systematic parameters (such as solvent, irradiation time, reaction temperature, power level, molar ratio of the reactants and reactant concentration) to optimise the reaction conditions for MOFs synthesis [86]. Basically, the synthesis of MOFs using MW irradiation can allow fast synthesis of products with small crystallite sizes. Besides fast crystallization, the other advantages with this technique include phase selectivity, narrow particle size distribution and also facile morphology control [51].

A few metal (III) carboxylate-based MOFs (M = Fe, Al, Cr, V, Ce) have been successfully synthesised via MW irradiation. For example, Cr-MIL-100 was synthesised using microwave at 200 °C over 4 h with a 44% yield which was comparable to conventional electric synthesis. Meanwhile, Cr-MIL-101 was successfully synthesised with increasing reaction time in the microwave which resulted in an increase in the dimensions of nanoparticle sizes from 40–80 to 70–90 nm and more homogeneity [39]. In another report, nanocrystals of Cr-MIL-101 was produced without using THF as solvent and lowering the reactant concentration and increasing the pH of the reaction mixture which resulted in ~50 nm crystal sizes and a yield of 37% [42]. Synthesis conditions such as temperature and method of heating greatly affected the crystal sizes obtained.

2.5 *Electrochemical Synthesis*

The first pioneers of synthesising MOFs using electrochemical techniques was by researchers at BASF (a German chemical company) in 2005 [3]. The main objective of this technique in the syntheses are the exclusion of anions, such as nitrate, perchlorate, or chloride. Instead of using metal salts, the metal ions in anodic dissolution are introduced to the reaction medium, which contains the dissolved linker molecules and a conducting salt. This is to avoid the formation of anions in the course of the reaction and to initiate a continuous process resulting in large amounts of product. To avoid deposition of metal on the cathode, protic solvents, or other compounds such as acrylic, acrylonitrile, or maleic esters were used during the synthesis [86]. Electrochemical techniques are advantageous for industrial processes due to their ability to run continuously and high yields of product was obtained compared to normal batch reactions.

The well-known MOF copper-based MOF, HKUST-1 consisting of Cu^{2+} ions constructed from 1,3,5-benzenetricarboxylic acid (BTC) was used to illustrate the formation of coated copper electrodes by electrochemical reaction. When an anodic voltage was applied to the copper electrode, Cu^{2+} ions were produced as a conducting salt into the synthesis solution containing BTC and methyl tributyl ammonium methyl sulfate (MTBS). Densely packed films of HKUST-1 crystals were found to be easily prepared by electrochemical methods and the chemical structures were in agreement with the conventionally synthesised HKUST-1 [3]. By altering the synthesis conditions, the crystal size can be tuned with a range of 2–50 μm. Therefore, the thickness of the thin film can be varied accordingly. When high voltage is applied from 2.5 V up to 25 V in the system, a high concentration of metal ions is produced thus yielding smaller sized crystals coating on the surface of the thin film. Therefore, if more water was added to the synthesis mixture, crystal formation would slow down thus yielding larger crystals [3].

Porous organic-based conducting materials have been on demand due to their good properties in applications such as photovoltaics, super capacitors, batteries, catalysis and chemical sensing. Therefore, the electrochemical technique is one of the common techniques used to prepare MOF thin films by depositing active substrates or materials into a parent MOF. For example, a study was carried out by employing electrochemical technique to synthesis microporous conductive polymers using MOF thin film.

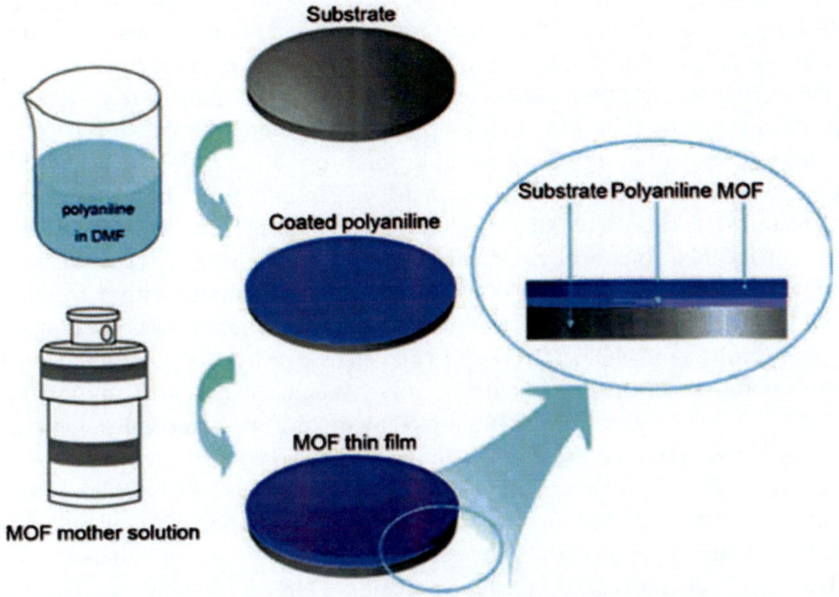

Fig. 6 Preparation of MOF thin film with polyaniline. Sourced from Lu et al. [62]

Three different parent MOFs were used which were HKUST-1, $Zn_2(BDC)_2DABCO$ and MIL-68 (In). The MOF thin film acted as a template to enhance the conductivity and formation of the porous polymer. In this case, polyaniline played a role to tighten the MOF layer through imino group bonding and transferring electrode under electrochemical conditions. As illustrated in Fig. 6, a layer of conducting polyaniline in DMF was coated on a Pt electrode (substrate). The polyaniline/Pt was then introduced into the MOF solution to allow heterogeneous nucleation and crystal intergrowth. The study found that the thickness of MOF/polyaniline/Pt thin films were in a range of 2–100 μm. At the end of the process, porous polyaniline was obtained by digesting the MOF thin film in HCl solution. The prepared porous polyaniline exhibited a high BET surface area of 986 $m^2\ g^{-1}$ and high electric conductivity of 0.125 S cm^{-1} [62].

2.6 Mechanochemical Synthesis

In mechanochemical synthesis, the mechanical breakage of intermolecular bonds followed by a chemical transformation takes place for example, by milling in ball mills. This technique can be carried out at room temperature under solvent-free conditions and short reaction times, normally in the range of 10–60 min, which results in high yields containing small-sized particles. In mechanochemical reactions, the presence of a liquid component can lead to an easy crystallization process, high crystallinity of product and yield. Therefore, the diffusion and permanent forms of new surfaces can be obtained by continuous milling of the components.

This technique was first performed on MOFs under solvent-free conditions using organic linkers with low melting point and hydrated metal salts containing basic anions. The common use of metal acetates or carbonates lead to crystalline compounds with the presence of acetic acid as a by-product. The by-product can be removed by thermal activation. In 2006, the synthesis of a Cu(INA) MOF was carried out by mixing copper acetate and isonicotinic acid (HINA) in a ball mill for a few minutes. This resulted in a well-crystallised product with a formula of $Cu(INA)_2.x\ H_2O.y$ AcOH [75]. Two years later, this technique was used again to compare two different MOFs (HKUST-1 and $Cu(INA)_2$) by investigating the metals used, the mode of introduction, the linker, solvents and also the synthesis conditions [74]. HKUST-1 was successfully synthesised by mixing copper acetate and benzene tricarboxylic acid (H_3BTC) under solvent-free conditions. However, the formation of HKUST-1 needed longer mixing times compared to $Cu(INA)_2$ which occurred spontaneously after 1 min of mixing. Interestingly, the addition of small amounts of acetic acid during milling sharply increased the formation rate of $Cu(INA)_2$ rather than HKUST-1. This was due to the high porosity of HKUST-1 which caused instant absorption of liquid components preventing participation in dissolution thus affecting the reaction mixture [13].

2.7 *Sonochemical Synthesis*

Sonochemical synthesis basically takes place with the supply of high-energy ultrasound energy to a reaction mixture. Ultrasound radiation is the cyclic mechanical vibration with a frequency between 20 and 10 kHz, which is the upper limit of human hearing. Cyclic alternating areas of compression (high pressure) and rarefaction (low pressure) will be formed when high-energy ultrasound waves interact with liquids. In the low pressure region, the pressure drops below the vapor pressure of the solvents and reactants lead to cavitation. Cavitation is a phenomenon in which small bubbles grow under the alternating pressure by the diffusion of solute vapor into the volume of the bubble, thus accumulating ultrasonic energy. When the bubbles reach their maximum size, they become unstable and collapse. This process of bubble formation leads to the rapid release of energy with heating and cooling rates of $>10^{10}\ K\ s^{-1}$, temperatures of ~5000 K, and pressures of ~1000 bar [86].

The synthesis of MOFs using sonochemical techniques is easy to carry out since it is very fast, energy-efficient, environmentally-friendly and can be done at room temperature. Interestingly, nanocrystals are often obtained by sonochemical synthesis which are advantageous for specific applications. High yields of product have been reported with short reaction times at ambient temperatures. The first sonochemical synthesis of $[Zn_3(BTC)_2]$ was carried out by mixing $Zn(OAc)_2$ and H_3BTC in ethanol and water in an ultrasonic bath at room temperature conditions. The only parameter tested was the reaction time, where short reaction times of 5–10 min led to spherical

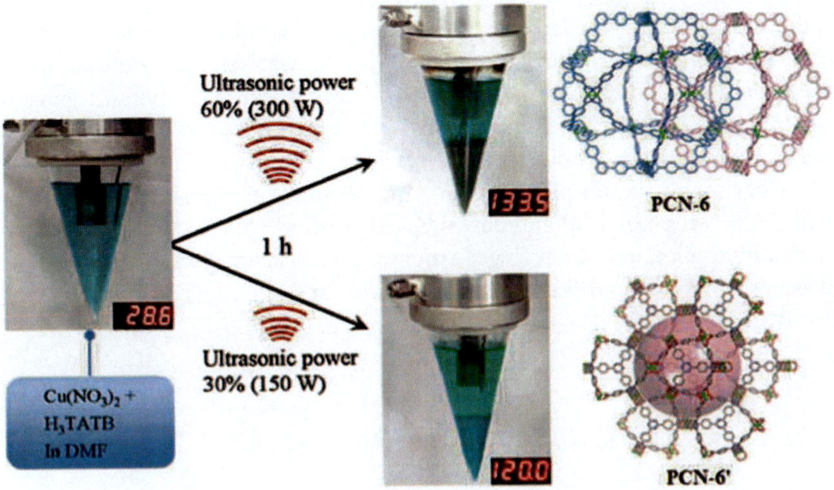

Fig. 7 Sonochemical synthesis of the catenation isomers PCN-6 and PCN-6′ at different ultrasonic power level. Sourced from Kim et al. [43]

particles in the range of 100–200 nm, while long reaction times at 30 and 90 min resulted in long needle-like shapes with diameters up to 900 nm [77].

Another study investigated the effect of ultrasonic power levels on two examples of interpenetrated and non-interpenetrated structures of PCN-6/PCN-6′ (MOF containing copper nitrate and 4,4′,4″-*S*-triazine-2,4,6-triyl-tribenzoic acid, H_3TATB) and IRMOF-9/IRMOF-10 (MOF containing zinc nitrate and 4,4′-biphenyldicarboxylic acid, BPDC) [43]. From the findings, lower power levels led to non-interpenetrated structures whereas higher power levels led to phase-pure interpenetrated structures of the MOFs as in Fig. 7. Thus, consistently small uniformly-shaped particles were obtained with sizes of 1.5–2 μm for PCN-6′, 4.5–6 μm for PCN-6 and ~5–20 μm for IRMOF-9 and -10 from the sonochemical synthesis [86].

Thus, MOFs can be synthesised by conventional synthesis (including solvothermal and non-solvothermal), microwave, electrochemical, mechanochemical and sonochemical synthesis but the most commonly used synthesis method for MOFs is the solvothermal method as it is possible to tune the nucleation process and enhance the crystal growth. However, this depends on the type of product required, for example, nanosized MOFs or MOF thin films as per their use in various applications.

3 Physical and Chemical Properties of MOFs

MOFs are advantageous compared to other well-known porous materials such as zeolites, activated carbon and mesoporous silica, due to their ability to tune the structure and functionality. The tunability in MOFs is significantly different from the traditional zeolites whose pores are confined by rigid tetrahedral oxide skeletons which are difficult to alter. Previously, MOFs research only focused on direct synthesis methods to design new structures. Different synthetic methods have been studied to develop different sizes, shapes and improved properties of MOFs for specific applications. The choice of appropriate metal clusters and organic linkers or the design of organic linkers play an important role in the synthesis of MOFs, in order to obtain interesting physical and chemical properties. This makes them promising candidates in applications such as gas storage, catalysis, drug delivery, sensors and luminescent materials. In this section, the physical and chemical properties of MOFs are briefly described.

3.1 Rigidity and Flexibility of the Frameworks

Construction of MOFs by organic and inorganic elements can lead to materials with unique properties. Therefore, the use of multidentate rigid ligands incorporated in MOFs can result in stable frameworks [23]. These rigid organic linkers allow robust and high structural stability. The concept of reticular design of MOFs (IRMOF series) was described by Yaghi et al. [100] where the use of long bridging ligands led to interpenetration and decrease in void space. To overcome these issues, Kitagawa et al. [44] reported CPL-*n* compounds as a "pillared layer construction" in the synthesis of MOFs [11]. A pillared layer construction is coordination of pillar linkers into metal nodes from 2D layers to construct 3D porous frameworks which are more stable structures. From the findings, the MOFs showed some structural transformations after guest molecule accommodation which meant that the MOFs could respond to external stimuli, making it special compared to rigid zeolites and activated carbon.

MOFs can be classified into three classes, where the 1st generation MOFs have frameworks that collapse irreversibly upon removal of guest molecules. The 2nd generation MOFs are non-conformationally changeable frameworks which maintain their robustness after the removal of guest molecule. Lastly, the 3rd generation of frameworks exhibit flexibility and dynamic functionalities, leading to guest responsive adsorption behaviour [11, 34].

When external stimuli such as removal of guest molecules, heating, magnetic and electric fields were applied to the flexible MOFs, some reversible structural transformations occurred whereas there was no change in the robust MOFs. Examples of robust MOFs are MOF-5 and HKUST-1, while MIL-53 and MIL-88 are examples of flexible MOFs [17, 20, 55, 81]. Robust MOFs with high porosity and surface area with permanent pore size and shape are assets in gas storage applications. Meanwhile, flexible MOFs are advantageous in gas separation applications [11].

3.2 Surface Area

In terms of pore sizes, specific pore volumes and surface area, MOF materials are much better than zeolite materials. These properties make them advantageous in selective catalytic reactions including bulky molecules which could not penetrate into the smaller zeolite pores. Microporous MOFs are defined as MOFs with pore sizes of the framework less than 2 nm which can selectively capture small molecules. The pore sizes for mesoporous MOFs are between 2 and 50 nm which can encapsulate large molecules [46].

Surface area dimensions of MOFs can be tailored accordingly where longer linkers with a number of phenyl rings can increase the surface area and build ultra-high porosity MOFs. Previous research showed the surface area of MOF-177 at 4,500 m^2/g, which was almost five times higher than zeolites with the narrowest dimension (10.8 Å) of the pores [16]. The construction of NU-1000 (NU is Northwestern University MOF-100) with a tetratopic linker, 1,3,6,8-(p-benzoate)pyrene and zirconium cluster was reported to form a mesoporous framework with pore sizes estimated at about 30 Å and microporous pores of around 12 Å (Fig. 8) [94]. Due to the high surface area possessed by NU-1000, Chen et al. [19] was able to encapsulate insulin into NU-1000 as an alternative to the traditional method of injection of insulin into type-2-diabetes patients. The mesoporous size of NU-1000 allowed insulin with the size of 13 Å × 13 Å to diffuse through the framework with a high

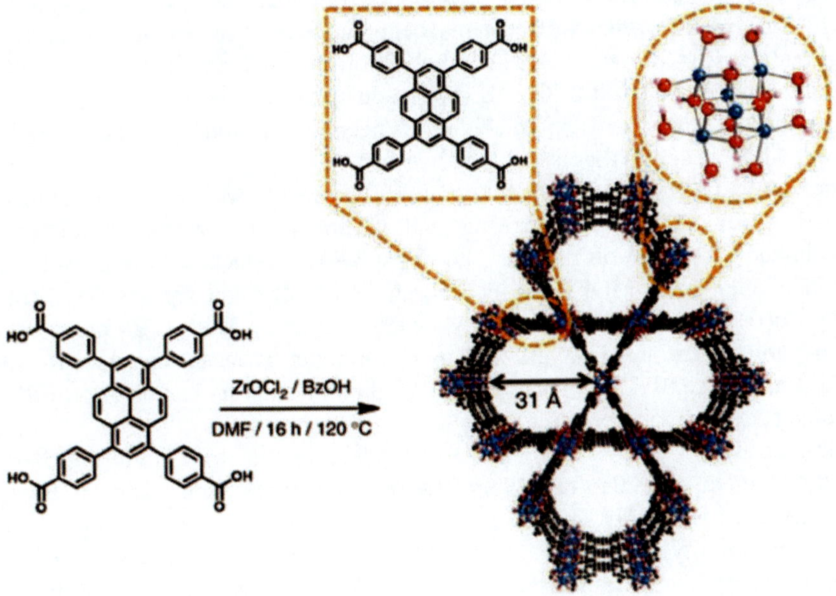

Fig. 8 The structure of NU-1000 synthesised from a tetratopic linker, 1,3,6,8-(p-benzoate)pyrene and zirconium cluster, $ZrOCl_2$ with benzoic acid. Sourced from Wang et al. [94]

loading of about 40% insulin in 30 min, therefore proving that the MOFs was able to capture large-sized molecules due to their high surface area [19].

MOFs have great potential in gas storage applications. Several MOFs were reported to have a high methane uptake of ~160 cm^3 (STP) storage [36]. Due to large surface area of MOFs, high adsorption capacity of methane were obtained. However, microporous MOFs are unable to encapsulate large molecules such as proteins and drugs. Therefore, mesoporous MOFs play a more important role to overcome the problems faced by microporous MOFs, due to their larger pore sizes.

3.3 Thermal Stability

The strong bond in Si–O and Al–O is difficult to break due to the high kinetic stability of zeolites that is consistent with their high energy barrier. The coordination bond strengths are lower in MOFs and therefore MOFs are expected to be less thermally stable. This usually happens and typically decomposition occurs at lower temperatures than zeolites. The stability of the framework is mainly controlled by the inorganic building blocks and the chemical bond strength between the inorganic building blocks and the connectors (linkers). Oxygen- and nitrogen-containing organic linkers-provide stability to the frameworks from the strong covalent interactions between metal and linkers. Identification of stable and flexible new inorganic building units is therefore of utmost importance for the development of new, usable, open hybrid structures. Generally, thermal degradation of MOFs involves node-linker bond breakage first followed by linker combustion and the thermal stability of MOFs is usually limited to 350–400 °C. MIL-53 (made up of scandium and oxygen (ScO_6) nodes with 1,4-benzodicarboxylic acid struts between the nodes) is one example of a porous MOF structure that is stable above 500 °C [14].

There are a few types of MOFs degradation processes which are amorphization [7], node-cluster dehydration, melting [90], hydrogenation or graphitization [25]. In the linker-graphitization process, the degradation products can be useful materials. Generally, most MOFs are incorporated from divalent cations (i.e., Zn(II), Cu(II), Co(II) and Cd(II)), and carboxylate or N-donating linkers. By using equivalent oxyanion-terminated linkers with a large number of vacant metal centre sites, such as Ln(III), Al(III) Zr(IV) and Ti(IV), the thermal stability can be improved due to stronger metal–ligand bonding.

Wang et al. [92, 93] reported an aerosol-based approach to study the thermal stability of metal–organic frameworks (MOFs) for the synthesis of MOF-based gas phases used for highly active catalysis. Temperature-programmed electrospray-differential mobility analysis (TP-ES-DMA) was used to characterize temperature-dependent morphological changes directly in the gas phase, and the results were shown to be highly related to the thermal stability of MOFs determined by traditional porosity measurements and crystallinity. Three different MOFs have been used in the studies which were Cr-based MIL-88B-NH_2, Fe-based MIL-88B-NH_2

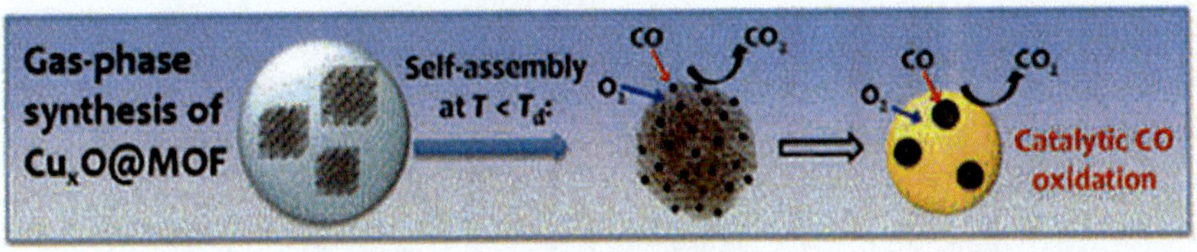

Fig. 9 The encapsulation of MOFs (Cr-based MIL-88B-NH_2, Fe-based MIL-88B-NH_2 and UiO-66) with copper oxide using gas-phase evaporation-induced self-assembly approach and catalytic CO Oxidation. Sourced from Wang et al. [92, 93]

and UiO-66. Studies show that Cr-based MIL-88B-NH_2 exhibited a higher decomposition temperature (T_d) at 350 °C compared to the Fe-based MIL-88B-NH_2 which was at 250 °C. This is because Cr provides a more stable binding with the amino-1,4-benzenediacarboxylic acid (NH_2-1,4-BDC) linkers. Meanwhile, there was a significant decrease of T_d by ~100 °C in UiO-66 [92, 93]. Interestingly, UiO-66 showed high catalytic activity and stability in carbon monoxide oxidation after encapsulation with copper oxide using a gas-phase evaporation-induced self-assembly approach (Fig. 9). This was because UiO-66 provided an active support template creating Cu–Zr–O interfaces which enhanced the catalytic CO oxidation.

Hence, thermal stability is very important in order for MOFs to be of use in specific applications especially in heterogeneous catalytic applications. The stability of the framework can be increased by ensuring good hard-hard or soft-soft interactions between ligand and metal to increase the bond strengths and high decomposition temperatures.

3.4 *Chirality*

Chirality is widespread in nature and chiral MOFs can be used in chiral recognition, chiral separation and chiral catalysis. Chirality is an important feature of many materials and has applications in chemistry, pharmacy, agriculture and medicine. Enantiomers of the same substance can show different functionalities from each in human bodies or the environment [61]. Chirality can be introduced into the MOF by simply choosing a chiral building blocks from the organic ligands.

Mallick et al. [63] was first reported on a chiral hexagonal MgMOF constructed from an achiral organic building unit named Mg-MOF-1 [Mg(3,5-PDC)(H_2O)] (Fig. 10), synthesized using a solvothermal method in dimethylformamide (DMF). Structure determination by single-crystal X-ray diffraction revealed that this chiral MOF was constructed by coupling Mg^{2+} ion helices with achiral 3,5-pyridine dicarboxylates, coordinated to water molecules which showed a 3D framework parallel to the hexagonal channels. Interestingly, Mg-MOF-1 remained stable and porous after evacuation of the coordinated water molecules. This MOF has been investigated in gas adsorption application and was found to have good selectivity towards hydrogen

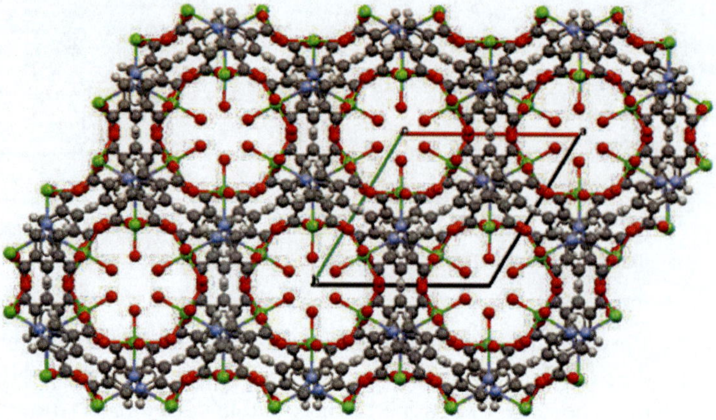

Fig. 10 Crystal Structure of Mg-MOF-1. Re-drawn from CCDC 772894

(H_2) gas (*ca.* 0.8 wt% at 77 K) and carbon dioxide (CO_2) uptake (*ca.* 0.7 mmol g^{-1} at 298 K) [63].

3.5 Luminescent Properties

One of the most interesting properties of MOFs is their luminescence behaviour. These materials are an important area of research because of their applications as small molecule sensors, pH sensors, light concentrators, photovoltaic devices, antennae in photosensitive bioinorganic compounds and high technology optics. MOFs can be considered excellent solid-state luminescence materials since they have structural predictability and well-defined environments for the chromophores in crystalline form. The combination of size or shape-selective absorption with luminescence can make MOFs good sensor materials. Luminescence supports the coordination of polymers with d^{10} elements and bridging ligand containing nitrogen or oxygen. Benzene di- and tricarboxylates are examples of bridges that may exhibit luminescence in coordination of polymers with metal nodes such as zinc or cadmium [38].

For examples, two stilbene-based MOFs were successfully synthesised by Bauer et al. [6] by using trans-4,4′-stilbene dicarboxylic acid (LH_2) and zinc nitrate in two different solvents. The use of dimethyformamide (DMF) has resulted in 2D network structure $Zn_3L_3(DMF)_2$ while a 3D porous framework structure of Zn_4OL_3 was obtained when diethylformamide (DEF) was used (Fig. 11). The optical properties of both MOFs indicated that the LH_2 organic ligand acted as a chromophore. In fact, the rigidity of the stilbene ligand increased after coordinated with the metal centre thus resulting in increased emission lifetimes of the MOFs crystals [6].

In particular, lanthanide MOFs also have unique luminescence properties such as high luminescence quantum yields, long-lived emission large Stokes shifts and

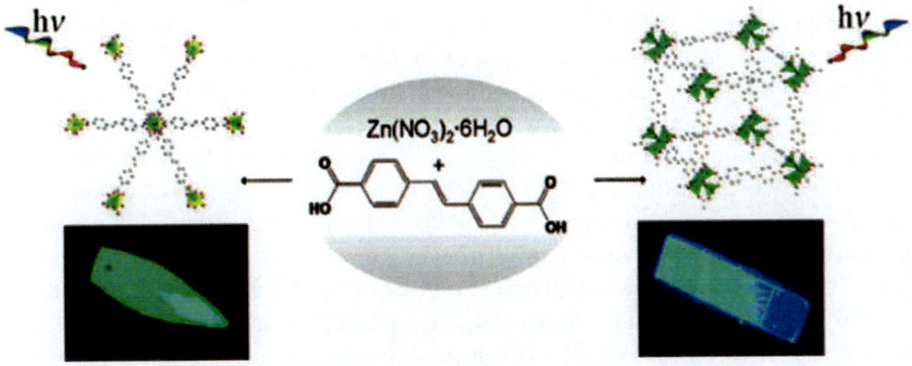

Fig. 11 Structure and luminescence properties of $Zn_3L_3(DMF)_2$ (left) and Zn_4OL_3 (right). Sourced from Bauer et al. [6]

characteristically sharp line emissions [21]. These materials produce an effective platform for chemical sensing since the luminescent intensity of lanthanide ions is very sensitive to the details of the coordination environment structure and the interaction of ligands or lanthanide ions with the analyte [21]. A MOF, Eu(BTC) was reported to have high selectivity luminescent sensing towards small molecules such as dimethyl formamide (DMF), acetone and ethanol [18]. When the activated Eu(BTC) was immersed into different solvents, they exhibited significant improvement due to the effects of quenching. From the report, a gradual decrease in fluorescence intensity was observed upon addition of acetone to the 1-propanol emulsion of Eu(BTC). This suggested that the binding interaction of the open metal sites with guest solvent molecules played an important role where the weakly coordinated 1-propanol was replaced by DMF and acetone molecules causing the addition and reduction of luminescence properties [21].

3.6 Magnetic Properties

Magnetic properties have become of interest in the construction of MOFs as magnetism can used to further enhance the potential applications of the MOFs. Generally, the linked with diamagnetic organic ligands paramagnetic metal ions are which can then be used efficiently in transmitting magnetic exchange [48]. Transition metals such as V, Cr, Mn, Fe, Co, Ni and Cu are common metal centres that have paramagnetic properties. The variable oxidation states that exist in these metal ions allow two important parameters which are spin quantum and magnetic anisotropy. Therefore, the use of polycarboxylic ligands with these metals can form MOFs with different flexible structures and different bridging modes which can affect the magnetic properties. In addition, the introduction of guest molecules in non-magnetic frameworks can enhance the magnetism of the MOFs thus these properties are advantageous for MOFs as sensors, lasers, non-linear optics and display devices [31, 49].

There are two common magnetic phenomena that can occur in the MOFs which are ferromagnetism and anti-ferromagnetism. Generally, ferromagnetism is created when the metal ions have unpaired electrons and thus the atom has a net magnetic moment. When an external magnetic field is applied, a parallel coupling spin exhibits strong attraction towards the magnetic field and their magnetic properties can be retained after the removal of the external field. Therefore, Magnetic Metal Organic Frameworks (MMOFs) can be designed by the choice of magnetic materials for their specific purposes [49]. An example of a chiral 3D Ni-Glurate, $[Ni_{20}(H_2O)_8(C_5H_6O_4)_{20}.40H_2O]$ was successfully synthesised showing ferromagnetic behaviour with a Curie temperature (*Tc*) of 4 K. Since nickel has four parallel neighbouring electrons, Ni–O–Ni bonds are formed between nickel clusters and oxygen from carboxylate ligands which occurred at edge-sharing nickel with octahedral geometry. However, the low *Tc* value indicated that the frameworks showed weak ferromagnetic interactions due to significant large Ni–O–Ni bridge angles (>90°) on the crystal structures [30].

Antiferromagnetic properties are generated when an antiparallel coupling of the metal ions and organic free radical entities were formed in the MOF structure. A cobalt MOF constructed from a 4,4′,4″-*S*-Triazine-2,4,6-triyl-tribenzoate (TATB) ligand with a formula of $Co_3(TATB)_2(H_2O)_2$ (Fig. 12a) formed a non-interpenetrated porous network where trimetallic cobalt clusters were extended by the TATB ligand pairs. This compound showed antiferromagnetic coupling between the neighbouring Co^{2+} (Fig. 12b). From the magnetic susceptibility measurements, the paramagnetic properties were found at a temperature range of 2–300 K [102].

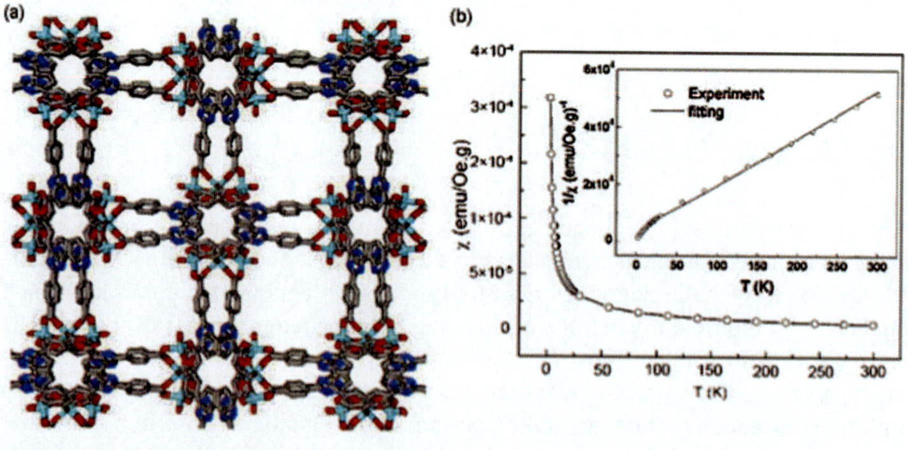

Fig. 12 **a** 3D framework structure of $Co_3(TATB)_2(H_2O)_2$. **b** Susceptibility as a function of temperature (inset shows the $1/\chi$ vs. T curve along with theoretical fitting). Sourced from Yu et al. [102]

Another study was carried out by Harbuzaru et al. [31] in the synthesis of Ln^{3+} based materials, ITQMOF-1 and ITQMOF-2 (ITQMOF = Instituto de Tecnologia Quimica Metal Organic Framework) with 4,4′-(hexafluoroisopropylidene)-bis(benzoic acid) (HFIPBB) ligand. In particular, both MOFs had similar properties but different structures since ITQMOF-2 was obtained from a secondary phase of the synthesised ITQMOF-1. The luminescent and magnetic properties of ITQMOF-1-Tb showed a strong green emission under UV light and a ferromagnetic interaction of Tb^{3+} ions. Meanwhile, the mixed Eu and Gd ITQMOF-1-(5Eu-95Gd) showed an antiferromagnetic interaction of Gd^{3+} and Eu^{3+} ions, thus exhibiting a red colored emission under UV light. Interestingly, ITQMOF-1-Eu was investigated for their use as sensor devices to detect ethanol in the presence of water under ambient conditions, and they were found to display excellent potential as sensors.

4 Metal Organic Frameworks as Composites

Over the last few decades, much efforts have intensified in making novel MOFs and studying their applications. Despite their versatility and functionality, MOFs also have several problems associated with them which is poor chemical stability that interferes with their full potential use [82]. Besides that, due to the high metal content of MOFs, some solvent molecules can still remain attached and bonded with metal atoms thus hindering the interaction of substrate and the MOFs. This may be cause the limitation of the MOFs applications [82]. MOFs also have several problems with their physical properties as most have brittle crystalline frameworks and typically exist as insoluble powders [70]. They also have poor physical strength, low process ability, hydrolytical instability and suffer from irreversible structural degradation after several hours of exposure to humid air.

Thus, studies on MOFs composites were initiated to form multifunctional composites that exhibit high performance with sophisticated architectures in order to rectify the issues associated with MOFs.

MOFs composites are materials that are made up of one MOF and one or more definite constituent materials where the character of each stays intact in any task. In general, there are dominant components that stands for a matrix material (continuous phase). Other components that make up a composite with matrix material represent a functional species or discontinuous phase [106]. Composite MOFs with other components have been reported to be excellent materials in many applications such as gas storage and catalysis. In this section, four types of MOF-composites will be discussed (Fig. 13).

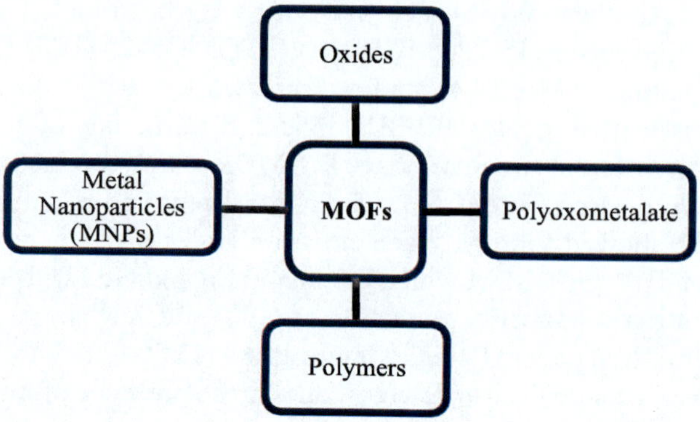

Fig. 13 MOFs composites with their functional materials

4.1 *Preparation of MOFs Composites*

There are several approaches for the synthesis of MOF composites. According to the composite structures, there are two basic types of MOF composites which are discontinuous phase and continuous phase. For the preparation of MOF composites where MOFs are in a continuous phase, several methods have been used. This method can be categorized into several main classes. In one of these methods, MOF precursors are mixed with pre-synthesised materials and then the synthesis procedure is performed.

This concept is known as 'bottle around ship' (BAS). In the BAS method, large particles of the composite materials are immobilized inside the cage of the MOF (Fig. 14a). This can only be done by in situ synthesis of the MOF. The MOF is built from the precursors around the composing materials, and when the MOF formation is complete, the composing materials are inserted into the pores [2, 98, 106]. This approach involves incorporating MNPs in a MOF matrix, which involves the control of metal precursor using a variety of techniques. For example, infiltration of solutions, deposition of vapors and solid grinding to introduce metal precursor, followed by reduction of the metal precursors to metal atoms.

Petit et al. [73] disclosed several articles on Graphite oxide/MOF (GO/MOF) composites prepared using this procedure and successfully applied it in the gas phase of adsorptive removal of toxic gas such as NH_3, H_2S, NO_2, and so on. In these composites, the GO layer was slightly separated and arranged between the MOF planar cage structures. In some cases, for this type of synthesis, the particles of the composing materials can be trapped in the pores of the MOF and thus can be stable or immobilized [73].

In another preparation method, the MOF composite was formed from the precursors of the composite materials and the preformed MOF. This method is sometimes called the 'ship in a bottle' (SIB) (Fig. 14b). In this case, the composing material is forming inside the MOF cage and the diffusion of the precursors is usually carried out in the cages by a solvent system. Due to the size increase, the particles remain stable inside the cages. In most cases, the precursors of the composing materials

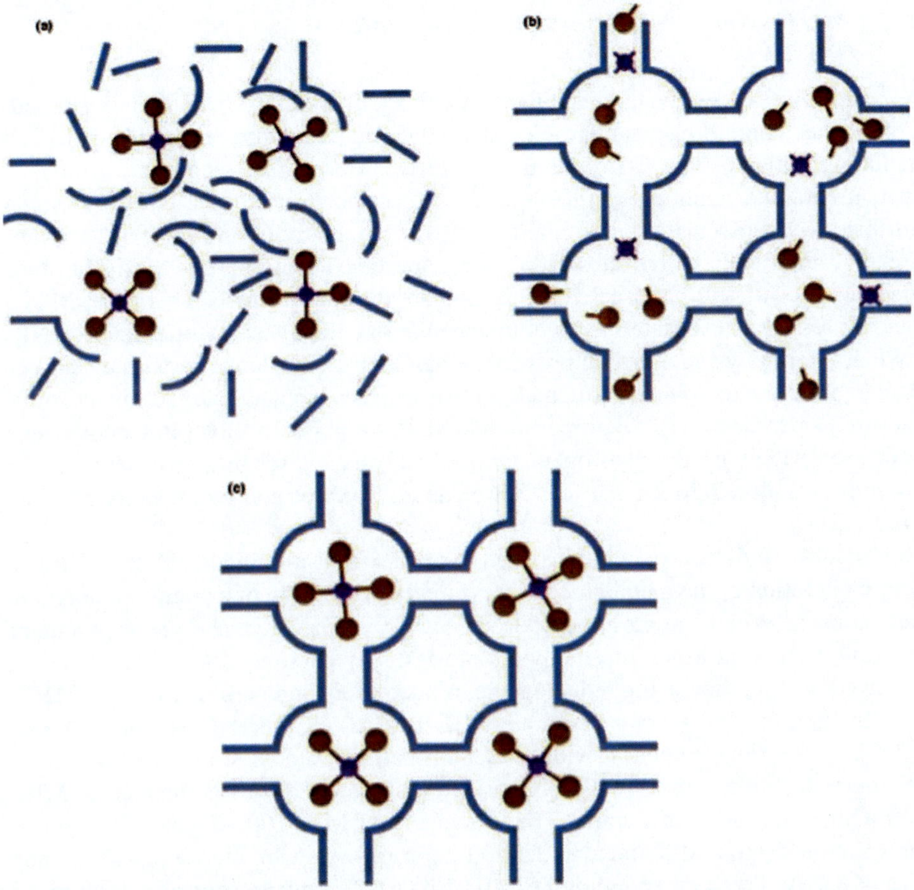

Fig. 14 Different strategies in the encapsulation of composite formation: **a** 'bottle around ship' (BAS) approach where the porous host materials are produced around the encapsulated materials; **b** 'ship in a bottle' (SIB) where guest moieties are prepared in the presence of porous host and; and **c** the composite prepared by either 'bottle around ship' (BAS) approach or 'ship in a bottle' (SIB) approach. Sourced from Ahmed and Jhung [2]

are added into the solution and disseminated into the pores. Then the composite is formed by stabilizing the complexity inside the pores of the MOFs by chemical or thermal methods (Fig. 14c) [2, 98, 106].

One of the best examples of this is the direct encapsulation of different porphyrines into zeolite-like Metal Organic Frameworks (ZMOFs). The resulting composites were successfully used in catalytic oxidation. Although SIB techniques are widely known for zeolite composites, their research in the field of MOFs has only recently begun and there are many opportunities to apply this encapsulation method to MOFs [73].

4.2 MOF-Metal Nanoparticle Composites

As a new class of catalysts, metal nanoparticles (MNPs) have been widely studied. Composites containing nanoparticles and MOFs have unique properties of MOFs including incisive crystal structures, extended surface area, single site catalysis, special confined nanopores, tunable and uniform pore structures, but avoid some intrinsic weaknesses such as limited electrical conductivity and lack of "conventional" catalytically active sites [106] in nanoparticles using porous MOFs because they have permanent nanoscale cavities and are thermally stable. Therefore, MOFs can be used as a host materials to hold and stabilize the guest nanoparticles. MNPs have different physicochemical properties than their bulk counterparts as a result of the large surface to volume ratio, making them attractive candidates for many applications. Depending on metal properties and MOF supports, MNP@MOF composites have been widely used in hydrogen storage, catalysis and sensing by utilizing MNP growth restriction in MOF matrices as well as surface area and size selectivity of the MOFs [106].

A solution infiltration method is used to prepare non-mobilized MNPs in MOFs, where a solution of the metal precursor, normally in the form of a common inorganic salt, is used. When the dissolved porous MOFs is soaked in the metal precursor solution, MNPs infiltrate into the pores of MOFs by capillary force, which is then followed by reduction using reducing agents such as H_2 and $NaBH_4$ to obtain MNPs accumulated in MOFs. As an example, MIL-100(Al) ($[Cr_3F(H_2O)_3O(btc)_2]$, btc = 1,3,5-benzenetricarboxylate), with tetrachloropalladinic acid solution was used in this purpose, Zlotea et al. [107] synthesized dispersed Pd NPs embedded in MIL-100 with a mean size of 2.5 nm. The framework of MIL-100(Al) has two types of large cavities (internal diameters: 2.1 and 2.5 nm) with a smaller window opening size of 8.7 Å. The corresponding Pd/MIL-100(Al) composite showed a high metal content of 10 wt% without degradation of the porous host. The significant decrease in BET surface area and the total pore volume of the composite determined that the insertion of Pd NPs altered the pore structure of MIL-100(Al) [107].

4.3 MOF–Metal Oxide Composites

Metal oxide nanomaterials with shape, size, crystallinity and functionality are widely used in applications such as electronics, optics, electrochemical energy consuming and storage, solar energy harvesting and also catalysis. To further improve the properties and introduce new functions, several attempts to integrate metal oxides, especially magnetic or semiconductor properties, and MOFs into core–shell nanostructures have been implemented [106]. The preparation methods for these composites are as follows:

1. The generation of metal oxides in the cavities of MOFs via oxidative annealing or decomposition in the preloaded precursors.

2. The encapsulation of pre-synthesised metal oxide NPs inside the MOFs matrices. In this method, the metal oxide cores that undergoes pre-treating with a material such as Polyvinyl pyrrolidone (PVP) has increased flexibility toward aMOFs and can facilitate the growth of the MOFs around metal oxide NPs. Alternatively, the use of metal oxide NPs as both the template and metal precursors for the formation of MOFs, namely self-template synthetic strategy, can also be used to obtain well-defined core–shell nanostructures.

As an example, Müller et al. [66] reported a solvent-free two step syntheses used in the synthesis of metal-oxide-NP@MOF nanocomposites by dispersing the metal oxide nanomaterials inside the MOFs. The two steps were using gas phase infiltration of the volatile metal precursors and followed by oxidative annealing of the loaded metal precursors. Nanosized ZnO and TiO_2 from $ZnEt_2$ and $Ti(O\text{-}iPr)_4$ were hosted inside MOF-5. Based on the results, doubly loaded Cu/ZnO@MOF-5 and Au/MO_x@MOF-5 (M = Zn, Ti; x = 1, 2) samples were prepared by gas phase loading of MO_x@MOF-5 with the corresponding volatile organometallic molecules and hydrogenolysis was thermally activated [66].

4.4 *MOF–Organic Polymer Composites*

Organic polymers own many unique characteristics, including ease of production, light weight, and good thermal and chemical stability, where the properties are desirable for integration with other functional materials to make composites. Besides that, compared to the polymers in the bulk state, confined polymers at the nanometer scale exhibit more fascinating and unexpected properties. Therefore, a new methodology for controlling polymer synthesis within porous MOFs has been developed to produce MOF–organic polymer composites formed from various combinations of MOFs and organic polymers that would constitute a class of composite materials with mixed properties.

Solution blending is the simplest and most effective technique to incorporate MOF crystals in the polymers. Polymers are usually soluble in organic solvents so this technique has been widely used to synthesize MOF/polymer composites. To obtain the maximum homogeneity of the MOFs within the solution, sonication or mechanical stirring is applied to a solution mixture containing the MOFs and dissolved polymer. Finally, evaporation of the solvent results in the formation of a MOF/polymer composite [106].

In 2005, the first polymerisation of styrene (St) in the nanochannels (7.5×7.5 Å^2) of $[M_2(bdc)_2(ted)]$ (M = Cu or Zn; ted = triethylenediamine) was reported by Uemura et al. Quantitative recovery of the stationary polystyrene (PSt) with a molecular weight of ~55000 was achieved by decomposing the host framework in NaOH solution. The recovered PSt showed a narrower distribution of molecular weight than those synthesized in bulk and solution [89].

4.5 *MOF-Polyoxometalate Composites*

Polyoxometalates (POMs), is a species of metal oxide clusters that has high negative charge properties and are a distinctive class of molecular inorganic compounds with different composition, sizes and shapes [106]. POMs play an important role in various fields, such as catalysis, medicine, electrochemistry, photochromism and magnetism. Particularly, the catalytic applications of POMs in oxidation reactions has been of interest to researchers. However, under catalytic conditions, POMs tend to have low specific surface area and decreased stability. Thus, immobilizing POMs in porous solid materials such as MOFs is a promising approach to solve their inherent problems and then optimize their catalytic performance. The dispersion of POMs in MOFs prevent POMs from conglomerating and deactivating, thereby their catalytic properties can be enhanced (Scheme 1).

In recent years, many guest structures of POM@MOF have been synthesized through self-assembly reactions mainly driven by the chemical coordination of each component. Buru et al. [12] prepared a series of POM@HKUST-1 composites based on heteropolyacids with a formula of $H_nXM_{12}O_{40}$ (X = Si, Ge, P, As; M = W, Mo), from simple one-pot hydrothermal reactions (Fig. 15). X-ray crystallographic analysis showed that the Keggin (heteroponic acid structure) of the polyanions were alternately arrayed as guests in HKUST-1 cage, and only about half of the pores in the host were occupied. The catalytic performance of $PW_{12}O_{40}$@HKUST-1 with accessible catalytically active sites was evaluated by the hydrolysis of esters in excess water. The result showed a high catalytic activity and size selectivity of the $PW_{12}O_{40}$@HKUST-1 [12].

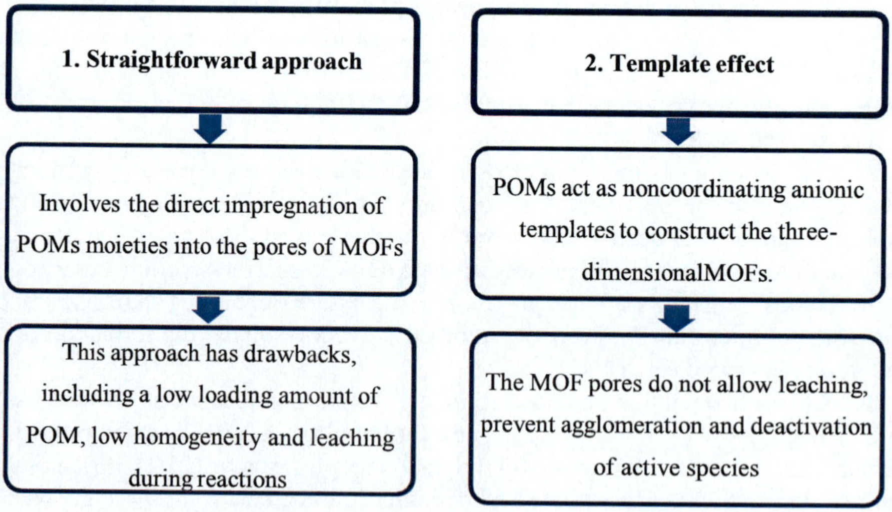

Scheme 1 Method of preparation of MOFs polyoxometalate composites

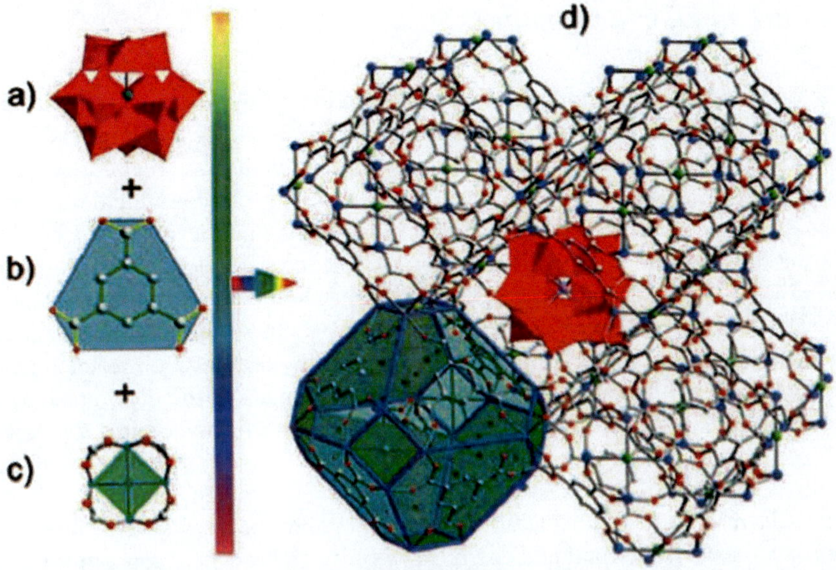

Fig. 15 The composite of POM @ HKUST-1 composite by simply mixing MOF and polyanion precursors. **a** Keggin polyanion. **b** Three-connected node and hexagonal face (blue) defined by a BTC ligand linked to six adjacent Cu^{2+} ions. **c** SBU and square face (green) defined by four Cu^{2+} ions. **d** Cube of eight sodalite-like truncated-octahedral cages sharing square faces. Sourced from Buru et al. [12]

5 Metal-Organic Frameworks in Emerging Engineering Applications

In recent years, MOFs have been regarded as one of the important porous materials. The combination of various types of metal nodes and unlimited choices of organic linkers has led to the expansion of the MOF family. MOFs possess advantages over zeolites and activated carbons as traditional porous materials are unable to tune the structures. The organic linkers in MOFs can be modified by incorporating active functional groups to tune the pore surfaces of MOFs and the construction of MOFs using unsaturated metal centre enable MOFs to fulfill some important and challenging applications. These examples of structural tunability with some properties of MOFs including large surface area, structural tunability, and high thermal and chemical stabilities have made MOFs ideal candidates for industrial applications. MOFs have been extensively used in gas storage, drugs delivery, chemical sensing, catalysis and semiconductors. This review intends to describe the important structural features of MOFs in industrial applications.

5.1 Carbon Dioxide Capture

The sharp rising level of atmospheric carbon dioxide, CO_2 has caused serious environmental concern as a result of anthropogenic emissions due to human activities. The major activities, predominantly from the combustion of fossil fuels, are expected to increase in the future due to the accelerating demand for economic development which causes global warming and deterioration such as land droughts, destruction of marine life and floods [50]. Hence, there is a need to develop materials that can remove the excessive amounts of CO_2 from atmosphere. Aqueous amine solutions have been used extensively to remove CO_2 from gas mixtures. However, aqueous amine solutions have several limitations including low regeneration as the aqueous amine solutions are unstable towards heating which leads to the decomposition of the solution, thus, decreasing the adsorbent material performance in CO_2 capture [79].

Zeolites are one of the most established adsorbents to remove CO_2 and these materials have attracted considerable attention as they possess several interesting features such as high thermal and chemical stability and low heat capacities [9, 41]. However, the performance of zeolites gradually decreased due to the exposure to water vapor in the flue gas stream which lead to low CO_2 adsorption over time [54].

In the past decade, metal-organic frameworks (MOFs) have been recognized as one of the most innovative materials in chemistry and serve as promising adsorbents for CO_2 capture [65]. In comparison with traditional adsorbent materials, MOFs demonstrated excellent properties in terms of extraordinary surface area, tunable pores and diverse active sites for industrial applications [88]. The design of MOFs for CO_2 capture is based on the pore sizes, active functional groups, pore volume and unsaturated metal sites. Mg-MOF-74 is an example of a MOF that exhibited excellent capacity in storing CO_2 at 298 K and 1 bar. Mg-MOF-74 is comprised of magnesium (Mg(II) in the metal cluster of the framework which forms one-dimensional secondary building units (SBUs) bonded to 2,5-dihydroxyterephtalate that play a role as inorganic nodes in Mg-MOF-74 (Fig. 16) leading to the formation of large hexagonal channels (~12 Å), thus, allowing CO_2 molecules to coordinate at vacant sites of the unsaturated metal centre of Mg(II) after the evacuation of guest molecules in the framework. It was observed from the uptake of CO_2 using Mg-MOF-74 that the unsaturated metal centres were a main factor as they exhibited high heats of adsorption (Q_{st}) at low pressure, hence, allowing strong electrostatic interactions and high selectivity toward CO_2 [9].

5.2 Methane Storage

An alternative cleaner fuel using natural gas has attracted considerable attraction as a new method to replace our dependency on petroleum. Methane gas is a component of natural gas which is environmentally clean and cheaper and it is considered as

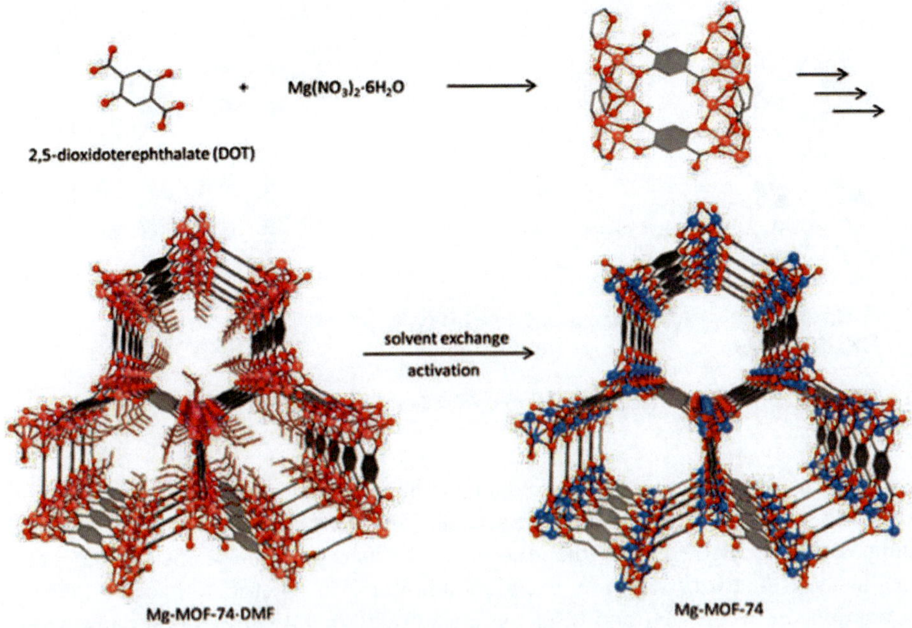

Fig. 16 The structure of Mg-MOF-74. C atoms are shown in gray, O atoms in red, 6-coordinate Mg atoms and terminal ligands in pink, and 5-coordinate Mg atoms in blue. H atoms and terminal ligands on the fragment at top right are omitted for clarity. Sourced from Britt et al. [9]

an alternative source of energy for automobile transport [15]. Methane also contains the highest hydrogen to carbon ratio in comparison to other hydrocarbons and the combustion of methane emits only small amounts of CO_2 [50]. Despite these benefits, the deployment of methane in the transportation sector has been quite challenging as the volumetric density of methane gas is very low at about 0.04 MJ/L in ambient conditions which constrains the utilization of methane gas in various applications [64]. Hence, it was deemed necessary to develop alternative ways to enhance energy density for large scale productions of natural gas.

Compressed Natural gas (CNG) and Liquefied Natural Gas (LNG) are the common methods used for natural gas storage (such as methane). Nevertheless, these methods are associated with several disadvantages which include expensive storage cost and efficiency methane storage at ambient conditions. Therefore, Adsorbed Natural Gas (ANG) has become the most prominent method to store methane by reducing the pressure at room temperature to enable the methane to be stored in lightweight and inexpensive fuel tanks. The use of adsorbent materials for methane storage has been evaluated using zeolites and activated carbons. However, zeolites exhibit low methane uptake at 100 cm^3 (STP) cm^{-3} (Standard temperature: 273.15 K and pressure: 101.325 kPa) due to their extreme hydrophilicity and low micropore volume. Activated carbon also recorded low methane uptake around 50–160 cm^3 at STP as it is quite difficult to tune the pore size of activated carbons which limits it's practical application as adsorbent materials for gas storage [32].

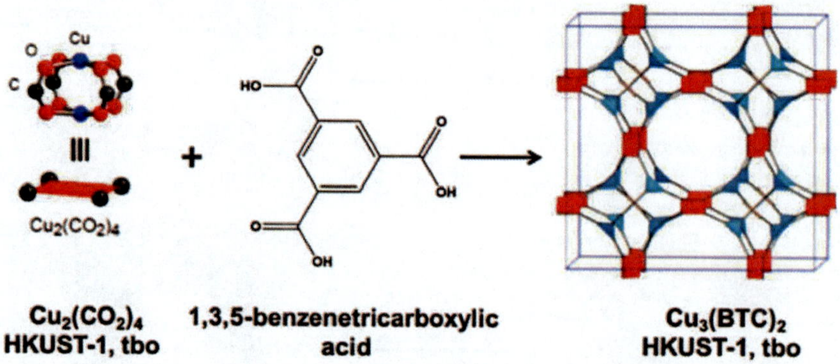

Fig. 17 The formation of HKUST-1. Sourced from Furukawa et al. [27]

MOFs have been recently studied in methane storage applications relying on electrostatic and van der Waals interactions [56]. Growing studies have reported using established MOFs for methane storage. The incorporation of unsaturated metal sites in the framework led to an excellent amount of CH_4 storage based on strong interactions between CH_4 and MOFs at unsaturated metal sites. Based on previous studies, HKUST-1 in Fig. 11 was utilized as adsorbent material for CH_4 storage and the framework exhibited remarkable performance on CH_4 uptake of about 276 cm^3 (STP) cm^{-3} at 298 K and 65 bar [71]. Recent studies have also reported high total volumetric and gravimetric CH_4 uptakes in a new MOF, Cu-**tbo**-MOF-5. This type of the framework was inspired from the building unit of HKUST-1 that was arranged in a **tbo** (The bold letter code represents network topologies from the Reticular Chemistry Structure Research (RCSR) which is an underlying topology constructed with 3,4-connected networks with square SBUs containing copper paddle wheels (Fig. 17).

Cu-**tbo**-MOF-5 is comprised of 3,6-dimethyl-1,2,4,5-tetra-(biphenyl-3′,5′-dicarboxylic acid) benzene, (H_8L) with Cu(II) synthesised under solvothermal conditions (Fig. 12). The presence of methyl groups at the centre of the organic linkers are key in increasing the electron density of the ring which lead to strong π–π interactions in the framework and CH_4 thus, increasing the maximum capability of Cu-**tbo**-MOF to capture CH_4. Consequently, Cu-**tbo**-MOF recorded a volumetric total uptake about 221 cm^3 (STP) cm^{-3} and a gravimetric total capacity of 372 cm^3 (STP) cm^{-3} at 85 bar and 298 K. Cu-**tbo**-MOF proved to be the best performing MOFs for CH_4 which was attributed to the structural features of the framework containing unsaturated metal sites, Cu(II) and expanded organic linkers that provided good interaction sites, thus, increasing the capacity of CH_4 capture [85] (Fig. 18).

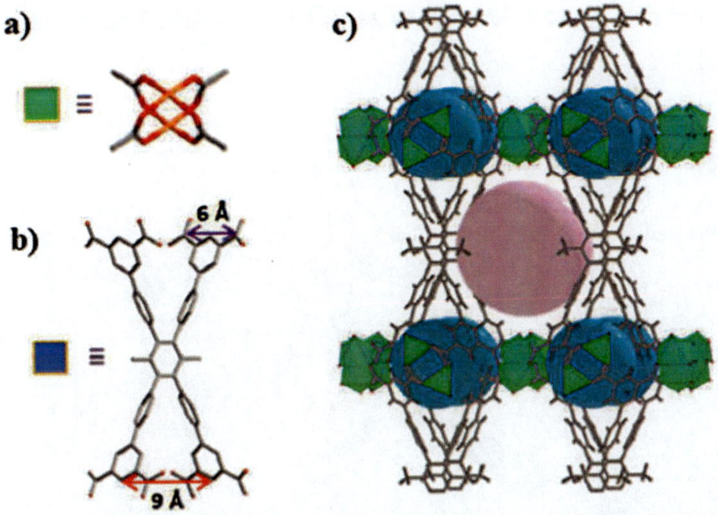

Fig. 18 **a** Four connected square-like paddle wheels, **b** Octatropic linker, 3,6-dimethyl-1,2,4,5-tetra-(biphenyl-3′,5′-dicarboxylic acid)benzene, (H_8L) and **c** Crystal structure of Cu-tbo-MOF-5. Sourced from Spanopoulos et al. [85]

5.3 Volatile Organic Compounds (VOCs)

Aromatic hydrocarbons such as toluene, xylenes, benzene and ethylbenzene are classified as components of volatile organic compounds (VOCs) which are one of the groups of toxic pollutants. These types of hydrocarbons are present in indoor/outdoor air which are emitted from cleaning products, paints, gasoline and chemical industries. Long exposure to these pollutants causes adverse effects on environmental and human health. The development of solid porous materials are highly desirable to capture VOCs from the environment, thus, minimizing the impact on health and environmental concerns from the emission of VOCs [4].

Vellingiri and colleagues [91] carried out an investigation of the interaction of VOCs (benzene, toluene, styrene and p-xylene) with several types of MOFs; MOF-199, MOF-5 and Eu-MOF. From the experimental data, MOF-199/HKUST-1 was found to have strong interactions with polar VOCs due to the presence of unsaturated metal sites, Cu(II), thus, showing a high amount of adsorption capacity of VOCs at 72.2 mg g^{-1} (Fig. 19). DFT simulation analysis also proved that the strong π–π interactions between 1,3,4-tricarboxylic acid and VOCs molecules played an important role in influencing the adsorption of VOCs molecules [91]. Another example was reported on the use of HKUST-1 as a pre-concentration material which had a good performance in the adsorption and desorption of benzene at low concentrations [52].

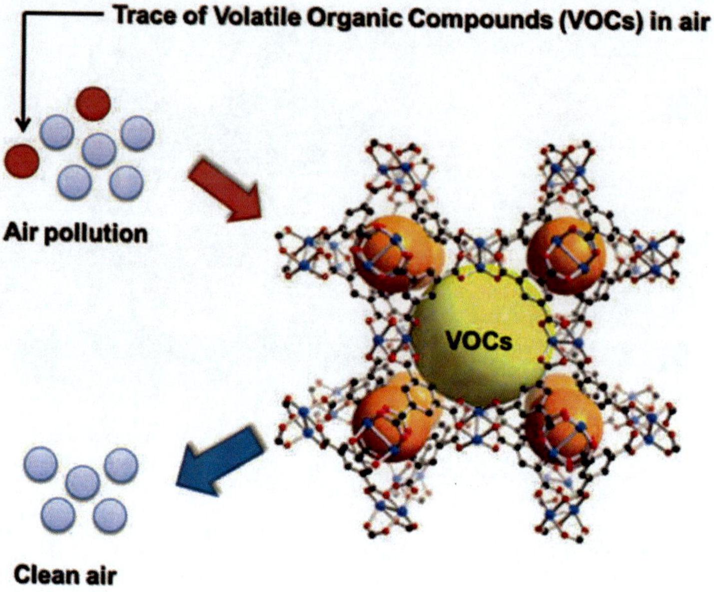

Fig. 19 Illustration of HKUST-1/MOF-199 capturing VOCs from the surrounding. Sourced from Leidinger et al. [52]

5.4 *Catalytic Applications in Knoevenagel Condensation*

Knoevenagel condensation is defined as the reaction between a carbonyl compound (aldehydes/ketones) and a compound containing methylene groups for the formation of new carbon-carbon bonds [45]. This condensation is important for the production of α,β-unsaturated compounds that act as intermediates in the formation of drugs, natural products and chemicals for industrial applications [68]. Solid base catalysts such as alkali or alkali earth metal oxides have been widely used in the condensation reaction in homogeneous systems. However, the separation and disposal of solid-based catalysts in homogeneous systems has been quite challenging, thus, affecting the recyclability of the catalyst and low catalyst loading [68].

In pursuit of developing solid base catalysts in heterogeneous systems, MOFs have attracted continuous attraction owing to their high thermal and chemical stability which allows the recyclability of the framework after being reused several times. Different types of basic sites on MOFs produces different yields and this was proven from the cooperative study of the catalytic activity of benzaldehyde and malonitrile in toluene at 23 °C for 2 h using a solid porous catalyst; an unfunctionalised MOFs, UiO-66 and a functionalized series of UiO-66, NH_2-UiO-66 and RNH_2-NH-UiO-66. The types of basic sites on MOFs structures play an important role in determining the yield of the product. From the experimental results, the unfunctionalised UiO-66 had yields of below 1% which was due to absence of amine groups in the structure. The catalytic activity of NH_2-UiO-66 and UiO-66 showed remarkable differences, where RNH_2-UiO-66 exhibited high yields of around 97% in contrast to the unfunctionalised UiO-66 and NH_2-UiO-66. This was due to the incorporation of primary aliphatic amines on RNH_2-UiO-66 rather on the aromatic amines of NH_2-UiO-66 which only recorded low yields of about 5% [105].

5.5 *Chemical Sensors*

MOFs have framework structures that possess interesting structural features for a wide range of applications. Luminescent MOFs (LMOFs) are constructed using d^{10} metal ions that often lead to ligand-ligand charge transfer. Taking advantage of the structural features of LMOFs, many researchers have used LMOF structures for chemical sensors.

This concept was applied in synthesising LMOF-311 and LMOF-312 using 1,1,2,2-tetrakis(4-(4-carboxyphenyl)phenyl) ethane (H_4TCBPE) and cadmium(II) nitrate tetrahydrate, $Cd(NO_3)_2.4H_2O$ for LMOF-311. LMOF-312 was synthesised using H_4tcbpe and zinc(II) nitrate hexahydrate. Both MOFs were used as chemical sensors for nitroaromatic compounds (nitrobenzene, 2,6-dinitrotoluene, *p*-nitrotoluene, *m*-nitrotoluene). The photoluminescence spectra of LMOF-311 and LMOF-312 indicated good luminescence, where LMOF-311 acquired an internal quantum yield about 66.8% and LMOF-312 achieved a yield of 65.7%. LMOF-311 was selected for chemical sensor application studies using fluorescence titration by adding the nitroaromatic compounds and the Stern-Volmer (SV) plot was applied to interpret the sensing efficiency (K_{SV}) of LMOF-311. From the range obtained for nitroaromatic compounds, the highest K_{SV} constant was for 2,6-dinitrotoluene at 2063 M^{-1} which proved that LMOF-311 detected to 2,6-dinitrotoluene more strongly than other nitroaromatic compounds [92, 93].

Another iridium(III)-based MOF was successfully developed as a chemical sensor for Fe^{3+} and $Cr_2O_7^{2-}$ and an optical sensor for selective sensing for adenosine triphosphate (ATP^{2-}) over adenosine diphosphate (ADP^{2-}) in aqueous media. The Iridium(III)-based MOF was constructed using the iridium complex, $Ir(ppy)_2(L)$(ppy = 2-phenylpyridine and L = organic dicarboxylates) and cadmium nitrate hexahydrate, $Cd(NO_3)_2$. $6H_2O$. The iridium complex is a significant complex for chemical sensing due to their long-lived phosphorescence and high quantum yields which possess distinct advantages for sensor applications. The sensor activity of the iridium-based MOF showed good data for Fe^{3+} where the K_{SV} value was recorded at 1.165 $\times$ 104 M^{-1} which was due to a strong quenching effect on the iridium-based MOF. For the detection of the anion, $Cr_2O_7^{2-}$, the iridium-based MOF clearly showed high selectivity towards $Cr_2O_7^{2-}$ and recorded a high K_{SV} value around at 3.475 $\times$ 104 M^{-1} [24].

5.6 Drugs Delivery Systems

The structural features of MOFs allow the encapsulation of drugs and the frameworks are promising materials to deliver drugs to the specific locations in the human body. The interior environment of the MOFs structures also play an important role for drug administration. This is based on research that was conducted using UiO-66 for the encapsulation of ibuprofen (IBU) where the building unit of the framework was based on zirconium(IV) oxoclusters, ($Zr_6O_4(OH)_4$) and 1,4-benzenedicarboxylic (BDC) arranged as a coordination polymer, forming tetrahedral (~8 Å) and octahedral (~11 Å) cavities [80]. UiO-66 was tested for its chemical nature using simulated water adsorption isotherms by Grand Canonical Monte Carlo (GCMC) simulations. From water adsorption isotherms, UiO-66 was classified as hydrophobic because of the water condensing at 80% relative humidity in the pores of the frameworks [28]. Ibuprofen (IBU) is a hydrophobic drug (log *P*: 3.97) and has functionalised benzene linkers with specific functional groups such as carboxylic acids and alkane chains which create weak interactions (π–π staking and hydrogen bonding) within the pores of UiO-66 that enhances the loading capacity of IBU into UiO-66. The loadings of IBU into UiO-66 achieved 97% for the first hour, which indicated the same compatible chemical nature (hydrophobicity) of the drug and MOFs. To prove the presence of IBU in UiO-66, several characterisations were carried out using Fourier transform infrared spectroscopy (FTIR) to observe the shift of the ν(C=O) group. That was indicative of interactions of IBU with the framework structure in comparison to the FTIR of pure IBU. For the drug release study, high performance liquid chromatography (HPLC) was conducted to measure the amount of IBU delivered, which recorded about $46 \pm 2\%$ as the influence of hydrophobicity of the framework decided the drug release performance [80].

Another example of a cationic drug carrier, MOF-74-Fe(III) was successfully developed by Yang et al. [97] through the oxidation of neutral MOF-74-Fe(III) [36]. The study showed that the cationic MOF exhibited around 15.9 wt% loading capacity of ibuprofen (IBU) [97] (Fig. 20). Interestingly, during the preparation process, two distinctive IBU^- anions were present in the MOF channel after the production of

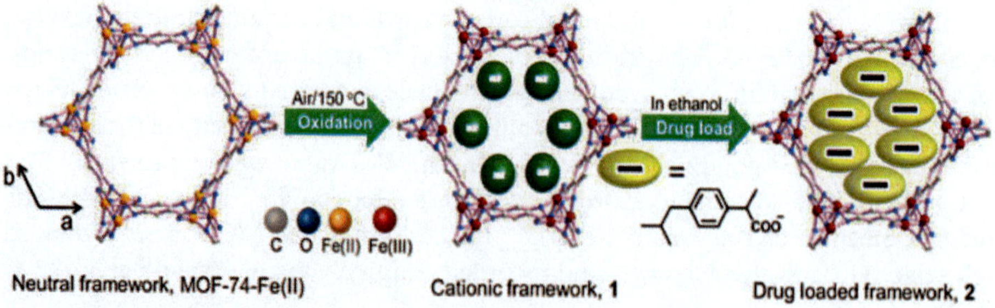

Fig. 20 Schematic presentation of the process of MOF-74-Fe(II) oxidation to a cationic framework and drug loading. Sourced from Hu et al. [36]

hydroxide, thus, two different release behaviors were observed. The first release was sodium ibuprofen and other coordinated free anions due to ion exchange between the encapsulated drug and PBS solution. The following release was the targeted drug through competitive adsorption with phosphate anions [36].

5.7 Semiconductors

In recent years, there has been increasing interest in the development of electronic and optoelectronic devices with high performance energy storage to replace traditional inorganic materials. MOFs are promising for electronic applications due to their unique properties which include structural tunability and large surface area [26]. However, the implementation of MOFs in device applications such as electrochemical capacitors [84], thermoelectric generators [10] fuel cells [8] and chemical sensors [47] remain challenging due to their low electrical conductivity [87].

The structural features of MOFs containing carboxylate linkers and metal ions that are connected by electronegative oxygen or nitrogen atoms require a high amount of voltage for electrons to pass through the organic linkers. As a result, poor overlap from *d*-orbital of metal to the electronegative oxygen atoms results in the low electrical conductivity of MOFs. Therefore, MOFs have been designed to improve electrical conductivity by constructing the framework using planar organic linkers that consist of benzene and triphenylene derivatives linkers with active functional groups (amine, thiol and hydroxyl) and square metal complexes to form two-dimensional (2D) MOFs with stacked honeycomb lattices to improve electrical conductivity. The stacked honeycomb lattice from benzene and triphenylene derivatives ligands allow full π-d conjugation which is important to improve the electrical conductivity of MOFs. In 2018, Dong and colleagues developed a 2D MOF based on polycyclic aromatic hydrocarbon, 1,2,3,4,5,6,7,8,9,10,11,12-perthiolated coronene (PTC) as an organic linker to coordinate with planar iron-bis(dithiolene) as shown in Fig. 21, forming a framework structure of hexagonal lattices and van der Waals layer to be utilized as the material for ferromagnetic semiconductors. From the experimental results, the coronene-based conjugated PTC-Fe 2D MOFs showed excellent electrical conductivity with ~10 S cm^{-1} at 300 K and the framework also showed semiconductor properties as the value of electrical conductivity gradually decreased upon cooling. The semiconductor behaviour of the MOFs was measured by variable-temperature studies where the bond between PTC and Fe displayed a ferromagnetic ground state at low temperature. This was due to the hybridization of *d*/*p* orbital of Fe and the coronene core with the iron-bis(dithiolene) nodes. The hybridization of the orbitals influenced the magnetic properties of the stacked honeycomb lattice. Incorporation of square metal complex and high number of benzene rings in the framework structure enables MOFs to be promising materials for device applications [22].

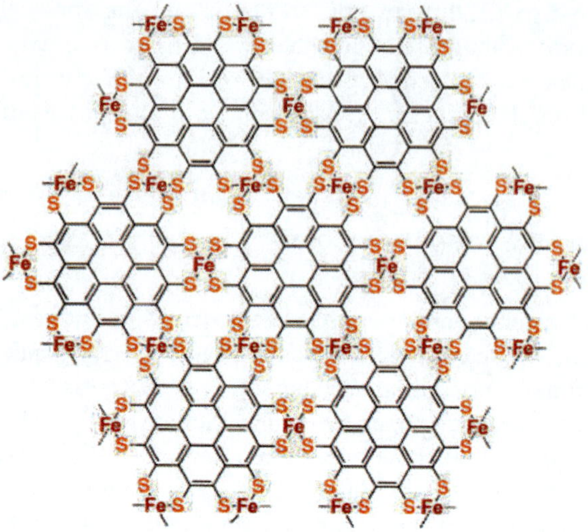

Fig. 21 The structure of PTC-Fe 2D MOF. Sourced from Dong et al. [22]

6 Conclusions, Limitations and Future Research

MOFs are porous materials constructed via self-assembly from inorganic and organic bridging which are constitute a relatively new and actively studied class of materials. In principle, the wide choice of metal clusters together with infinite choices of ligands allows the formation of materials that have a large range of structural, catalytic, magnetic, electrical, and optical properties. Thus, the rational design and functionalisation on the structures can be achieved by using rigid and flexible organic bridging ligands which enable the fine-tuning of their properties.

As shown in the synthesis of MOFs, the variety of possible organic linkers and positively and negatively charged frameworks makes MOFs are more advantageous materials compared to zeolites, activated carbon and mesoporous silica. The various synthesis methods have proved that MOFs can be simply synthesised and can give different types of crystals according to their purposed. Therefore, in situ and ex situ methods are necessary to study the different stages of crystallization because these techniques will comprise of various analytical tools with the combination of theoretical methods. These methods are suitable to synthesise new or improved metastable polymorphs and MOFs [76]. Not only the synthesis has to be carefully planned and strategized but the activation and characterisation of these new MOFs need to be considered. Synthesis of functional metal organic frameworks need to be well-planned to improve the properties for specific applications.

The study of MOFs as platforms to prepare composites is still at their early stages but very rapid progress is being made. When designing the composites, the primary consideration is the structure and property of the parent MOFs. The fact is that the current studies are often focused on MOF-5, MIL series, ZIF family and HKUST-1. Thus, alternatives to synthesise chemically and thermally stable porous MOFs on a large-scale synthesis are necessary. For the time being, the nature of the functional materials added to MOFs is very clear which is critical to construct composite structures and improve the composite properties. The nature of the interface between the MOFs and the secondary composites affects the composite performance. Therefore, an understanding of the interactions that occur at the interface is highly recommended in order to investigate structure–property relationships and to design high-performance MOF-composites for the next generation. The combination of different guest moieties within one MOF and the use of different functional material can open up new approaches in developing multifunctional materials.

From the engineering applications of MOFs, we can conclude that MOFs constructed from flexible and rigid ligands are found to be useful in selective gas adsorption. Therefore, MOFs has become one of the most promising applications in size-, shape-, and enantio-selective catalysis. Since microporous MOFs have limited pore sizes for the encapsulation of drugs, much research is necessary to understand the potential of MOFs for drug delivery applications. Many studies have reported on the toxicity of MOFs and their bio-degradation, thus, more efforts should be made to design nontoxic or low toxicity and good biocompatible MOFs. These can allow prolonged stay in blood circulation and ensure the decomposition of products can be processed through the body's metabolic system.

Future work may be focused on more discussion on the activation processes, post synthetic modification and Solvent Assisted Linker Exchange (SALE) in the synthesis of MOFs, in order to improve their physical and chemical properties. Silica nanochemistry and quantum dots (QDs) have multiple functionalities, and unique electrical and optical properties. Additionally, the immobilization of enzymes on a solid support can enhance their reactivity and selectivity, Hence, MOF-silica, MOF-quantum dots and MOF-enzyme composites are future promising leads in MOF-composites research.

In conclusion, MOF research has achieved great advances in the last decade from their structural diversity till their applications. Many new types of applications have emerged and will continue to do so, as real-world solutions are needed. Therefore, the future of this field of research is indeed very bright.

References

1. Abrahams, B.F., Hoskins, B.F., Michail, D.M., Robson, R.: Assembly of porphyrin building blocks into network structures with large channels. Nature **369**(6483), 727–729 (1994)
2. Ahmed, I., Jhung, S.H.: Composites of metal–organic frameworks: preparation and application in adsorption. Mater. Today **17**(3), 136–146 (2014)
3. Ameloot, R., Stappers, L., Fransaer, J., Alaerts, L., Sels, B.F., De Vos, D.E.: Patterned growth of metal-organic framework coatings by electrochemical synthesis. Chem. Mater. **21**(13), 2580–2582 (2009)
4. Barea, E., Montoro, C., Navarro, J.A.R.: Toxic gas removal metal–organic frameworks for the capture and degradation of toxic gases and vapours. Chem. Soc. Rev. **43**(16), 5419–5430 (2014)
5. Batten, S.R., Hoskins, B.F., Moubaraki, B., Murray, K.S., Robson, R.: Crystal structures and magnetic properties of the interpenetrating rutile-related compounds $M(tcm)_2$ [M =

octahedral, divalent metal; tcm = tricyanomethanide, $C(CN)_3-$] and the Sheet Structures of [$M(tcm)_2(EtOH)_2$] (M = Co or Ni). J. Chem. Soc. Dalton Trans. (17), 2977–2986 (1999)
6. Bauer, C.A., Timofeeva, T.V., Settersten, T.B., Patterson, B.D., Liu, V.H., Simmons, B.A., et al.: Influence of connectivity and porosity on ligand-based luminescence in zinc metal-organic frameworks. J. Am. Chem. Soc. **129**(22), 7136–7144 (2007)
7. Bennett, T.D., Goodwin, A.L., Dove, M.T., Keen, D.A., Tucker, M.G., Barney, E.R., et al.: Structure and properties of an amorphous metal-organic framework. Phys. Rev. Lett. **104**(11), 115503 (2010)
8. Brandon, N.P., Brett, D.J.: Engineering porous materials for fuel cell applications. Philos. Trans. R. Soc. A: Math. Phys. Eng. Sci. **364**(1838), 147–159 (2006)
9. Britt, D., Furukawa, H., Wang, B., Glover, T.G., Yaghi, O.M.: Highly efficient separation of carbon dioxide by a metal-organic framework replete with open metal sites. Proc. Natl. Acad. Sci. **106**(49), 20637–20640 (2009)
10. Bubnova, O., Crispin, X.: Towards polymer-based organic thermoelectric generators. Energy Environ. Sci. **5**(11), 9345–9362 (2012)
11. Bureekaew, S., Shimomura, S., Kitagawa, S.: Chemistry and application of flexible porous coordination polymers. Sci. Technol. Adv. Mater. **9**(1), 014108 (2008)
12. Buru, C.T., Li, P., Mehdi, B.L., Dohnalkova, A., Platero-Prats, A.E., Browning, N.D., et al.: Adsorption of a catalytically accessible polyoxometalate in a mesoporous channel-type metal–organic framework. Chem. Mater. **29**(12), 5174–5181 (2017)
13. Butova, V.V., Soldatov, M.A., Guda, A.A., Lomachenko, K.A., Lamberti, C.: Metal-organic frameworks: structure, properties, methods of synthesis and characterization. Russ. Chem. Rev. **85**(3), 280–307 (2016)
14. Cavka, J.H., Jakobsen, S., Olsbye, U., Guillou, N., Lamberti, C., Bordiga, S., et al.: A new zirconium inorganic building brick forming metal organic frameworks with exceptional stability. J. Am. Chem. Soc. **130**(42), 13850–13851 (2008)
15. Celzard, A., Fierro, V.: Preparing a suitable material designed for methane storage: a comprehensive report. Energy Fuels **19**(2), 573–583 (2005)
16. Chae, H.K., Siberio-Pérez, D.Y., Kim, J., Go, Y., Eddaoudi, M., Matzger, A.J., et al.: A route to high surface area, porosity and inclusion of large molecules in crystals. Nature **427**(6974), 523–527 (2004)
17. Chalati, T., Horcajada, P., Gref, R., Couvreur, P., Serre, C.: Optimisation of the synthesis of MOF nanoparticles made of flexible porous iron fumarate MIL-88A. J. Mater. Chem. **21**(7), 2220–2227 (2011)
18. Chen, B., Yang, Y., Zapata, F., Lin, G., Qian, G., Lobkovsky, E.B.: Luminescent open metal sites within a metal–organic framework for sensing small molecules. Adv. Mater. **19**(13), 1693–1696 (2007)
19. Chen, Y., Li, P., Modica, J.A., Drout, R.J., Farha, O.K.: Acid-resistant mesoporous metal–organic framework toward oral insulin delivery: protein encapsulation, protection, and release. J. Am. Chem. Soc. **140**(17), 5678–5681 (2018)
20. Chui, S.S.Y., Lo, S.M.F., Charmant, J.P.H., Orpen, A.G., Williams, I.D.: A chemically functionalizable nanoporous material. Science **283**(5405), 1148–1150 (1999)
21. Cui, Y., Chen, B., Qian, G.: Lanthanide metal-organic frameworks for luminescent sensing and light-emitting applications. Coord. Chem. Rev. **273–274**, 76–86 (2014)
22. Dong, R., Zhang, Z., Tranca, D.C., Zhou, S., Wang, M., Adler, P., et al.: A coronene-based semiconducting two-dimensional metal-organic framework with ferromagnetic behavior. Nat. Commun. **9**(1), 2637 (2018)
23. Eddaoudi, M., Moler, D.B., Li, H., Chen, B., Reineke, T.M., O'Keeffe, M., et al.: Modular chemistry: secondary building units as a basis for the design of highly porous and robust metal–organic carboxylate frameworks. Acc. Chem. Res. **34**(4), 319–330 (2001)
24. Fan, K., Bao, S.-S., Nie, W.-X., Liao, C.-H., Zheng, L.-M.: Iridium(III)-based metal–organic frameworks as multiresponsive luminescent sensors for Fe^{3+}, $Cr_2O_7{}^{2-}$, and ATP^{2-} in aqueous media. Inorg. Chem. **57**(3), 1079–1089 (2018)

25. Fujiwara, Y.-I., Horike, S., Kongpatpanich, K., Sugiyama, T., Tobori, N., Nishihara, H., et al.: Control of pore distribution of porous carbons derived from Mg^{2+} porous coordination polymers. Inorgan. Chem. Front. **2**(5), 473–476 (2015)
26. Furukawa, H., Cordova, K.E., O'Keeffe, M., Yaghi, O.M.: The chemistry and applications of metal-organic frameworks. Science **341**(6149), 1230444 (2013)
27. Furukawa, H., Go, Y.B., Ko, N., Park, Y.K., Uribe-Romo, F.J., Kim, J., et al.: Isoreticular expansion of metal–organic frameworks with triangular and square building units and the lowest calculated density for porous crystals. Inorg. Chem. **50**(18), 9147–9152 (2011)
28. Ghosh, P., Colón, Y.J., Snurr, R.Q.: Water adsorption in UiO-66: the importance of defects. Chem. Commun. **50**(77), 11329–11331 (2014)
29. Gonzalez-Nelson, A., Coudert, F.-X., van der Veen, A.M.: Rotational dynamics of linkers in metal–organic frameworks. Nanomaterials **9**(3), 330 (2019)
30. Guillou, N., Livage, C., Drillon, M., Férey, G.: The chirality, porosity, and ferromagnetism of a 3D nickel glutarate with intersecting 20-membered ring channels. Angew. Chem. Int. Ed. **42**(43), 5314–5317 (2003)
31. Harbuzaru, B.V., Corma, A., Rey, F., Atienzar, P., Jordá, J.L., García, H., et al.: Metal–organic nanoporous structures with anisotropic photoluminescence and magnetic properties and their use as sensors. Angew. Chem. Int. Ed. **47**(6), 1080–1083 (2008)
32. He, Y., Zhou, W., Qian, G., Chen, B.: Methane storage in metal–organic frameworks. Chem. Soc. Rev. **43**(16), 5657–5678 (2014)
33. Horcajada, P., Chalati, T., Serre, C., Gillet, B., Sebrie, C., Baati, T., et al.: Porous metal–organic-framework nanoscale carriers as a potential platform for drug delivery and imaging. Nat. Mater. **9**, 172–178 (2009)
34. Horike, S., Shimomura, S., Kitagawa, S.: Soft porous crystals. Nat. Chem. **1**, 695–704 (2009)
35. Hoskins, B.F., Robson, R.: Design and construction of a new class of scaffolding-like materials comprising infinite polymeric frameworks of 3D-linked molecular rods. A reappraisal of the zinc cyanide and cadmium cyanide structures and the synthesis and structure of the diamond-related frameworks $[N(CH_3)_4][CuIZnII(CN)_4]$ and CuI[4,4′,4″,4‴-tetracyanotetraphenylmethane] $BF_4.xC_6H_5NO_2$. J. Am. Chem. Soc. **112**(4), 1546–1554 (1990)
36. Hu, Q., Yu, J., Liu, M., Liu, A., Dou, Z., Yang, Y.: A low cytotoxic cationic metal–organic framework carrier for controllable drug release. J. Med. Chem. **57**(13), 5679–5685 (2014)
37. James, S.L.: Metal-organic frameworks. Chem. Soc. Rev. **32**(5), 276–288 (2003)
38. Janiak, C., Vieth, J.K.: MOFs, MILs and more: concepts, properties and applications for Porous Coordination Networks (PCNs). New J. Chem. **34**(11), 2366–2388 (2010)
39. Jhung, S.H., Lee, J.H., Yoon, J.W., Serre, C., Férey, G., Chang, J.S.: Microwave synthesis of chromium terephthalate MIL-101 and its benzene sorption ability. Adv. Mater. **19**(1), 121–124 (2007)
40. Kalmutzki, M.J., Hanikel, N., Yaghi, O.M.: Secondary Building Units as the Turning Point in the Development of the Reticular Chemistry of MOFs. Sci. Adv. **4**(10), eaat9180 (2018)
41. Kanezashi, M., O'Brien-Abraham, J., Lin, Y.S., Suzuki, K.: Gas permeation through DDR-type zeolite membranes at high temperatures. AIChE J. **54**(6), 1478–1486 (2008)
42. Khan, N.A., Kang, I.J., Seok, H.Y., Jhung, S.H.: Facile synthesis of nano-sized metal-organic frameworks, chromium-benzenedicarboxylate, MIL-101. Chem. Eng. J. **166**(3), 1152–1157 (2011)
43. Kim, J., Yang, S.-T., Choi, S.B., Sim, J., Kim, J., Ahn, W.-S.: Control of catenation in CuTATB-n metal–organic frameworks by sonochemical synthesis and its effect on CO_2 adsorption. J. Mater. Chem. **21**(9), 3070–3076 (2011)
44. Kitagawa, S., Kitaura, R., Noro, S.-I.: Functional porous coordination polymers. Angew. Chem. Int. Ed. **43**(18), 2334–2375 (2004)
45. Knoevenagel, E.: Condensationen zwischen Malonester und Aldehyden unter dem Einfluss von Ammoniak und organischen Aminen. Ber. Dtsch. Chem. Ges. **31**(3), 2585–2595 (1898)
46. Koo, J., Hwang, I.-C., Yu, X., Saha, S., Kim, Y., Kim, K.: Hollowing out MOFs: hierarchical micro- and mesoporous MOFs with tailorable porosity via selective acid etching. Chem. Sci. **8**(10), 6799–6803 (2017)

47. Kreno, L.E., Leong, K., Farha, O.K., Allendorf, M., Van Duyne, R.P., Hupp, J.T.: Metal–organic framework materials as chemical sensors. Chem. Rev. **112**(2), 1105–1125 (2012)
48. Kuppler, R.J., Timmons, D.J., Fang, Q.-R., Li, J.-R., Makal, T.A., Young, M.D., et al.: Potential applications of metal-organic frameworks. Coord. Chem. Rev. **253**(23), 3042–3066 (2009)
49. Kurmoo, M.: Magnetic metal–organic frameworks. Chem. Soc. Rev. **38**(5), 1353–1379 (2009)
50. Lee, S.-Y., Park, S.-J.: A review on solid adsorbents for carbon dioxide capture. J. Ind. Eng. Chem. **23**, 1–11 (2015)
51. Lee, Y.-R., Kim, J., Ahn, W.-S.: Synthesis of metal-organic frameworks: a mini review. Korean J. Chem. Eng. **30**(9), 1667–1680 (2013)
52. Leidinger, M., Rieger, M., Weishaupt, D., Sauerwald, T., Nägele, M., Hürttlen, J., et al.: Trace gas VOC detection using metal-organic frameworks as pre-concentrators and semiconductor gas sensors. Procedia Eng. **120**, 1042–1045 (2015)
53. Li, B., Wen, H.-M., Zhou, W., Chen, B.: Porous metal–organic frameworks for gas storage and separation: what, how, and why? J. Phys. Chem. Lett. **5**(20), 3468–3479 (2014)
54. Li, G., Xiao, P., Webley, P., Zhang, J., Singh, R., Marshall, M.: Capture of CO_2 from high humidity flue gas by vacuum swing adsorption with Zeolite 13X. Adsorption **14**(2), 415–422 (2008)
55. Li, H., Eddaoudi, M., O'Keeffe, M., Yaghi, O.M.: Design and synthesis of an exceptionally stable and highly porous metal-organic framework. Nature **402**, 276 (1999)
56. Li, H., Wang, K., Sun, Y., Lollar, C.T., Li, J., Zhou, H.-C.: Recent advances in gas storage and separation using metal–organic frameworks. Mater. Today **21**(2), 108–121 (2018)
57. Li, J.-R., Kuppler, R.J., Zhou, H.-C.: Selective gas adsorption and separation in metal–organic frameworks. Chem. Soc. Rev. **38**(5), 1477–1504 (2009)
58. Lin, J.-D., Cheng, J.-W., Du, S.-W.: Five d^{10} 3D metal–organic frameworks constructed from aromatic polycarboxylate acids and flexible imidazole-based ligands. Cryst. Growth Des. **8**(9), 3345–3353 (2008)
59. Lin, W., Rieter, W.J., Taylor, K.M.L.: Modular synthesis of functional nanoscale coordination polymers. Angew. Chem. Int. Ed. **48**(4), 650–658 (2009)
60. Liu, J., Chen, L., Cui, H., Zhang, J., Zhang, L., Su, C.Y.: Applications of metal-organic frameworks in heterogeneous supramolecular catalysis. Chem. Soc. Rev. **43**(16), 6011–6061 (2014)
61. Liu, Y., Xuan, W., Cui, Y.: Engineering homochiral metal-organic frameworks for heterogeneous asymmetric catalysis and enantioselective separation. Adv. Mater. **22**(37), 4112–4135 (2010)
62. Lu, C., Ben, T., Xu, S., Qiu, S.: Electrochemical synthesis of a microporous conductive polymer based on a metal–organic framework thin film. Angew. Chem. Int. Ed. **53**(25), 6454–6458 (2014)
63. Mallick, A., Saha, S., Pachfule, P., Roy, S., Banerjee, R.: Selective CO_2 and H_2 adsorption in a chiral magnesium-based metal organic framework (Mg-MOF) with open metal sites. J. Mater. Chem. **20**(41), 9073–9080 (2010)
64. Mason, J.A., Veenstra, M., Long, J.R.: Evaluating metal–organic frameworks for natural gas storage. Chem. Sci. **5**(1), 32–51 (2014)
65. Millward, A.R., Yaghi, O.M.: Metal–organic frameworks with exceptionally high capacity for storage of carbon dioxide at room temperature. J. Am. Chem. Soc. **127**(51), 17998–17999 (2005)
66. Müller, M., Hermes, S., Kähler, K., van den Berg, M.W.E., Muhler, M., Fischer, R.A.: Loading of MOF-5 with Cu and ZnO nanoparticles by gas-phase infiltration with organometallic precursors: properties of Cu/ZnO@MOF-5 as catalyst for methanol synthesis. Chem. Mater. **20**(14), 4576–4587 (2008)
67. Nishiyabu, K.: 15-Powder Space Holder Metal Injection Molding (PSH-MIM) of microporous metals. In: Heaney, D.F. (ed.) Handbook of Metal Injection Molding, pp. 349–390. Woodhead Publishing (2012)
68. Opanasenko, M., Dhakshinamoorthy, A., Shamzhy, M., Nachtigall, P., Horáček, M., Garcia, H., et al.: Comparison of the catalytic activity of MOFs and zeolites in Knoevenagel condensation. Catal. Sci. Technol. **3**(2), 500–507 (2013)

69. Park, J., Jiang, Q., Feng, D., Mao, L., Zhou, H.-C.: Size-controlled synthesis of porphyrinic metal–organic framework and functionalization for targeted photodynamic therapy. J. Am. Chem. Soc. **138**(10), 3518–3525 (2016)
70. Pastore, V.J., Cook, T.R., Rzayev, J.: Polymer–MOF hybrid composites with high porosity and stability through surface-selective ligand exchange. Chem. Mater. **30**(23), 8639–8649 (2018)
71. Peng, Y., Krungleviciute, V., Eryazici, I., Hupp, J.T., Farha, O.K., Yildirim, T.: Methane storage in metal–organic frameworks: current records, surprise findings, and challenges. J. Am. Chem. Soc. **135**(32), 11887–11894 (2013)
72. Perry Iv, J.J., Perman, J.A., Zaworotko, M.J.: Design and synthesis of metal–organic frameworks using metal–organic polyhedra as supermolecular building blocks. Chem. Soc. Rev. **38**(5), 1400–1417 (2009)
73. Petit, C., Levasseur, B., Mendoza, B., Bandosz, T.J.: Reactive adsorption of acidic gases on MOF/graphite oxide composites. Microporous Mesoporous Mater. **154**, 107–112 (2012)
74. Pichon, A., James, S.L.: An array-based study of reactivity under solvent-free mechanochemical conditions-insights and trends. CrystEngComm **10**(12), 1839–1847 (2008)
75. Pichon, A., Lazuen-Garay, A., James, S.L.: Solvent-free synthesis of a microporous metal–organic framework. CrystEngComm **8**(3), 211–214 (2006)
76. Pienack, N., Bensch, W.: In-situ monitoring of the formation of crystalline solids. Angew. Chem. Int. Ed. **50**(9), 2014–2034 (2011)
77. Qiu, L.-G., Li, Z.-Q., Wu, Y., Wang, W., Xu, T., Jiang, X.: Facile synthesis of nanocrystals of a microporous metal–organic framework by an ultrasonic method and selective sensing of organoamines. Chem. Commun. (31), 3642–3644 (2008)
78. Robson, R.: Design and its limitations in the construction of bi- and poly-nuclear coordination complexes and coordination polymers (aka Mofs): a personal view. Dalton Trans. (38), 5113–5131 (2008)
79. Rochelle, G.T.: Amine scrubbing for CO_2 capture. Science **325**(5948), 1652–1654 (2009)
80. Rojas, S., Colinet, I., Cunha, D., Hidalgo, T., Salles, F., Serre, C., et al.: Toward understanding drug incorporation and delivery from biocompatible metal–organic frameworks in view of cutaneous administration. ACS Omega **3**(3), 2994–3003 (2018)
81. Serre, C., Millange, F., Thouvenot, C., Noguès, M., Marsolier, G., Louër, D., et al.: Very large breathing effect in the first nanoporous chromium(III)-based solids: MIL-53 or CrIII(OH)·$\{O_2C–C_6H_4–CO_2\}$·$\{HO_2C–C_6H_4–CO_2H\}$x.H_2Oy. J. Am. Chem. Soc. **124**(45), 13519–13526 (2002)
82. Shih, Y.-H., Wang, K.-Y., Singco, B., Lin, C.-H., Huang, H.-Y.: Metal–organic framework–polymer composite as a highly efficient sorbent for sulfonamide adsorption and desorption: effect of coordinatively unsaturated metal site and topology. Langmuir **32**(44), 11465–11473 (2016)
83. Shono, T., Mingos, D.M.P., Baghurst, D.R., Lickiss, P.D.: Novel Energy Sources for Reactions, vol. Chapter 4. The Press Syndicate of the University of Cambridge, Cambridge (2000)
84. Simon, P., Gogotsi, Y.: Materials for electrochemical capacitors. In: Nanoscience and Technology, pp. 320–329. Co-Published with Macmillan Publishers Ltd., UK (2009)
85. Spanopoulos, I., Tsangarakis, C., Klontzas, E., Tylianakis, E., Froudakis, G., Adil, K., et al.: Reticular synthesis of HKUST-like tbo-MOFs with enhanced CH_4 storage. J. Am. Chem. Soc. **138**(5), 1568–1574 (2016)
86. Stock, N., Biswas, S.: Synthesis of Metal-Organic Frameworks (MOFs): routes to various MOF topologies, morphologies, and composites. Chem. Rev. **112**(2), 933–969 (2012)
87. Sun, L., Campbell, M.G., Dincă, M.: Electrically conductive porous metal–organic frameworks. Angew. Chem. Int. Ed. **55**(11), 3566–3579 (2016)
88. Trickett, C.A., Helal, A., Al-Maythalony, B.A., Yamani, Z.H., Cordova, K.E., Yaghi, O.M.: The chemistry of metal–organic frameworks for CO_2 capture, regeneration and conversion. Nat. Rev. Mater. **2**, 17045 (2017)
89. Uemura, T., Kitagawa, K., Horike, S., Kawamura, T., Kitagawa, S., Mizuno, M., et al.: Radical polymerisation of styrene in porous coordination polymers. Chem. Commun. (48), 5968–5970 (2005)

90. Umeyama, D., Horike, S., Inukai, M., Itakura, T., Kitagawa, S.: Reversible solid-to-liquid phase transition of coordination polymer crystals. J. Am. Chem. Soc. **137**(2), 864–870 (2015)
91. Vellingiri, K., Szulejko, J.E., Kumar, P., Kwon, E.E., Kim, K.-H., Deep, A., et al.: Metal organic frameworks as sorption media for volatile and semi-volatile organic compounds at ambient conditions. Sci. Rep. **6**, 27813 (2016)
92. Wang, F.-M., Zhou, L., Lustig, W.P., Hu, Z., Li, J.-F., Hu, B.-X., et al.: Highly luminescent metal–organic frameworks based on an aggregation-induced emission ligand as chemical sensors for nitroaromatic compounds. Cryst. Growth Des. **18**(9), 5166–5173 (2018)
93. Wang, H.-L., Yeh, H., Chen, Y.-C., Lai, Y.-C., Lin, C.-Y., Lu, K.-Y., et al.: Thermal stability of metal–organic frameworks and encapsulation of CuO nanocrystals for highly active catalysis. ACS Appl. Mater. Interfaces **10**(11), 9332–9341 (2018)
94. Wang, T.C., Vermeulen, N.A., Kim, I.S., Martinson, A.B.F., Stoddart, J.F., Hupp, J.T., et al.: Scalable synthesis and post-modification of a mesoporous metal-organic framework called NU-1000. Nat. Protoc. **11**, 149–162 (2015)
95. Wang, Y.B., Liu, D.S., Pan, T.H., Liang, Q., Huang, X.H., Wu, S.T., et al.: Structural variation from 1D to 3D: effect of metal centres on the construction of metal–organic coordination polymers with N-(1H-tetrazol-5-yl)benzamide ligand. CrystEngComm **12**(11), 3886–3893 (2010)
96. Work, W.J., Horie, K., Hess, M., Stepto, R.F.T.: Definition of terms related to polymer blends, composites, and multiphase polymeric materials (IUPAC Recommendations 2004). Pure Appl. Chem. **76**(11), 1985–2007 (2004)
97. Wu, M.-X., Yang, Y.-W.: Metal–Organic Framework (MOF)-based drug/cargo delivery and cancer therapy. Adv. Mater. **29**(23), 1606134 (2017)
98. Xiang, W., Zhang, Y., Lin, H., Liu, C.-J.: Nanoparticle/metal–organic framework composites for catalytic applications: current status and perspective. Molecules **22**(12), 2103 (2017)
99. Xuan, W., Zhu, C., Liu, Y., Cui, Y.: Mesoporous metal–organic framework materials. Chem. Soc. Rev. **41**(5), 1677–1695 (2012)
100. Yaghi, O.M., O'Keeffe, M., Ockwig, N.W., Chae, H.K., Eddaoudi, M., Kim, J.: Reticular synthesis and the design of new materials. Nature **423**(6941), 705–714 (2003)
101. Yi, F.-Y., Chen, D., Wu, M.-K., Han, L., Jiang, H.-L.: Chemical sensors based on metal–organic frameworks. ChemPlusChem **81**(8), 675–690 (2016)
102. Yu, C., Ma, S., Pechan, M.J., Zhou, H.-C.: Magnetic properties of a noninterpenetrating chiral porous cobalt metal-organic framewok. J. Appl. Phys. **101**(9), 09E108 (2007)
103. Yu, J., Mu, C., Yan, B., Qin, X., Shen, C., Xue, H., et al.: Nanoparticle/MOF composites: preparations and applications. Mater. Horiz. **4**(4), 557–569 (2017)
104. Zhang, J., Shreeve, J.N.M.: 3D nitrogen-rich metal–organic frameworks: opportunities for safer energetics. Dalton Trans. **45**(6), 2363–2368 (2016)
105. Zhu, L., Liu, X.-Q., Jiang, H.-L., Sun, L.-B.: Metal–organic frameworks for heterogeneous basic catalysis. Chem. Rev. **117**(12), 8129–8176 (2017)
106. Zhu, Q.-L., Xu, Q.: Metal–organic framework composites. Chem. Soc. Rev. **43**(16), 5468–5512 (2014)
107. Zlotea, C., Campesi, R., Cuevas, F., Leroy, E., Dibandjo, P., Volkringer, C., et al.: Pd nanoparticles embedded into a metal-organic framework: synthesis, structural characteristics, and hydrogen sorption properties. J. Am. Chem. Soc. **132**(9), 2991–2997 (2010)

Conducting Polymers and Their Composites

Ankit Jadhav, Sundus Saeed Qureshi, Harshit Jadhav, Sabzoi Nizamuddin, Abdul Sattar Jatoi, Shaukat Ali Mazari, Israr Ahmed, Humair Ahmed Baloch, M. T. H. Siddiqui, and Nabisab Mujawar Mubarak

Abstract The on-going shrinkage in the size of electronic devices and the high-power generation of electrical appliances have led to new challenges in insulating polymers for packaging applications. The main aim of this review is to study fabrication of polymer composites coupled with improved electrical properties and thermal resistivity, with lower dielectric constant, lowers thermal expansion and the cost associated with the production of such composites. Polymeric materials are an excellent choice for production of electronic materials as they can be processed with ease and are economical. The disadvantage of using polymers for processing of electrically conductive materials is that they possess low thermal conductivity. However, the properties can be enhanced by incorporating appropriate fillers that have high thermal conductivity; these include boron nitride, aluminium oxide and silicon nitride. This

A. Jadhav
Department of Mechanical Engineering, Ahmedabad Institute of Technology, Ahmedabad Gujarat 380060, India

S. S. Qureshi
Institute of Environmental Engineering and Management, Mehran University of Engineering and Technology Jamshoro, Sindh, Pakistan

H. Jadhav
Department of Chemical Engineering, Vishwakarma Government Engineering College, Ahmedabad, Gujarat 382424, India

S. Nizamuddin (✉) · H. A. Baloch · M. T. H. Siddiqui
School of Engineering, RMIT University, Melbourne 3000, Australia
e-mail: nizamuddin.nizamuddin@rmit.edu.au

A. S. Jatoi · S. A. Mazari
Department of Chemical Engineering, Dawood University of Engineering and Technology, Karachi, Sindh, Pakistan

I. Ahmed
School of Chemical Engineering, The University of Faisalabad, Faisalabad, Punjab, Pakistan

N. M. Mubarak
Department of Chemical Engineering, Faculty of Engineering and Science, Curtin University, 98009 Miri, Sarawak, Malaysia

N. M. Mubarak et al. (eds.), *Contemporary Nanomaterials in Material Engineering Applications*, Engineering Materials,
https://doi.org/10.1007/978-3-030-62761-4_6

chapter aims to review the available conducting polymers, fillers, processing techniques, final properties, and the end use applications of such conducting polymer composites.

Keywords Polymers · Conductivity · Conduction mechanism · Composite · Electrode

1 Introduction

Conducting polymers were discovered by chemical polymerization of pyrrole and aniline about a century ago [120]. After their discovery, these materials gained increasing interest and popularity [142] due to their outstanding processability and low cost [60]. The insulating nature of such polymers is the basis for their application in electrical and electronic fields [189]. However, material scientists are interested in imparting conduction to polymers by mixing polymers with conductive materials to form conductive polymer composites [109]. The conductive polymer composites provide several advantages, including low density, redox properties, ease of shaping, unique morphologies, corrosion resistance and a wide range of electrical conductivities [189, 7]. Due to these extraordinary characteristics of conductive polymer composites, they have several applications in electric devices, drug delivery, chemosensory, and actuators, etc. [1, 96]. The conducting polymers are also being tested for their applications in solar cells, which shows the potential of their use within that field [65]. Similarly, due to suitable physical and electrical characteristics, low cost, ease of manufacturing, conducting composites are being tested, and used for several applications like, supercapacitors, electrochromic devices, batteries, solar cells biomedical applications, and sensors [110].

These are polymers with conjugated chain structures, which are doped with epoxies to increase their conductivity through redox reaction and protonation. This happens on the cost of elasticity of the polymer as when any electron shifts from one orbit to another which causes deformation in shape, which ultimately determines the size of lattice deformation [176]. Some of the well-known conducting polymers are polypyrrole (PPy), polythiophene, polyacetylene, poly(3,4-ethylenedioxythiophene) (PEDOT), and polyaniline (PANI), etc. [118]. Chemical preparation of conducting polymers through oxidative coupling enables the formation of conducting polymers to the sizes of micrometer (mm) and nanometer (nm) ranges [125].

Shape memory polymers (SMPs) have also been introduced. These polymers come with better desired properties, ease in manufacturing and low cost [88]. The SMPs and blends provide with improved mechanical, thermal, electrical, magnetic, shape, light and water properties [103]. The fabricated desired properties expand the application horizon of SMPs. These are potential candidates for uses in aerospace, biomedical devices, energy, textiles, civil engineering, bionics engineering, electronics and household appliances and products [102].

The most commonly used conductive components for blending with polymers to fabricate conductive polymer composites to impart conduction in polymers are carbon fibers, conducting polymers, e.g., polyaniline, carbon black and metal particles [109]. Different conductive fillers show different behaviour in imparting conduction into polymers because of their diverse intrinsic nature. For example, threshold percolation of fibre blended nitrile rubber composite lies between 10 and 20 per hundred rubber (phr), whereas it falls in the range of 20 to 35 phr for carbon black incorporated nitrile rubber composites [128].

Vapor grown carbon nanofibers (VGCNF)/polymers have been reported as robust conductive polymers and are cost effective compared to CNTs, multi-walled carbon nanotubes (MWNTs), single wall carbon nanotubes (SWNTs) [3]. This provides an option of replacement of CNTs, MWNTs and SWNTs with VGCNF. Strong inter-fibre bonding and preparation of these fibres with polymeric resins sometimes reduce electrical conductivity and tensile strength. A fibre loading of 15% has been suggested to keep resistivity's below 0.15 Ω cm and conductivity has been reported to increase up to ten times by addition of epoxy composites [129, 159]. The percolation threshold of below 1 wt.% fibre has been found to provide better conductivity for VGCNFs [186]. Boron nitride (BN) helps in electrical insulation when added to VGCNFs for thermal conduction applications [133].

2 Polymers and Conducting Polymers

Polymers have several macroscopic and microscopic characteristics, including electrical transport, semiconducting and optical properties [179]. The invention of conductive polyacetylene has attracted a lot of attention to conductive polymers. Doping at maintained thermal stability and mechanical flexibility makes these conducting polymers for several extended applications. There are several techniques and methods for polymers synthesis, including chemical and electrochemical polymerization. Several methods for synthesizing new conducting polymer composites have been reported involving the use of template-directed synthesis, gas phase polymerization, chemical functionalization, formation of conductive polymer compounds. The morphology, thickness, chemical state, and conductivity of the compound can be easily controlled. Various materials and composites are presented in Fig. 1. The conducting polymer composites use materials like metal ions, nanoparticles, metal nanostructures and metal oxides, carbon materials, molecular species, i.e., metal phthalocyanines and biologically active components such as enzymes, antibodies, antigens, and proteins. The purpose of preparing new composites is to derive new unique properties not observed in the individual components [20, 36, 40, 77]. These polymers conduct electricity because of substitution of the compound that may have metal conductivity or maybe a semiconductor [6].

The conducting nanomaterial polymers are of interest due to their enhanced and superior characteristics than their host counterparts. Conductive polymers have received considerable attention due to their range of excellent properties such as high

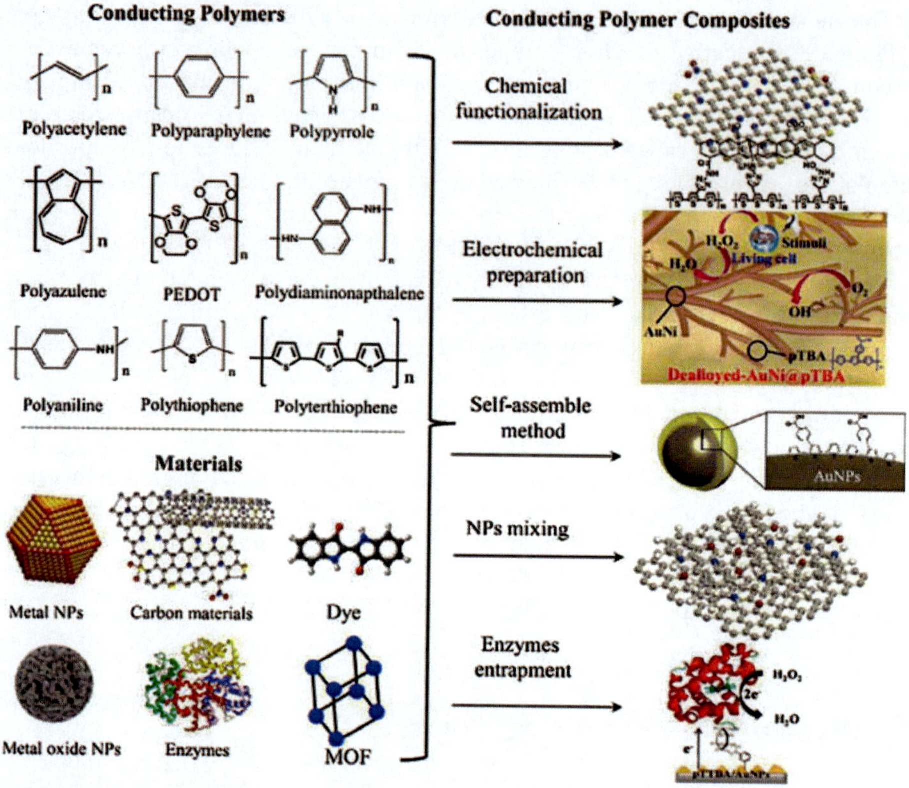

Fig. 1 Conducting polymers and their composite materials [110]

conductivity, high specific strength, high specific modulus, high thermal stability, corrosion resistance, fatigue resistance, eco-friendly and economical. Some of the most commonly used applications for conductive polymers are conductive adhesives, electrostatic materials, electromagnetic shielded electromagnetic interference (EMI), artificial nerves, aircraft structures, diodes and transistors [72]. Figure 2 represents an outline of a broad conductivity range for conducting polymers and their composites.

2.1 History and Conduction Mechanism of Conducting Polymers

In 1975, the innovation of poly sulphur nitride intensified the research on conducting polymers, which acts as a superconductor in low temperatures [43]. The electrical conductivity of polyamide was enhanced from 10^{-9} to 10^{5} S cm^{-1} by doping with oxidizing agents (AsF_5, I_2, $NOPF_6$) or reducing agents (sodium naphthalide) [146]. This triggered interest in the scientific community to study and develop new systems

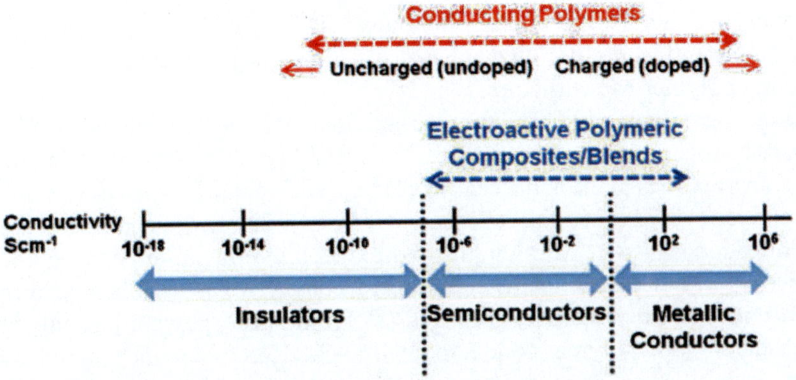

Fig. 2 Conductivity ranges for conducting polymers and their composites [67]

of conducting polymers. The synthesis of poly-paraphenylene was carried out by Ivory et al. [63], they developed transfer complexes of highly conducting charge with both p and n types dopants. The conductivity value was increased by several magnitudes i.e., from 10^{-5} to 500 cm^{-1}. Treatment with aqueous H_2SO_4 through electrochemical oxidation on the platinum electrode can lead to the production of conducting polymer known as 'Pyrrole Black' [66].

The conductive mechanism in these polymers is quite complicated as these materials display conductivity along a spectrum of about fifteen magnitude orders, yet many distinct processes within separate systems. The conducting polymers demonstrate increased electrical conductivity through several orders of doping magnitude. To understand the electronic phenomena in such systems, the principle of solitons, polarons and bipolarons was used [17].

2.2 *Conducting Polymers*

The π- electron backbone present in conducting polymers is liable for unusual behaviour in terms of electrical conductivity. This elongated π-conjugated arrangement in conducting polymers comprises of single and double bonds which alternate throughout the polymer chain. Such organic polymers with high electrical conductive values are named as 'synthetic metals' [41]. Many researchers studied conducting polymers contributing in various applications such as analytical chemistry, biosensing devices [48], supercapacitors [168], memories [9], transistors [9], lithium ionic batteries [112], sensors [163], and actuators/artificial muscles [8].

Conducting polymer nanostructures have quickly become a growing field of studies since the last two decades as they show unique characteristics linked to their nanoscale size and have significantly enhanced device efficiency [78, 89]. Several techniques like hard/soft-template techniques [114, 165], electrospinning technology [182], and well-controlled solution synthesis [57] can be used to synthesize conducting polymer nanostructures.

Conducting polymer nanostructure are anticipated to show an enhanced efficiency than bulk conducting polymers in terms of technological applications due to their distinctive properties of the nanoscale size [188]. The unique characteristics linked to this nanoscale provide better electrical conductivity [172], wide specific surface area [122], brief ion transport path lengths and reduced interfacial impedance between the electrolyte and electrodes, due to mixed conductive mechanism in terms of ionic and electronic conductivity [119, 172]. Polythiophene, polyaniline, polypyrrole are considered good polymer conductors with better electrochemical activity. They are used in lithium batteries and pseudocapacitors as essential electrode materials [111, 112, 136].

Fused deposition modelling commonly referred to as additive manufacturing technology is widely used in three-dimensional (3D) printing. Additive manufacturing technologies have a broad range of applications in the field of automotive, tissue engineering, aerospace, nanotechnology and fashion industry [37, 61, 94]. Fused deposition modelling 3D printing is a rapid, simple, reliable and affordable technique for the production of plastic structures and prototypes with a refined resolution allowing complex designs [166, 170]. High functionality and excellent electrical properties can be achieved by incorporating particle and nanomaterial reinforcements into polymers. The electrical conductivity of manufactured parts by 3D printing enables the in-built sensors and signal processing functionalities of structure [28]. Polymers can be made electrically conductive with the addition of graphene, carbon nanotubes and carbon black or by their combination [126]. The three main processes for production of conductive polymer nanocomposites are solution processing, in-situ polymerization, and melt mixing. Wei et al. [169] used graphene reinforced ABS composite for fused deposition modelling and observed improved electrical conductivity. A 5.6 wt.% graphene loading resulted in an improvement of four instructions of magnitude in the electrical conductivity [169].

By incorporating the thermally reduced graphene nanofiller, the potential to impart conductivity to polymers is considered superior to other competing nanofillers like carbon nanotubes [137]. The conductive fillers in this technique having preferential affinity are added as a co-continuous structure into the immiscible polymer blend, resulting in the selective localization of nanofiller at the interface or inside one of the phases [58]. This selective localization is controlled mainly by joined action of thermodynamics and kinetic factors [173, 180]. During melt-mixing, the nanoparticles selectively locate in a more favourable phase to reduce free energy of the system. In contrast, the kinetic factors define the localization of nanoparticles in the phase having lower viscosity during melt mixing.

Carbon nanomaterials have been proven to be very beneficial in material and polymer industries for years after their development. Carbon nanomaterials like fullerene, carbon nanofiber, graphene, etc. and their hybrids have proven to be revolutionary in producing astonishing composites used in the field of electronics, automobile, showbiz, and sports industry. Due to their fascinating multifunctional properties, they are also used in aerospace and defense industries. The electrical properties of these composite materials are one of the most important properties in determining the effectiveness of polymer matrices materials [81]. The study of electrical properties of numerous carbon nanofiller composites proves that the performance is greatly influenced by using different fabricating techniques, types of polymer matrices used and the materials involved in a process [91].

The composites containing carbon black, even with a least amount of fillers, can have electrical conductivity of 101 S/m due to filler's zero-dimensional shape. A zero-dimensional shape cannot be constructed easily as it requires efficient conductivity and stronger networks between the matrices [115]. Contrary to the carbon black-reinforced composites, the one-dimensional shaped carbon nanofillers and carbon nanotubes can have a better electrical conductivity of 10^4 S/m [107]. It is recommended that the nanofillers should be handled carefully otherwise high percolation can occur as well as agglomeration of nanofillers can be expected [95, 101].

Carbon nanotubes decorated with metal helps in increasing the electrical conductivity of composites by 10^6 S/m. Among all nanocomposites, this has the best electrical conductivity that can be achieved. Moreover, the plating of metal particles also limits agglomeration affecting a significant decrease in loading level and percolation threshold [23]. Polymer composites, which are graphene based, include both reduced graphene oxides and graphites. It also has an excellent electrical conductivity of 10^1 S/m, though the percolation threshold is not satisfactory. The graphene-based nanocomposites potential can be further explored if graphene layers from graphite are detached. Moreover, efficient techniques are required to prevent aggregate formation [81]. Among others, the 3D carbon reinforced nanocomposites display a balanced type of composites. It shows significant improvements in decreasing percolation threshold, loading level and enhanced electrical properties. In situ fabrication designs impart high electrical conductivity in the nano-foam reinforced 3D composites [70].

Recent developments in the electrical properties of polymeric materials help towards production of dielectric polymer composites. Mostly polymers have low thermal conductivities around 0.1 to 0.5 W/mK instead, they exhibit significant thermal insulation. Thermal conductivity of polymeric materials can be increased by introducing fillers like boron nitride, aluminium oxide, silicon nitride, etc. [60]. Several factors are involved in determining the thermal and electrical conductivity of the polymers, such as intrinsic thermal conductance of the filler and matrix, filler type/size/shape/aspect ratio/loading/dispersion, etc. Specifically, thermal conductivity cannot be readily determined by only figuring out one of the following factors. All factors must be accounted whenever the property, such as thermal conductivity, is of concern. For example, the researcher may expect that with increasing filler size, the loading must increase as well. However, it has been proven wrong many times that by increasing filler size, a decrease in loading has been observed [55].

Surface treatment methods are very important in increasing the thermal conductivity of filler particles. This technique not only improves the filler dispersion properties but also enhances its adhesion between the polymer matrix and filler. This decreases the thermal resistance of the polymeric material significantly and enhances its thermal conductivity. The effectiveness of thermal conductivity of the composite is also dependent upon the thickness of surface treatment [60].

A composite such as silica–carbon-epoxy resin was prepared to attempt higher conductivity of composites. Such kind of composites improves not only electrical conductivity but also present enhanced service properties. It was proved that the content of silica also helps to enhance the conductivity of composites [190]. Conductive strong networks made up by carbon black-particles determine the conductivity of a material. By introducing silica into composites facilitates and resists the formation of conductive networks. There is still some ongoing research in this area of study [190].

2.3 *Classification of Conducting Polymers*

In general, polymers are insulating materials; however, four major types of semiconducting polymers are established to date, including charge transfer polymers, conjugated conductive polymers, ion-conductive polymers and electrically conductive filled polymers. Conductive filled polymers were first produced in 1930 to prevent corona discharge. Inventions of new substances with superior properties often produce the latest technology. The main discovery of conductive polymers growth was observed in 1960. The polysulfide has an ambient temperature conductivity of about 103 $(cm.cm)^{-1}$. These innovations are important because they indicate the presence of high polymer conductivities and encourage the increased quantities of work needed to make other polymer conductors. Electronic polymer conductors are important candidates for many applications because many of their features have common problems in conventional random access memories, including corrosion, matrix inconsistency, weight, and ecological veracity. In addition to corrosion resistance and lightweight, several important features of the conductive polymers can be adapted to different applications. The work of conductive polymers has given rise to new scientific concepts and the potential of novel technologies. The significance of conductive polymers is reflected in the 2000 Nobel Prize in Chemistry, MacDiarmid, Shirakawa and Heeger, who invented and developed the conductive polymers [92]. Due to its unique properties such as electrical properties, reversible doping and doping procedures, controlled chemical and electrochemical properties, and simple processability, several conductive polymers and their derivatives have attracted special attention in nanoscience and nanotechnology. These include polyacetylene, polypyrrole, polyaniline, poly(p-phenylene), poly(phenylene)))s, poly(p-phenylene vinylene), poly(3,4- Ethylene dioxythiophene), polyfuran and other polythiophene [86, 90, 151]. The structure of some of these conductive polymers is shown in Fig. 3.

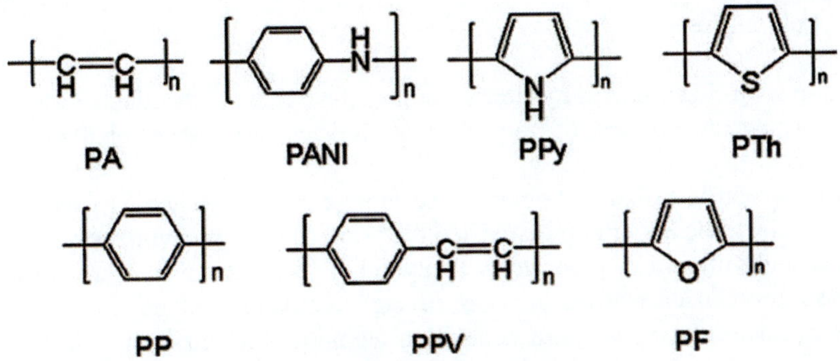

Fig. 3 Structures of some common conducting polymers [27]

2.3.1 Polyacetylene

In the mid-1970s, polyacetylene, which is an organic polymer having a repeating unit (C_2H_2) was the first conductive polymer produced by Heeger [53]. The presence of highly conductive polyacetylene resulted in a rapid injection of research activities to discover new conductive polymers, which enabled the use of organic compounds in microelectronics. This discovery found that a rapid combustion around polymer with oxidative doping increases conductivity [90]. In 1976, Alan MacDiarmid et al. reported that mixing with iodine improves conductivity of polyacetylene raises six times, i.e. because of load carriers. Furthermore, it has been reported that polymers produced at varying folding levels exhibit a wide variety of electrical properties ranging from insulator to metal. The polyacetylene itself is too unstable to have any real value, and its structure includes the core of all conjugated conductive polymers. Very few people believed that a correctly substituted polyacetylene molecule would exhibit superconductivity at ambient temperature [52]. Yamamoto et al. are the first to shape the electrical conductivity of the order of 105 S/cm for trans polyacetylene sample [177].

2.3.2 Poly (thiophene)s

Poly(thiophene) is an important conductive polymer for the production of environmentally and thermally stable materials. In 1980, a new Poly(thiophene) derivative was developed at the Bayer Research Laboratory in Germany [83]. Poly(thiophene) is like many other linear polyachromatic compounds and is insoluble in organic solvents [152, 135] studied the electrochemical synthesis and explained the conducting abilities of substituted Poly(thiophene). Conventional synthesis methods [98], general methods [134], Conjugated-polymer-based chemical sensors [100] for conducting polymers are well reported in the literature.

2.3.3 Polyaniline

Polyaniline was first reported by Henry Lethe in the middle of the nineteenth century while working on products of polyaniline in acidic media under electrochemical and chemical oxidation processes [75, 74] reported the impact of electrode material on non-volatile resistive-memory apparatus (electrical-switching type) finding that lower electrode material transformed the properties of the living polymer layer. Athawale and Kulkarni [5] reported that polyaniline responded well to ethanol during the substitution of derivatives such as ethanol, methanol, butanol, propanol, and heptanol sensors. Using vacuum deposition techniques Misra et al. [106] synthesized a high quality doped polyaniline film. Crowley et al. [24] studied the manufacturing and operation of polyaniline/$CuCl_2$ as an H_2S sensor. Banerjee discussed on the manufacturing of Polyaniline/nanofiber reinforced nanocomposite microbalance sensor as hydrochloric acid (HCl) sensor [12]. The authors found that low concentrations of HCl was quickly detected in natural water systems. De Surville et al. [31] reported high conductivity in polyaniline by electrochemical polymerization. Likewise, Diaz and Logan [34] synthesized electroactive films of polyaniline which can be utilized as an electrode.

2.3.4 Polypyrrole

Polypyrrole, which is an organic polymer produced through the polymerization of pyrrole, was found to be a conducting polymer in 1968. Several studies exist on polypyrrole for ease of preparation and excellent redox performance [44], capacity to give immense conductivity, stabilized oxidized form [171], water solubility, useful electrical and optical properties. Pyrrole black is obtained as a powder through chemical polymerization of pyrrole. Those are well-known to be pyrrole polymers in which the bonding is mainly through alpha, alpha' carbon [99]. described the pyrolysis of tetraiodopyrrole to make astonishingly conductive materials, an independent film with sufficient mechanical properties was synthesized using electrochemical techniques, to study the system as a conductive polymer [33]. Dallolio et al. [25] manufactured polypyrrole via oxidation of pyrrole in H_2SO_4 as a black powder couple with ambient temperature. Their work was later extended by IBM personnel, who found that the film of this polymer could be obtained through electrochemical polymerization [147].

2.3.5 Poly (3,4-Ethylenedioxythiophene)

Poly (3,4-ethylenedioxythiophene) is a conductive polymer based on 3,4-ethylenedioxythiophene monomer. Poly (3,4-ethylenedioxythiophene) transparent thin oxide films, highly stable and acceptable bandwidth, and short redox potential [68]. Antistatic coatings for polymers and glass, multi-conductor bodies, organic LED screens, nanofiber electrodes for unit stimulation, cathode materials, solar cells in electrolytic capacitors, printed circuit boards, color textiles in rapidly growing organic semiconductor areas, transparent electrodes for thick film electroluminescence, welding passage and unloading [14, 45, 73]. Vanadium pentoxide nanofibers are used to synthesize poly (3,4-ethylenedioxythiophene) by nano-fiber grafting methods. Add 3,4-ethylene glycol dithiophene to an aqueous camphor H_2SO_4 and V_2O_5 nanofiber sol–gel solution and initiate polymerization with $(NH_4)_2S_2O_8$ [191]. Preparation of alkyl-substituted ethylenedioxythiophene derivatives (EDT-C1, EDT-C6, and EDT-C10) was reported by Heywang and Jonas in 1992 [54].

2.3.6 Poly (Phenylene Vinylene)

Poly (phenylene vinylene) is a diamagnetic material, having a structure between polyaniline and polypyrrole, and having a very low conductivity of 10–13 S/cm [18]. Oriented poly (phenylene vinylene) has high crystallinity, mechanical strength and environmental stability. When iodine, iron chloride, alkali metal or acid are added, the conductivity increases but the stability decreases. Typically, the unsubstituted misaligned poly (phenylene vinylene) shows reasonable conductivity in the doping range of 10–3–100 S/cm synthesized by the popular Gilch pathway. The first PLED luminescent layer was prepared in 1990 using poly (phenylene vinylene) [139]. Although poly (phenylene vinylene) based devices have poor absorption and lens degradation, their derivatives encompass many applications in research cells [80].

2.3.7 Polyphenylene and Polyparaphenylene

Polyparaphenylene is a precursor of the conductive polymer of the rigid rod polymer host family made from repeating p-phenylene units and converted into conductive form using an oxidizing agent or additive. In 1980, polyaniline was doped with polyparaphenylene to achieve comparable conductivity [11]. It is an example of a non-alkyne polymer which can be folded by an electron donor or electron acceptor to provide electrical conductivity. Ballard et al. studied homogenization using benzene as transformation substrate for synthesizing of polyparaphenylene to obtain a cis-dihydrofuran derivative from bacterial fermentation [71]. Very high crystalline polyparaphenylene film was synthesized via electrochemical oxidation of C_6H_6/96% H_2SO_4 solution.

3 Physical and Chemical Properties of Conductive Polymers

Conductive polymers exhibit conductive, magnetic, electronic, optical, humectant, mechanical, and microwave absorption properties. A number of the properties of conductive polymers are shown in Fig. 4.

3.1 Chemical and Electrochemical Oxidation or Reduction

Poly-acetylene, through conversion of chemical oxidation or reduction reaction to metallic conductor, is considered the first intrinsic insulating polymer $(CH)_x$. Ozaki et al. suggested a list of oxidizing and reducing agents responsible for the effective modification for the conductivity of $(CH)_x$ [113]. Many oxidizing agents show effectiveness, while alkali metals were the only reducing agents. The following reactions show a typical oxidation and reduction process during a charge transfer.

$$(CH)_x + 3/2yI_2 \rightarrow (CH)_x^{y+}\left(I_3^-\right)_y \tag{1}$$

$$(CH)_x + yNa^0 \rightarrow (CH)_x^{y-}\left(Na^+\right)_y \tag{2}$$

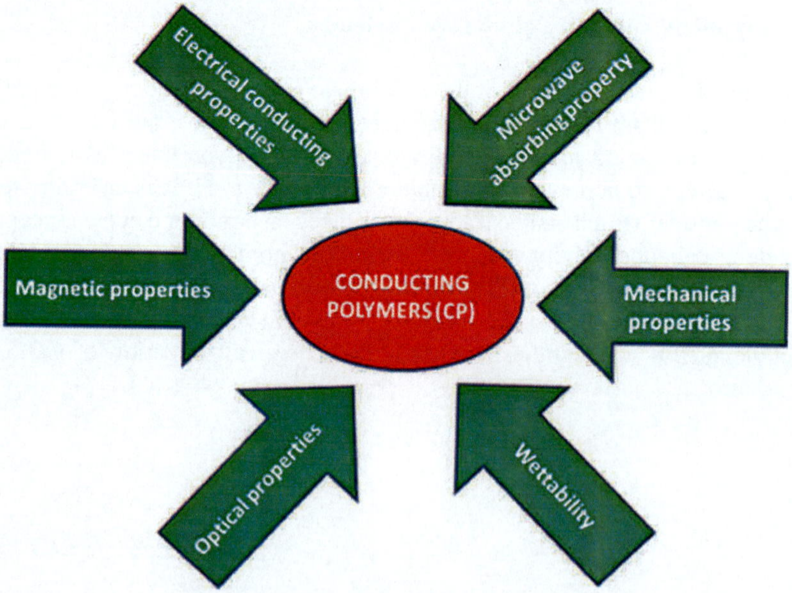

Fig. 4 Properties of conducting polymers [27]

Polyolefinic ions are presented apparently in the above equations of $(CH)_x$ conductive derivatives produced from oxidation/reduction agents through charge neutralized by counter ions [152]. These counterions intercalate through the planes. However, poly-acetylene remains an interesting member of this category of polymers. In some cases, the monomer was reacted with oxidizing agents and oxidation and polymerization was performed simultaneously in a single step, e.g. conductive poly-acetylenes were obtained by reacting AsF_5 with solid acetylene [174].

Electrochemical oxidation or reduction can also lead to the formation of poly-acetylene. The counter ions show up in the electrolyte solution. Electrochemical oxidation and polymerization are simultaneously carried on for chemical modification. This method produces flexible and stable polypyrrole films by electrochemical oxidation and polymerization of pyrrole [174].

3.2 Magnetic Properties

Conducting polymers have extraordinary magnetic properties and technical applications. In addition, the structural and magnetic properties of nanomaterials are incorporated into the polymer matrix [108]. The electrical conductivity of polyaniline/nickel oxide nanocomposites were investigated. As the nickel oxide ratio increased in polyaniline, magnetization increased. Conductive polymer/ferromagnetic thin film was prepared with side anodizing technology [178]. In a process where sodium dodecylbenzene sulphonate was used as a surfactant and an additive, a polypyrrole composite with ferromagnetic behaviour is prepared by a chemical method. The magnetic properties of the final composite exhibit ferromagnetic properties, such as high saturation magnetization and achieved the maximum conductivity of 10 S/cm in the most suitable synthetic medium [87]. Composite particles containing different proportions of $NiFe_{0.95}Gd_{0.05}O_4$ ferrite/polyaniline were prepared by in situ polymerization. Aphesteguyet al. chose different ferrite or aniline ratios to polyaniline and found that the magnetic properties were increased [4]. Electrical and Magnetic properties of polyaniline coatings were investigated.

The coercive force is largely unaffected, and the saturation magnetization is considerably reduced. Zaidi et al. studied the preparation and characterization of a new polymer polyanilinecyanokinodimethane, and receptor molecule tetracyanokinodimethane, prepared by polyaniline [185]. The magnetic calculations concluded that the polymer was ferromagnetic or ferromagnetic with a Curie temperature above 350 K and with a maximum saturation magnetization of 0.1 JT^{-1} kg^{-1}. Gosk et al. described the magnetic properties of polyaniline doped with $FeCl_3$ [42]. Ferromagnetism in polyaniline was observed by Trivedi [161]. Billas and his group studied the magnetism in iron, cobalt and nickel clusters from atomic to bulk level [16]. Ferromagnetic polyaniline with conductivity in an applied magnetic field has also been synthesized and characterized [192].

Owing to the well-known technological properties besides their remarkable magnetic properties, magnetism of conductive polymers exhibits great significance.

Polymer matrix manifests the structural characteristics along with other magnetic attributes; thus, making the transition metal oxide nanoparticles is quite distinct in their nature. The magnetic and transport features of conducting polyaniline/nickel oxides have already been evaluated by Nandapure et al. [108]. It was also assessed that with the enhancement of weight percentage of nickel oxide nanoparticles in polyaniline, the conductivity of polyaniline/nickel oxide nanocomposites gets diminished, while ameliorating the magnetization. Anodic oxidation technique was employed by Yan et al. to compose the conductive polymer/ferromagnet films [178]. The chemical technique in the presence of p-dodecyl benzenesulfonic acid sodium salt as a dopant and surfactant was deployed to fashion a composite of polypyrrole with ferromagnetic behaviour. Ferromagnetic behaviour as in the high saturated magnetization was illustrated through magnetic attributes of the ending composites. In the presence of utmost conducive circumstances, the maximized value of conductivity procured was of 10 S/cm [87]. In situ polymerization approach was successfully exploited to synthesize particles of $NiFe_{1.95}Gd_{0.05}O_4$ composites with varying fractions of ferrite and polyaniline.

Magnetic aspects were also enhanced with the usage of the enhanced ratio of polyaniline, while individual proportions of ferrite/aniline were analysed by Aphesteguy et al. [4]. It was also analysed that polyaniline coating gave varying magnetic properties along with conductive properties. The coercivity nearly remains unaltered, while the saturation magnetization remarkably curtailed. The preparation and characterization of a unique type of polymer, namely, polyanilinecyanoquinodimethane was delineated by using polyaniline along with an acceptor molecule tetracyanoquinodimethane [185]. The magnetic features of polyaniline doped with $FeCl_3$ were also reported by Gosk and co-workers [42].

3.3 *Optical Properties*

Das and Prusty [27] reported the remarkable optical properties such as optical refractive index for conductive polymers. The innovative properties of conductive polymers coupled with molecular dopants such as fullerenes have been detected by [181]. The optical reflectance polyacetylene coupled with enhanced accumulation of potassium along with the optical features analysed by Kramers–Kronig study, were well evaluated by Tanner et al. [157].

Akagi et al. [2] produced composition of helical PA $[(CH)_x]$ with the usage of left and right-handed screw structures under the chiral nematic reaction field. The concomitant effect on the optical and electrical properties of indium tin oxide/CP electrodes after treatment with ammonium sulfide was evaluated by Lin et al. [85].

Extensive reports on unique optical properties are also available in the literature [27, 181] reported novel optical features in CPs doped with molecular dopants like fullerenes. Tanner et al. [157] investigated optical reflectivity of doped high concentration potassium and the optical properties predicted by the Kramers–Kronig study. For the first time in the chiral nematic reaction field, Piao synthesized the polyacetylene $[(CH)_x]$ helix with the left and right helical structure [2].

3.4 Mechanical Properties

Conducting polymers magnetic properties have been analysed in detail in this section. The significant amount of information on unpaired spins and groups carrying charges are obtained from investigating this property. The mechanical properties for a conducting polymer based composite (CPC) of CNTs, polypropylene and graphite nanocomposites were investigated by Sulong et al. [154]. They also studied the results of different concentrations of fillers and functionality of chemicals on the mechanical properties of CPC. CPC, when functionalized, showed better results in strength and elongation as compared to the simple produced CPC from flexural and tensile tests. The highest flexural strength was found to be 80 MPa and tensile strength was 35 MPa. Functionalized CPC also has higher hardness than the simple produced CPC. The mechanical properties were calculated on machine and in transverse direction of uniaxially oriented polyphenylene vinylene thin film. Uniaxial extension and thermal exclusion of the poly sulphonium was used to synthesize this uniaxially thin film polyphenylene vinylene [93]. The properties resulted from it are anisotropic in nature and they are influenced by the degree of molecular orientation. In machine direction, the modulus was between 2.3 and 37 GPa and in the transverse direction, it was found to be between 2.3 and 0.5 GPa. Somanathan studied the tensile properties of 3-cyclohexyl thiophene conducting polymer. The mechanical properties of hybrid proton and the combination of sulfonated polyether ketones base was investigated by [144].

The mechanical properties of CP were thoroughly researched as they facilitate with valuable information on load bearing groups and unpaired rotation. Sulong et al. [154] discussed the mechanical properties of graphite, carbon nanotube, polypropylene nanocomposite (CPC) conductive polymer composites, etc. Better strength and elongation have been demonstrated by functionalized CPC than the CPC generated in tensile and bending tests. The maximum bending and tensile strength are 80 and 35 MPa, respectively. These polymers also show higher rigidity than CPC production. These characteristics are anisotropic and depend on the degree of molecular orientation. The Young modulus ranges from 2.3 to 37 GPa and lateral 2.3 and 0.5 GPa, as a function of the longitudinal plot ratio. Somanathan and Wegner [149] examined the tensile properties of conductive poly (3-cyclohexylthiophene) films at different temperatures.

3.5 Microwave Absorbing Properties

Conducting polymers are also investigated for their unique microwave absorbing properties. Ting studied these properties in detail through free space method. They determined the complex permeability/permittivity and loss due to reflection in the microwave range [160]. They also investigated that the addition of PANI over a wide frequency range was useful for achieving high absorption. TiO_2 nanoparticles, along with PANI nanocomposites for microwave absorptivity was studied by Phang et al. [124]. Initiated polymerization using a facile surface method was employed for synthesizing nanocomposites or nanorods of PANI with Titanium oxide. Various studies have been reported, one of which is by Guo which shows in a 3.5% solution of NaCl the absorption of microwave on magnesium alloy via electrochemical impedance spectroscope [46]. Polystyrene and polyvinyl pyrrolidone magnetic nanofiber properties of microwaves absorption were investigated by Guo et al. [46] At ambient conditions the nanofibers with diameters between 30 to 40 nm were found to have coercive field of 25 kV and 1.1 emu/g saturation magnetization [72].

Conductive polymer, an innovative material of interest to microwave is changing because of its low density, simple workability, and low cost. Ting determines the microwave absorption characteristics through determination of the complex permeability, and loss of reflection in the microwave frequency range by the free field method [144]. They also show that the addition of PANI is valuable for achieving excellent absorption over a wide frequency range. Phang et al. [123, 124] investigated microwave absorption properties of PANI nanocomposites containing TiO_2 nanoparticles [194] demonstrated the synthesis of PANI nanocomposites reinforced with tungsten oxide nanoparticles and nanorods using surface-initiated polymerization techniques. It was demonstrated that the electrochemical impedance spectroscopy study of the magnesium alloy microwave absorption shell in 3.5% by weight NaCl solution. Hosseini and Sadeghi [56] investigated the microwave absorption properties of polystyrene polyvinylpyrrolidone magnetic nanofibers and found that the magnetic nanofiber showed extraordinay microwave absorbing properties.

3.6 Electrical Conducting Properties

The electrical conductivity with respect to polymer mainly depends on doping percentage of, the alignment of the polymer chains, the length of the conjugate, and the purity of the sample. ECP is a molecule and lacks a long-range mode. The molecular structure of the polymer produces electron motion around each macromolecule. For polymers and inorganic semiconductors, the method of producing high conductivity is different. The increased conductivity depends on the additives in the polymer and involves the generation of self-aligned excitons such as solitons, polarons and dipoles. These particles create a strong interaction between the chain loads generated using the exciter. These polarons combine to produce a pair of non-electrode polarizers for use as charge carriers. Their conductive applications are eased through molecular engineering of the required properties [22]. Leung et al. calculated the optical as well as electrical properties of copolymers PA [76]. Tanaka et al. studied the optical and electrical properties of PA doped and its dependence on temperature [72]. The author also discussed the effects of thermal isomerization and impurities on the electrical properties of PA Tanaka and Danno [156] In 1989, Begin et al. studied the electrical properties of PAs that were highly prone to p-doping [13]. The conductivity of PA dopped is maximum achieved conductivity to date for any polymer material [162]. However, commercialization has not yet been achieved due to poor stability and processability issues. Matsushita explained the effects of printing on the resistance of doped PA and iodine [97]. Long et al. synthesized properties of electrical properties for carbon nanotube PANI composites [97]. Huang prepared highly self-assembled PANI gold nanowires and PANI CNTs [59]. Sarma et al. studied the synthesis of PANI compounds that result in Au nanoparticles as oxidation and reducing agents [59]. Chapman et al. discussed the PF-doped PPy with respect to its low energy conductivity [19]. Factors affecting conductivity are the density of the loads, their mobility, direction, presence of dopant materials, and the temperature.

3.7 Wettability

The performance of polymeric materials heavily depends on wettability. Other applications include anti-reflective properties for mobile phones or optical equipment such as cameras etc. Antibacterial adhesion is also one of the important applications along with other factors. A lot has been researched in this area of polymeric properties, Xu et al. investigated surface wettability of titanium oxide nanorod films combined and modified with triethoxyoctysilane [175].

Surface wetting exhibits a very imperative role for material properties, liquid distribution and separation, membrane cells and antibacterial adhesion are some of the applications where wettability is considered prime property. Xu et al. [175] reported synthesis and surface wettability of TiO_2 nanorod films altered with triethoxocytilsilane. Darmanin and Guittard [26] examined fibre structure, and wettability of the poly (3-alkyl-3,4-propylenedioxythiophene) for nanoporous, microporous or micro/nanostructure. Lin et al. [84] studied the development of the adjustable wettability characteristics of poly(3-alkylthiophene) films. Teh et al. [158] studied the effect of redox-induced PPy recombination on wettability and surface morphology. San and Isik-Gulsac [138] reported the effect of fluoropolymer-based additives on the wettability of polymer composite bipolar sheets with different additives/binders and additive/filler ratios was investigated.

4 Polymer Composites and Their Synthesis Methods

Conductive polymers can be useful in various commercial applications such as solar energy conversion [148], high-power fuel cells [130], supercapacitors [120], and energy storage [142], are gaining popularity since last decade. The polymers with electrical conductivity mainly includes polypyrrole [32], polyaniline [35] and polyacetylene [22]. These polymers are considered important due to their exceptional optical, chemical and electrical properties [116]. The main aim of the research in this area is to improve the applicability of conducting polymers [164]. This can be done by introducing appropriate organic and inorganic materials to form conducting polymer composites having superior properties than the polymer itself. The hybrid composite can be prepared by combining various materials, including CeO_2 [39], TiO_2 [153], SnO_2, and ZrO_2 [15, 29]. There are three main methods for preparing electrically conductive polymers: solution casting, chemical and electrochemical polymerization.

4.1 *Synthesis Methods*

4.1.1 Solution Casting

Solution casting method for synthesizing conducting composites involves the use of a solvent suitable for dissolving both polymers in order to attain a homogeneous mix. The polymers are mixed by stirring and finally cast to obtain composite films. Electrically conductive polyaniline (PANI) composite with acrylonitrile-butadine-styrene (ABS) as matrix was developed using solution casting technique, for use as a sensor material by [69]. Chloroform was selected as a suitable solvent for synthesizing the PANI-ABS composite, as it has polarity similar to that of PANI and ABS. Other organic and water-based solvents have also been used to fabricate conductive composites using the solution casting technique. Electrically conductive PANI-ABS composite was synthesized by adding different concentrations of PANI in ABS and 100 ml of chloroform. The resistance values at various loadings of PANI were reported to range between 100 and 330 F/g [38].

Solution casting is generally employed for carbon nanotubes (CNT) and graphene based polymer composites [50], as this method allows the even dispersion of nanotubes. In the solvent casting method, a polymer solution is prepared and nanofiller are dispersed separately in a solvent using sonication Fig. 5. In the case of polymer/CNT based nanocomposites, the CNT must be surface modified to achieve a stable and homogeneous dispersion. The polymer solution and nanofiller solution prepared using the same solvents are mixed, to allow the polymer to adsorb to the surface of the filler [67]. The solvent is finally removed through evaporation, to obtain the composite.

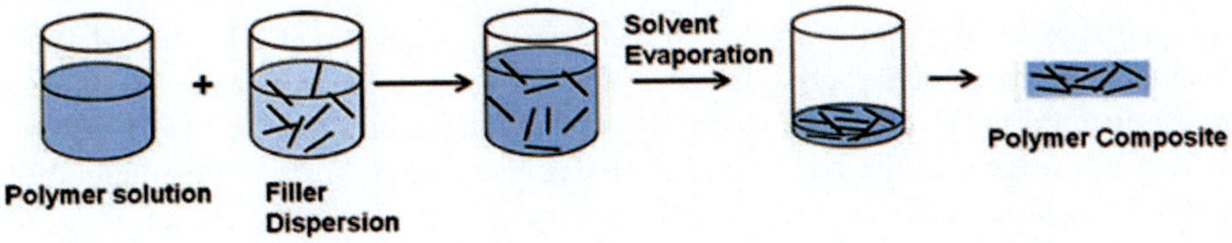

Fig. 5 Fabrication of conductive polymer composites through solution casting method. Figure taken with permission from [67]

4.1.2 Chemical Polymerization

Polymer composites having high stability and conductivity can be synthesized using chemical polymerization method [120]. This method includes synthesis of polymers with high conductivity by simple oxidations (with an oxidizing agent) of particular monomers. Ammonium persulfate and ferric chloride are the commonly used oxidizing agents. There are advantages and disadvantages to employing this method on a commercial scale. This method is favourable for large-scale commercial production because it delivers an inexpensive, relatively simple and effective method to accomplish polymerisation. When the oxidizing agent is added to the monomer solution, it leads to the formation of a thick layer or film of the polymer,this permits its bulk manufacture.

In addition, this method is suitable to synthesize all kinds of conductive polymers [10]. However, with this method, high homogeneity cannot be achieved as compared to other methods like electrochemical polymerization. Also, the final conductivity of the polymers after preparing by the chemical methods has at all times remained lesser than their electrochemically produced counterparts. Furthermore, the final properties of the synthesized polymer largely depend on the oxidant, the purity and choice of selected solvent, reaction temperature and time [10]. This in turn, affects the repeatability and reliability of the synthesis method.

4.1.3 Electrochemical Polymerization

Electrochemical polymerization is conducted by passing a current through the electrodes which are placed in a solution. This solution contains the solvent, monomer of the desired polymer and a doping material [10]. Through this, a thin film of desired polymer can be accumulated on the electrode. The polymer chains are formed as the current deposits the monomer, which oxidises the electrode having positive charge on the surface. The final properties of the polymer synthesized using electrochemical polymerization depend on the deposition charge, doping material, electrode system, temperature, time and the solvent used [10]. Additionally, this method of synthesis can only be used with some particular monomers which can easily endure oxidation, when in contact with electrical current,these include PANI and polypyrrole (PPY). The conductivity with respect to electrical of poly(vinyl chloride) (PVC) based composite with uniformly distributed PPY was studied by De Paoli et al. [30]. The main advantage of preparing the composites using the electrochemical polymerization method is that the final properties of the composite films can be altered by altering or controlling the conditions of electrolysis [131]. Further, this method of synthesizing electrically conductive polymer composites eradicates the necessity of harmful dopants and strong oxidizing materials [30]. A unique method of synthesising PVC/PPY composites was reported in this study. Firstly, a thin layer of PVC was formed on the outer surface of an electrode. Secondly, this surface-treated electrode was employed to make PPY in the pores of PVC. The solvent employed for electrochemical polymerization was Acetonitrile, as it has similar polarity to that of PVC [30]. The quantity of PPY in the PVC/PPY composite was altered by altering the time of electro polymerization. The final results showed that a PVC/PPY composite with superior conductivity, high mechanical strength and excellent stability was achieved using electro polymerization method.

Another study reported by Peng et al. [120], showed high homogeneity of carbon nanotube and good interaction among the polymer matrix and carbon nanotubes, when prepared using the electrochemical deposition method. The homogeneity of carbon nanotubes led to conjugation and strengthened electron delocalisation among the chains of the polymer [120]. This ultimately resulted in exceptional electrochemical storing capacities and rapid charge/discharge converting, making them favourable conductor materials aimed at high power supercapacitors.

Electrochemical polymerization allows the fast deposition of conductive polymers, however the number of bioactive particles that can be incapacitated into the polymer using this method is limited [10]. Also, the quantity of the polymer that can be produced is constrained by the electrode surface area and shape. This requisite of an electrode creates additional difficulty of creating the composites using electrochemical polymerization [145].

4.1.4 Melt Blending Method

This method includes the mixing of the polymer and filler in molten state [150]. This method is particularly of interest as it does not involve the use of toxic solvents. The filler is mixed in polymer matrix (thermoplastic) mechanically through extrusion at high temperatures, closer to the polymer matrix melting point. The resultant composite has the polymer chains intercalated among the filler fragments. The speediness and ease of this method has made it the most favoured one for large scale production. This method is also suitable for the polymers which cannot be synthesized using solution casting or any other processing technique [67].

5 Applications of Conducting Polymers and Composites

Several applications of conducting materials and composites have been discovered so far. Due to their nature of electronic and ionic conductivity, reversible doping, adjustable chemical and electrochemical characteristics, low cost and convenient processability they introduce due application in a variety of fields. Some of the well-researched areas for these materials have been batteries, light emitting diodes (LEDs), sensors/bio-sensors, electro-optic and optical devices, membranes/films microwave and conduction-based devices, electrochromic devices, electrochemical and mechanical devices, energy storage devices, fuel cells, supercapacitors, corrosion protection, electrorheological fluids, tissue engineering, neural interfaces, actuators semiconductors, lithography, artificial muscles, catalysis, drug delivery, biomedical, analytical applications etc. [140, 132, 27].

The batteries are based on three fundamental components, known as the anode, cathode, and electrolyte. The anodes and cathodes are selected based on their conducting potential so that the transfer of positive and negative ions should be facilitated with minimal dissipation of energy. Moreover, capacity, life, battery voltage and environmental implications play a major role selection of material for anodes, cathodes and electrolytes. Due to their nature, conducting polymers have witnessed their applications in batteries, especially lithium-ion batteries. The conventional liquid electrolyte batteries introduce a risk to electric vehicles, which can be mitigated using solid conducting polymers and composites. The solid conducting polymer electrolytes can be produced by crosslinking of mobile chains to develop networks or by the development of solid polymer electrolytes (SPEs) with supramolecular architectures [104]. Single-ion conducting polymer electrolytes (SIPEs) prepared by lithium sulfonamide methacrylic monomer, bifunctional poly(ethylene glycol) methyl ether dimethacrylate (PEGDM) and poly(ethylene glycol) methyl ether methacrylate (PEGM) in the presence of propylene carbonate (PC) shown lithium transference number approaching to unity and ionic conductivity $\sigma \approx 10^{-4}$ S cm^{-1} at ambient temperature [127]. Ion-solvation-site connectivity plays a vital role in polymer electrolytes for high lithium-ion conductance [105]. Poly(ε-caprolactone) (PCL)/Succinonitrile (SN) blends integrated with polyacrylonitrile(PAN)-skeleton with hierarchical architectures has exhibited high ionic conductivity, degree flexibility, large electrochemical windows, thermal stability and good flame-retardance strength [187].

Core/shell particles of poly (3,4-ethylenedioxythiophene) (PEDOT) and sulphur showed great electrochemical performance when they were as used as the cathode material for Li/S battery [21]. The polyaniline/TiO_2, while used a cathode in rechargeable battery, has shown charge–discharge properties up to 50 cycles than its conducting polymer alone [49]. Polypyrrole and polythiophenes doped with redox-active $Fe(CN)_6{}^{4-}$ anions are reported to enhance 56 to 300% redox capacity [193]. Attaching physical or chemically Fe^{III}/Fe^{II} redox couple to the Skelton of a conducting polymer enhances stability of the charge–discharge properties and increases electrode capacities as witnessed for chemically attaching ferrocene to the pyrrole Skelton [117]. Polypyrrole is reported to lower the initial irreversible capacity loss of the silicon anode due to a decrease in the thickness of the solid electrolyte-interface (SEI) layer, which leads to the higher coulombic efficiency and improved cycle life [47].

This highlights that conducting polymers and composites used in all three basic components as anodes, cathodes and electrolytes improve the efficiencies of batteries, their life and charge–discharge cycle, reduce environmental concerns, and lower the overall cost of the battery.

Different metals have been doped for the development of conducting polymers and composites for light emitting diode development. Due the deformative nature bistable curvatures shape-memory polymer light-emitting diodes (PLEDs) by silver nanowire/polymer electrode result in a minimal loss of electroluminescence [184]. Silver nanowire-polymer composite electrodes are used as electrophosphorescent polymer light-emitting diodes in different colours like blue, green, and red enhanced 20–50% efficiency than control devices on indium-doped tin oxide ITO/glass [79]. Single-walled carbon nanotube (SWNT)-polymer composite electrodes have been reported to increase the intrinsic stretch-ability [183]. SWNTs can increase Conductivity of more than 100 S cm^{-1} and stretch-ability of more than 100%, which can lead to a display stretch-ability up 50% without mechanical or electrical damage [141]. Elastomericpolymer light-emitting devices (EPLEDs) with rubbery elasticity can emit light when exposed to strains up to 120% [82]. In the presence of hydrogenated styrene ethylene butylene styrene (SEBS) block copolymers, the poly(3,4-ethylenedioxythiophene):poly(styrenesulfonate) (PEDOT:PSS) has depicted 3100 S/cm under 0% strain and over 4100 S/cm under 100% strain, which lasted even 3600 S/cm after 1000 cycles to 100% strain [167]. Modification of PEDOT:PSS with sodium-poly(styrenesulfonate) enhances work function and photoluminescence intensity of the overlying perovskite layer [121].

The conductive polymers have a potential application in the medical sciences also, for instance, intrinsically conducting polymers have been proven to be promising in medical field as well. These conducting polymers are used in wound healing and engineering skin tissues [143]. They have been found to be novel for fast wound care. Due to their enhanced antibacterial property, they have the potential in controlled drug usage/delivery as well. Hence the enhanced electrical conductivity property helps simulating the wound area directly and results in accelerated recovery of a wound. The electrical conductivity is modulated through an electrical stimuli which help in controlled release of drugs to the wound area [62]. The study mainly summarizes the data available in healing the wound and engineering skin tissues. Most commonly used conductive polymers for these purposes are polypyrrole, polyaniline, polythiophene etc. and their derivatives. Two of these conductive polymers (polyaniline and polypyrrole) have been examined and studied and provided potential outcpmes in this field. They are used frequently in wound care and hydrogels. Polypyrrole has been applied in skin tissue engineering. Conducting polymers are also used in combination with biomolecules or sometimes they are doped to make biocompatible polymers. Sometimes co-polymerization and blending is also done to enhance the biocompatibility of the polymers and their physical nature. Overall, conducting polymers are one of the major applications used in area of engineering of skin tissue and wound healing [155].

The incorporation of intrinsically/inherently conducting polymer with conductive fillers produce conductive polymer-based composites. These composites have great electromagnetic interference (EMI) shielding. They also show an increase in electromagnetic waves absorption rather than reflection which makes strong conductive networks [51]. Thus, the incorporation from recent advancements in the multi structural composites from intrinsically conducting polymer and conductive polymer-based composites have been proven to be promising techniques for electromagnetic interference (EMI) shielding applications. Enhanced electromagnetic interference shielding was achieved by presenting processing parameters, characteristics dependent on the frequency and morphological structures adjustments. The relevant mechanisms from the following applications were also comprehended. The study has helped understanding all the anomalies/nonuniformities involved for enhancing electromagnetic absorption inside the EMI shielding materials. For instance, in blends of biphasic polymers the localization and morphology of the filler has improved electromagnetic absorption. In polymer based materials either the lightweight or the one's with high EM absorptivity the feature that was found fascinating was EMI shielding potential of a composite [64].

6 Conclusion

The final properties of electrically conductive polymer composites, including the thermal conductivity, is dependent on the inherent thermal conductivity of the materials used, i.e. the matric and the filler. Further, the filler concentration added in the matrix, aspect ratio, shape, size and extent of miscibility with the matrix also affect the properties of conductive polymer composites. The overall properties of the prepared composites are reliant one or all of the above factors. Additionally, the filler surface treatments have been studied extensively to enhance the even distribution and miscibility of the filler within the matrix. Surface modifications of fillers have reduced the thermal resistance at the interface and ultimately this led to increased thermal conductivity. Filler must be selected appropriately, so that it improves the overall properties of the prepared composite, including the dielectric properties. Depending on the filler and polymer type the processing method can be selected. The fillers are added to improve the properties of the composite, as the polymer is not suitable by itself for commercial applications. Adding surface modified fillers have certainly increased the potential of conductive polymer composites for future applications.

References

1. Abidian, M.R., Kim, D.H., Martin, D.C.: Conducting-polymer nanotubes for controlled drug release. Adv. Mater. **18**(4), 405–409 (2006)
2. Akagi, K., Piao, G., Kaneko, S., Higuchi, I., Shirakawa, H., Kyotani, M.: Helical polyacetylene synthesized under chiral nematic liquid crystals. Synth. Met. **102**(1–3), 1406–1409 (1999)
3. Al-Saleh, M.H., Sundararaj, U.: A review of vapor grown carbon nanofiber/polymer conductive composites. Carbon **47**(1), 2–22 (2009)
4. Aphesteguy, J., Bercoff, P., Jacobo, S.: Preparation of magnetic and conductive Ni–Gd ferrite-polyaniline composite. Phys. B **398**(2), 200–203 (2007)
5. Athawale, A.A., Kulkarni, M.V.: Polyaniline and its substituted derivatives as sensor for aliphatic alcohols. Sens. Actuators B: Chem **67**(1–2), 173–177 (2000)
6. Awuzie, C.I.: Conducting polymers. Mater. Today: Proc. **4**(4, Part E), 5721–5726 (2017)
7. Bai, H., Zhao, L., Lu, C., Li, C., Shi, G.: Composite nanofibers of conducting polymers and hydrophobic insulating polymers: preparation and sensing applications. Polymer **50**(14), 3292–3301 (2009)
8. Baker, C.O., Shedd, B., Innis, P.C., Whitten, P.G., Spinks, G.M., Wallace, G.G., Kaner, R.B.: Monolithic actuators from flash-welded polyaniline nanofibers. Adv. Mater. **20**(1), 155–158 (2008)
9. Baker, C.O., Shedd, B., Tseng, R.J., Martinez-Morales, A.A., Ozkan, C.S., Ozkan, M., Yang, Y., Kaner, R.B.: Size control of gold nanoparticles grown on polyaniline nanofibers for bistable memory devices. ACS Nano **5**(5), 3469–3474 (2011)
10. Balint, R., Cassidy, N.J., Cartmell, S.H.: Conductive polymers: towards a smart biomaterial for tissue engineering. Acta Biomater. **10**(6), 2341–2353 (2014)
11. Ballard, D., Courtis, A., Shirley, I., Taylor, S.: Synthesis of polyphenylene from a cis-dihydrocatechol biologically produced monomer. Macromolecules **21**(2), 294–304 (1988)
12. Banerjee, S., Konwar, D., Kumar, A.: Polyaniline nanofiber reinforced nanocomposite based highly sensitive piezoelectric sensors for selective detection of hydrochloric acid: analysis of response mechanism. Sens. Actuators B: Chem. **190**, 199–207 (2014)
13. Begin, D., Marêché, J., Billaud, D.: Electrical properties of p-doped highly-oriented polyacetylene. Synth. Met. **34**(1–3), 671–676 (1989)
14. Bello, A., Giannetto, M., Mori, G., Seeber, R., Terzi, F., Zanardi, C.: Optimization of the DPV potential waveform for determination of ascorbic acid on PEDOT-modified electrodes. Sens. Actuators B: Chem. **121**(2), 430–435 (2007)
15. Bhattacharya, A., Ganguly, K., De, A., Sarkar, S.: A new conducting nanocomposite—PPy-zirconium (IV) oxide. Mater. Res. Bull. **31**(5), 527–530 (1996)
16. Billas, I.M., Chatelain, A., de Heer, W.A.: Magnetism from the atom to the bulk in iron, cobalt, and nickel clusters. Science **265**(5179), 1682–1684 (1994)
17. Bott, D., Skotheim, T.: Handbook of Conducting Polymers. Marcel Dekker, New York (1986)
18. Burroughes, J.H., Bradley, D.D., Brown, A., Marks, R., Mackay, K., Friend, R.H., Burns, P., Holmes, A.: Light-emitting diodes based on conjugated polymers. Nature **347**(6293), 539 (1990)
19. Chapman, B., Buckley, R., Kemp, N., Kaiser, A., Beaglehole, D., Trodahl, H.: Low-energy conductivity of PF 6-doped polypyrrole. Phys. Rev. B **60**(19), 13479 (1999)
20. Chen, G.Z., Shaffer, M.S., Coleby, D., Dixon, G., Zhou, W., Fray, D.J., Windle, A.H.: Carbon nanotube and polypyrrole composites: coating and doping. Adv. Mater. **12**(7), 522–526 (2000)
21. Chen, H., Dong, W., Ge, J., Wang, C., Wu, X., Lu, W., Chen, L.: Ultrafine sulfur nanoparticles in conducting polymer shell as cathode materials for high performance lithium/sulfur batteries. Sci. Rep. **3**, 1910 (2013)
22. Chiang, C.K., Fincher, C., Jr., Park, Y.W., Heeger, A.J., Shirakawa, H., Louis, E.J., Gau, S.C., MacDiarmid, A.G.: Electrical conductivity in doped polyacetylene. Phys. Rev. Lett. **39**(17), 1098 (1977)

23. Chun, K.Y., Oh, Y., Rho, J., Ahn, J.H., Kim, Y.J., Choi, H.R., Baik, S.: Highly conductive, printable and stretchable composite films of carbon nanotubes and silver. Nat. Nanotechnol. **5**(12), 853 (2010)
24. Crowley, K., Morrin, A., Shepherd, R.L., M. in het Panhuis, G. G. Wallace, M. R. Smyth and A. J. Killard, : Fabrication of polyaniline-based gas sensors using piezoelectric inkjet and screen printing for the detection of hydrogen sulfide. IEEE Sens. J. **10**(9), 1419–1426 (2010)
25. Dallolio, A., Dascola, G., Varacca, V., Bocchi, V.: Electronic paramagnetic resonance and conductivity of a black electrolytic oxypyrrole. Comptes Rendus Hebdomadaires Des Seances De L Academie Des Sciences Serie C **267**(6), 433–440 (1968)
26. Darmanin, T., Guittard, F.: Wettability of poly (3-alkyl-3, 4-propylenedioxythiophene) fibrous structures forming nanoporous, microporous or micro/nanostructured networks. Mater. Chem. Phys. **146**(1–2), 6–11 (2014)
27. Das, T.K., Prusty, S.: Review on conducting polymers and their applications. Polym.-Plast. Technol. Eng. **51**(14), 1487–1500 (2012)
28. Daver, F., Baez, E., Shanks, R.A., Brandt, M.: Conductive polyolefin–rubber nanocomposites with carbon nanotubes. Compos. A Appl. Sci. Manuf. **80**, 13–20 (2016)
29. De, A., Das, A., Lahiri, S.: Heavy ion irradiation on conducting polypyrrole and ZrO2–polypyrrole nanocomposites. Synth. Met. **144**(3), 303–307 (2004)
30. De Paoli, M.A., Waltman, R., Diaz, A., Bargon, J.: An electrically conductive plastic composite derived from polypyrrole and poly (vinyl chloride). Journal of Polymer Science: Polymer Chemistry Edition **23**(6), 1687–1698 (1985)
31. De Surville, R., Jozefowicz, M., Yu, L., Pepichon, J., Buvet, R.: Electrochemical chains using protolytic organic semiconductors. Electrochim. Acta **13**(6), 1451–1458 (1968)
32. Diaz, A., Kanazawa, K., Gardini, G.: Electrochemically conducting polypyrrole . J. Chem. Soc. Chem. Commun **14**, 635–636 (1979)
33. Diaz, A., Kanazawa, K.K., Gardini, G.P.: Electrochemical polymerization of pyrrole. J. Chem. Soc., Chem. Commun. (14), 635–636 (1979)
34. Diaz, A., Logan, J.: Electroactive polyaniline films. J. Electroanal. Chem. Interfacial Electrochem. **111**(1), 111–114 (1980)
35. Dogan, S., Akbulut, U., Toppare, L.: Conducting polymers of aniline I. Electrochemical synthesis of a conducting composite. Synth. Met. **53**(1), 29–35 (1992)
36. Downs, C., Nugent, J., Ajayan, P.M., Duquette, D.J., Santhanam, K.S.: Efficient polymerization of aniline at carbon nanotube electrodes. Adv. Mater. **11**(12), 1028–1031 (1999)
37. Dul, S., Fambri, L., Pegoretti, A.: Fused deposition modelling with ABS–graphene nanocomposites. Compos. A Appl. Sci. Manuf. **85**, 181–191 (2016)
38. Frackowiak, E., Khomenko, V., Jurewicz, K., Lota, K., Béguin, S.: Supercapacitors based on conducting polymers/nanotubes composites. **153**(2), 413–418 (2006)
39. Galembeck, A., Alves, O.L.: Chemical polymerization of pyrrole on CeO2 films. Synth. Met. **84**(1–3), 151–152 (1997)
40. Gao, M., Huang, S., Dai, L., Wallace, G., Gao, R., Wang, Z.: Aligned coaxial nanowires of carbon nanotubes sheathed with conducting polymers. Angew. Chem. **112**(20), 3810–3813 (2000)
41. Gerard, M., Chaubey, A., Malhotra, B.: Application of conducting polymers to biosensors. Biosens. Bioelectron. **17**(5), 345–359 (2002)
42. Gosk, J., Kulszewicz-Bajer, I., Twardowski, A.: Magnetic properties of polyaniline doped with FeCl3. Synth. Met. **156**(11–13), 773–778 (2006)
43. Greene, R.L., Street, G.B., Suter, L.: Superconductivity in polysulfur nitride (SN) x. Phys. Rev. Lett. **34**(10), 577 (1975)
44. Greenham, N.C., Friend, R.H., Brown, A.R., Bradley, D.D., Pichler, K., Burn, P.L., Kraft, A., Holmes, A.B.: Electroluminescent devices made with conjugated polymers. In: Electroluminescent Materials, Devices, and Large-Screen Displays. International Society for Optics and Photonics (1993)

45. Groenendaal, L., Jonas, F., Freitag, D., Pielartzik, H., Reynolds, J.R.: Poly (3, 4-ethylenedioxythiophene) and its derivatives: past, present, and future. Adv. Mater. **12**(7), 481–494 (2000)
46. Guo, Q., Du, K., Guo, X., Wang, F.: Electrochemical impedance spectroscope analysis of microwave absorbing coatings on magnesium alloy in 3.5 wt.% NaCl solution. Electrochim. Acta **98**, 190–198 (2013)
47. Guo, Z.P., Wang, J.Z., Liu, H.K., Dou, S.X.: Study of silicon/polypyrrole composite as anode materials for Li-ion batteries. J. Power Sources **146**(1), 448–451 (2005)
48. Gupta, N., Sharma, S., Mir, I.A., Kumar, D.: Advances in sensors based on conducting polymers (2006)
49. Gurunathan, K., Amalnerkar, D., Trivedi, D.: Synthesis and characterization of conducting polymer composite (PAn/TiO2) for cathode material in rechargeable battery. Matér. Lett. **57**(9–10), 1642–1648 (2003)
50. Gurunathan, T., Rao, C.R., Narayan, R., Raju, K.: Polyurethane conductive blends and composites: synthesis and applications perspective. J. Mater. Sci. **48**(1), 67–80 (2013)
51. Gurusiddesh, M., Madhu, B., Shankaramurthy, G.: Structural, dielectric, magnetic and electromagnetic interference shielding investigations of polyaniline decorated Co 0.5 Ni 0.5 Fe 2 O 4 nanoferrites. J. Mater. Sci.: Mater. Electron. **29**(4), 3502–3509 (2018)
52. Hatano, M., Kambara, S., Okamoto, S.: Paramagnetic and electric properties of polyacetylene. J. Polym. Sci. **51**(156), S26–S29 (1961)
53. Heeger, A.J.: Semiconducting and metallic polymers: the fourth generation of polymeric materials (Nobel lecture). Angew. Chem. Int. Ed. **40**(14), 2591–2611 (2001)
54. Heywang, G., Jonas, F.: Poly (alkylenedioxythiophene) s—new, very stable conducting polymers. Adv. Mater. **4**(2), 116–118 (1992)
55. Hill, R.F., Supancic, P.H.: Thermal conductivity of platelet-filled polymer composites. J. Am. Ceram. Soc. **85**(4), 851–857 (2002)
56. Hosseini, S.H., Sadeghi, M.: Investigation of microwave absorbing properties for magnetic nanofiber of polystyrene–polyvinylpyrrolidone. Curr. Appl. Phys. **14**(7), 928–931 (2014)
57. Huang, J., Kaner, R.B.: Nanofiber formation in the chemical polymerization of aniline: a mechanistic study. Angew. Chem. Int. Ed. **43**(43), 5817–5821 (2004)
58. Huang, J., Mao, C., Zhu, Y., Jiang, W., Yang, X.: Control of carbon nanotubes at the interface of a co-continuous immiscible polymer blend to fabricate conductive composites with ultralow percolation thresholds. Carbon **73**, 267–274 (2014)
59. Huang, K., Zhang, Y., Long, Y., Yuan, J., Han, D., Wang, Z., Niu, L., Chen, Z.: Preparation of highly conductive, self-assembled gold/polyaniline nanocables and polyaniline nanotubes. Chem.–A Eur. J. **12**(20), 5314–5319 (2006)
60. Huang, X., Jiang, P., Tanaka, T.: A review of dielectric polymer composites with high thermal conductivity. IEEE Electr. Insul. Mag. **27**(4), 8–16 (2011)
61. Hwang, S., Reyes, E.I., Moon, K.S., Rumpf, R.C., Kim, N.S.: Thermo-mechanical characterization of metal/polymer composite filaments and printing parameter study for fused deposition modeling in the 3D printing process. J. Electron. Mater. **44**(3), 771–777 (2015)
62. Isseroff, R.R., Dahle, S.E.: Electrical stimulation therapy and wound healing: where are we now? Adv. Wound Care **1**(6), 238–243 (2012)
63. Ivory, D., Miller, G., Sowa, J., Shacklette, L., Chance, R., Baughman, R.: Highly conducting charge-transfer complexes of poly (p-phenylene). J. Chem. Phys. **71**(3), 1506–1507 (1979)
64. Jiang, D., Murugadoss, V., Wang, Y., Lin, J., Ding, T., Wang, Z., Shao, Q., Wang, C., Liu, H., Lu, N.: Electromagnetic interference shielding polymers and nanocomposites-a review. Polym. Rev. **59**(2), 280–337 (2019)
65. Kalita, G., Umeno, M., Tanemura, M.: Blend of Silicon Nanostructures and Conducting Polymers for Solar Cells, pp. 495–508. Elsevier, Nanostructured Polymer Blends (2014)
66. Kanazawa, K.K., Diaz, A., Geiss, R.H., Gill, W.D., Kwak, J.F., Logan, J.A., Rabolt, J.F., Street, G.B.: 'Organic metals': polypyrrole, a stable synthetic 'metallic'polymer. J. Chem. Soc., Chem. Commun. (19), 854–855 (1979)

67. Kaur, G., Adhikari, R., Cass, P., Bown, M., Gunatillake, P.: Electrically conductive polymers and composites for biomedical applications. RSC Adv. **5**(47), 37553–37567 (2015)
68. Kirchmeyer, S., Reuter, K.: Scientific importance, properties and growing applications of poly (3, 4-ethylenedioxythiophene). J. Mater. Chem. **15**(21), 2077–2088 (2005)
69. Koul, S., Chandra, R., Dhawan, S.: Conducting polyaniline composite: a reusable sensor material for aqueous ammonia. Sens. Actuators B: Chem. **75**(3), 151–159 (2001)
70. Kuang, T., Chang, L., Chen, F., Sheng, Y., Fu, D., Peng, X.: Facile preparation of lightweight high-strength biodegradable polymer/multi-walled carbon nanotubes nanocomposite foams for electromagnetic interference shielding. Carbon **105**, 305–313 (2016)
71. Kumar, D., Sharma, R.: Advances in conductive polymers. Eur. Polym. J. **34**(8), 1053–1060 (1998)
72. Kumar, R., Singh, S., Yadav, B.: Conducting polymers: synthesis, properties and applications. Int. Adv. Res. J. Sci., Eng. Technol. **2**(11), 110–124 (2015)
73. Kumar, S.S., Mathiyarasu, J., Phani, K., Yegnaraman, V.: Simultaneous determination of dopamine and ascorbic acid on poly (3, 4-ethylenedioxythiophene) modified glassy carbon electrode. J. Solid State Electrochem. **10**(11), 905–913 (2006)
74. Lee, D., Baek, S., Ree, M., Kim, O.: Effect of the electrode material on the electrical-switching characteristics of nonvolatile memory devices based on poly ($ o $-anthranilic acid) thin films. IEEE Electron Device Lett. **29**(7), 694–697 (2008)
75. Letheby, H.: XXIX.—on the production of a blue substance by the electrolysis of sulphate of aniline. J. Chem. Soc. **15**, 161–163 (1862)
76. Leung, L.M., Tan, K.H., Lam, T., WeiDong, H.: Electrical and optical properties of polyacetylene copolymers. React. Funct. Polym. **50**(2), 173–179 (2002)
77. Li, D., Del Rio Castillo, A.E., Jussila, H., Ye, G., Ren, Z., Bai, J., Chen, X., Lipsanen, H., Sun, Z., Bonaccorso, F.: Black phosphorus polycarbonate polymer composite for pulsed fibre lasers. Appl. Mater. Today **4**, 17–23 (2016)
78. Li, D., Huang, J., Kaner, R.B.: Polyaniline nanofibers: a unique polymer nanostructure for versatile applications. Acc. Chem. Res. **42**(1), 135–145 (2008)
79. Li, L., Yu, Z., Hu, W., Chang, C.H., Chen, Q., Pei, Q.: Efficient flexible phosphorescent polymer light-emitting diodes based on silver nanowire-polymer composite electrode. Adv. Mater. **23**(46), 5563–5567 (2011)
80. Li, S., Li, Z., Fang, X., Chen, G.Q., Huang, Y., Xu, K.: Synthesis and characterization of polyparaphenylene from cis-dihydrocatechol. J. Appl. Polym. Sci. **110**(4), 2085–2093 (2008)
81. Li, Y., Huang, X., Zeng, L., Li, R., Tian, H., Fu, X., Wang, Y., Zhong, W.H.: A review of the electrical and mechanical properties of carbon nanofiller-reinforced polymer composites. J. Mater. Sci. **54**(2), 1036–1076 (2019)
82. Liang, J., Li, L., Niu, X., Yu, Z., Pei, Q.: Elastomeric polymer light-emitting devices and displays. Nat. Photonics **7**(10), 817 (2013)
83. Lin, J.W.P., Dudek, L.P.: Synthesis and properties of poly (2, 5-thienylene). J. Polym. Sci.: Polym. Chem. Edition **18**(9), 2869–2873 (1980)
84. Lin, P., Yan, F., Chan, H.L.: Improvement of the tunable wettability property of poly (3-alkylthiophene) films. Langmuir **25**(13), 7465–7470 (2009)
85. Lin, Y.J., Liu, B.Y., Chin, Y.M.: Effects of (NH4) 2Sx treatment on the electrical and optical properties of indium tin oxide/conducting polymer electrodes. Thin Solid Films **517**(18), 5508–5511 (2009)
86. Little, W.: Possibility of synthesizing an organic superconductor. Phys. Rev. **134**(6A), A1416 (1964)
87. Liu, J., Wan, M.: Composites of polypyrrole with conducting and ferromagnetic behaviors. J. Polym. Sci., Part a: Polym. Chem. **38**(15), 2734–2739 (2000)
88. Liu, Y., Lv, H., Lan, X., Leng, J., Du, S.: Review of electro-active shape-memory polymer composite. Compos. Sci. Technol. **69**(13), 2064–2068 (2009)
89. Long, Y., Duvail, J., Li, M., Gu, C., Liu, Z., Ringer, S.P.: Electrical conductivity studies on individual conjugated polymer nanowires: two-probe and four-probe results. Nanoscale Res. Lett. **5**(1), 237 (2010)

90. Lu, X., Zhang, W., Wang, C., Wen, T.C., Wei, Y.: One-dimensional conducting polymer nanocomposites: synthesis, properties and applications. Prog. Polym. Sci. **36**(5), 671–712 (2011)
91. Ma, P.C., Siddiqui, N.A., Marom, G., Kim, J.K.: Dispersion and functionalization of carbon nanotubes for polymer-based nanocomposites: a review. Compos. A Appl. Sci. Manuf. **41**(10), 1345–1367 (2010)
92. MacDiarmid, A.G.: "Synthetic metals": a novel role for organic polymers (Nobel lecture). Angew. Chem. Int. Ed. **40**(14), 2581–2590 (2001)
93. Machado, J., Masse, M., Karasz, F.: Anisotropic mechanical properties of uniaxially oriented electrically conducting poly (p-phenylene vinylene). Polymer **30**(11), 1992–1996 (1989)
94. Mannoor, M.S., Jiang, Z., James, T., Kong, Y.L., Malatesta, K.A., Soboyejo, W.O., Verma, N., Gracias, D.H., McAlpine, M.C.: 3D printed bionic ears. Nano Lett. **13**(6), 2634–2639 (2013)
95. Mapkar, J.A., Belashi, A., Berhan, L.M., Coleman, M.R.: Formation of high loading flexible carbon nanofiber network composites. Compos. Sci. Technol. **75**, 1–6 (2013)
96. Martin, C.R., Kohli, P.: The emerging field of nanotube biotechnology. Nat. Rev. Drug Discovery **2**(1), 29 (2003)
97. Matsushita, A., Akagi, K., Liang, T.S., Shirakawa, H.: Effects of pressure on the electrical resistivity of iodine-doped polyacetylene. Synth. Met. **101**(1–3), 447–448 (1999)
98. McCullough, R.D.: The chemistry of conducting polythiophenes. Adv. Mater. **10**(2), 93–116 (1998)
99. McNeill, R., Siudak, R., Wardlaw, J., Weiss, D.: Electronic conduction in polymers. I. The chemical structure of polypyrrole. Aust. J. Chem. **16**(6), 1056–1075 (1963)
100. McQuade, D.T., Pullen, A.E., Swager, T.M.: Conjugated polymer-based chemical sensors. Chem. Rev. **100**(7), 2537–2574 (2000)
101. Mecklenburg, M., Mizushima, D., Ohtake, N., Bauhofer, W., Fiedler, B., Schulte, K.: On the manufacturing and electrical and mechanical properties of ultra-high wt.% fraction aligned MWCNT and randomly oriented CNT epoxy composites. Carbon **91**, 275–290 (2015)
102. Meng, H., Li, G.: A review of stimuli-responsive shape memory polymer composites. Polymer **54**(9), 2199–2221 (2013)
103. Meng, Q., Hu, J.: A review of shape memory polymer composites and blends. Compos. A Appl. Sci. Manuf. **40**(11), 1661–1672 (2009)
104. Meyer, W.H.: Polymer electrolytes for lithium-ion batteries. Adv. Mater. **10**(6), 439–448 (1998)
105. Miller, T.F., III., Wang, Z.G., Coates, G.W., Balsara, N.P.: Designing polymer electrolytes for safe and high capacity rechargeable lithium batteries. Acc. Chem. Res. **50**(3), 590–593 (2017)
106. Misra, S., Mathur, P., Srivastava, B.: Vacuum-deposited nanocrystalline polyaniline thin film sensors for detection of carbon monoxide. Sens. Actuators, A **114**(1), 30–35 (2004)
107. Mittal, G., Dhand, V., Rhee, K.Y., Park, S.J., Lee, W.R.: A review on carbon nanotubes and graphene as fillers in reinforced polymer nanocomposites. J. Ind. Eng. Chem. **21**, 11–25 (2015)
108. Nandapure, B., Kondawar, S., Salunkhe, M., Nandapure, A.: Magnetic and transport properties of conducting polyaniline/nickel oxide nanocomposites. Adv. Mater. Lett. **4**(2), 134–140 (2013)
109. Narkis, M., Zilberman, M., Siegmann, A.: On the "curiosity" of electrically conductive melt processed doped-polyaniline/polymer blends versus carbon-black/polymer compounds. Polym. Adv. Technol. **8**(8), 525–528 (1997)
110. Naveen, M.H., Gurudatt, N.G., Shim, Y.B.: Applications of conducting polymer composites to electrochemical sensors: a review. Appl. Mater. Today **9**, 419–433 (2017)
111. Novák, P., Müller, K., Santhanam, K., Haas, O.: Electrochemically active polymers for rechargeable batteries. Chem. Rev. **97**(1), 207–282 (1997)
112. Oyama, N., Tatsuma, T., Sato, T., Sotomura, T.: Dimercaptan–polyaniline composite electrodes for lithium batteries with high energy density. Nature **373**(6515), 598 (1995)
113. Ozaki, M., Peebles, D., Weinberger, B., Heeger, A., MacDiarmid, A.: Semiconductor properties of polyacetylene p-(CH) x: n-CdS heterojunctions. J. Appl. Phys. **51**(8), 4252–4256 (1980)

114. Pan, L., Pu, L., Shi, Y., Song, S., Xu, Z., Zhang, R., Zheng, Y.: Synthesis of polyaniline nanotubes with a reactive template of manganese oxide. Adv. Mater. **19**(3), 461–464 (2007)
115. Pang, H., Xu, L., Yan, D.X., Li, Z.M.: Conductive polymer composites with segregated structures. Prog. Polym. Sci. **39**(11), 1908–1933 (2014)
116. Pant, H., Patra, M., Negi, S., Bhatia, A., Vadera, S., Kumar, N.: Studies on conductivity and dielectric properties of polyaniline-zinc sulphide composites. Bull. Mater. Sci. **29**(4), 379–384 (2006)
117. Park, K.S., Schougaard, S.B., Goodenough, J.B.: Conducting-polymer/iron-redox- couple composite cathodes for lithium secondary batteries. Adv. Mater. **19**(6), 848–851 (2007)
118. Park, S.-J., Son, Y.-R., Heo, Y.-J.: Chapter 6—Prospective synthesis approaches to emerging materials for supercapacitor. In: Cheong, K.Y., Impellizzeri, G., Fraga, M.A. (eds.) Emerging Materials for Energy Conversion and Storage, pp. 185–208. Elsevier
119. Pei, Q., Inganäs, O.: Electrochemical applications of the bending beam method. 2. Electroshrinking and slow relaxation in polypyrrole. The J. Phys. Chem. **97**(22), 6034–6041 (1993)
120. Peng, C., Zhang, S., Jewell, D., Chen, G.Z.: Carbon nanotube and conducting polymer composites for supercapacitors. Prog. Nat. Sci. **18**(7), 777–788 (2008)
121. Peng, X.F., Wu, X.Y., Ji, X.X., Ren, J., Wang, Q., Li, G.Q., Yang, X.H.: Modified conducting polymer hole injection layer for high-efficiency perovskite light-emitting devices: enhanced hole injection and reduced luminescence quenching. The J. Phys. Chem. Lett. **8**(19), 4691–4697 (2017)
122. Penner, R., Martin, C.: Microporous membrane-modified electrodes for preparation of chemical microstructures on electrode surfaces. J. Electrochem. Soc., Electrochemical Soc Inc 10 South Main Street, Pennington, NJ 08534 (1987)
123. Phang, S.W., Hino, T., Abdullah, M., Kuramoto, N.: Applications of polyaniline doubly doped with p-toluene sulphonic acid and dichloroacetic acid as microwave absorbing and shielding materials. Mater. Chem. Phys. **104**(2–3), 327–335 (2007)
124. Phang, S.W., Tadokoro, M., Watanabe, J., Kuramoto, N.: Microwave absorption behaviors of polyaniline nanocomposites containing TiO_2 nanoparticles. Curr. Appl. Phys. **8**(3–4), 391–394 (2008)
125. Plieth, W.: Electrochemistry for Materials Science. Elsevier (2008)
126. Pokharel, P., Xiao, D., Erogbogbo, F., Keles, O.: A hierarchical approach for creating electrically conductive network structure in polyurethane nanocomposites using a hybrid of graphene nanoplatelets, carbon black and multi-walled carbon nanotubes. Compos. B Eng. **161**, 169–182 (2019)
127. Porcarelli, L., Shaplov, A.S., Bella, F., Nair, J.R., Mecerreyes, D., Gerbaldi, C.: Single-ion conducting polymer electrolytes for lithium metal polymer batteries that operate at ambient temperature. ACS Energy Lett. **1**(4), 678–682 (2016)
128. Pramanik, P., Khastgir, D., Saha, T.: Conductive nitrile rubber composite containing carbon fillers: studies on mechanical properties and electrical conductivity. Composites **23**(3), 183–191 (1992)
129. Qian, C., Zhu, Y., Dong, Y., Fu, Y.: Vapor-grown carbon nanofiber/poly (ethylene-co-vinyl acetate) composites with electrical-active two-way shape memory behavior. J. Intell. Mater. Syst. Struct. **28**(19), 2749–2756 (2017)
130. Qiao, Y., Li, C.M., Bao, S.J., Bao, Q.L.: Carbon nanotube/polyaniline composite as anode material for microbial fuel cells. J. Power Sources **170**(1), 79–84 (2007)
131. Ramya, R., Sivasubramanian, R., Sangaranarayanan, M.: Conducting polymers-based electrochemical supercapacitors—progress and prospects. Electrochim. Acta **101**, 109–129 (2013)
132. Ravichandran, R., Sundarrajan, S., Venugopal, J.R., Mukherjee, S., Ramakrishna, S.: Applications of conducting polymers and their issues in biomedical engineering. J. Roy. Soc. Interface **7**(suppl_5), S559–S579 (2010)
133. Raza, M.A., Westwood, A.V.K., Stirling, C., Ahmad, R.: Effect of boron nitride addition on properties of vapour grown carbon nanofiber/rubbery epoxy composites for thermal interface applications. Compos. Sci. Technol. **120**, 9–16 (2015)

134. Reddinger, J.L., Reynolds, J.R.: Molecular engineering of π-conjugated polymers. In: Radical Polymerisation Polyelectrolytes, pp. 57-122. Springer (1999)
135. Roncali, J.: Conjugated poly (thiophenes): synthesis, functionalization, and applications. Chem. Rev. **92**(4), 711–738 (1992)
136. Ryu, K.S., Kim, K.M., Kang, S.G., Lee, G.J., Joo, J., Chang, S.H.: Electrochemical and physical characterization of lithium ionic salt doped polyaniline as a polymer electrode of lithium secondary battery. Synth. Met. **110**(3), 213–217 (2000)
137. Sadeghi, A., Moeini, R., Yeganeh, J.K.: Highly conductive PP/PET polymer blends with high electromagnetic interference shielding performances in the presence of thermally reduced graphene nanosheets prepared through melt compounding. Polym. Compos. (2018)
138. San, F.G.B., Isik-Gulsac, I.: Effect of surface wettability of polymer composite bipolar plates on polymer electrolyte membrane fuel cell performances. Int. J. Hydrogen Energy **38**(10), 4089–4098 (2013)
139. Sariciftci, N., Braun, D., Zhang, C., Srdanov, V., Heeger, A., Stucky, G., Wudl, F.: Semiconducting polymer-buckminsterfullerene heterojunctions: diodes, photodiodes, and photovoltaic cells. Appl. Phys. Lett. **62**(6), 585–587 (1993)
140. Scrosati, B.: Applications of Electroactive Polymers. Springer (1993)
141. Sekitani, T., Nakajima, H., Maeda, H., Fukushima, T., Aida, T., Hata, K., Someya, T.: Stretchable active-matrix organic light-emitting diode display using printable elastic conductors. Nat. Mater. **8**(6), 494 (2009)
142. Selampinar, F., Toppare, L., Akbulut, U., Yalçin, T., Süzer, Ş.: A conducting composite of polypyrrole II. As a gas sensor. Synth. Met. **68**(2), 109–116 (1995)
143. Sen, C.K., Gordillo, G.M., Roy, S., Kirsner, R., Lambert, L., Hunt, T.K., Gottrup, F., Gurtner, G.C., Longaker, M.T.: Human skin wounds: a major and snowballing threat to public health and the economy. Wound Repair and Regeneration **17**(6), 763–771 (2009)
144. Sgreccia, E., Khadhraoui, M., De Bonis, C., Licoccia, S., Di Vona, M., Knauth, P.: Mechanical properties of hybrid proton conducting polymer blends based on sulfonated polyetheretherketones. J. Power Sources **178**(2), 667–670 (2008)
145. Shi, G., Rouabhia, M., Wang, Z., Dao, L.H., Zhang, Z.: A novel electrically conductive and biodegradable composite made of polypyrrole nanoparticles and polylactide. Biomaterials **25**(13), 2477–2488 (2004)
146. Shirakawa, H., Louis, E.J., MacDiarmid, A.G., Chiang, C.K., Heeger, J.: Synthesis of electrically conducting organic polymers: halogen derivatives of polyacetylene, (CH) x. J. Chem. Soc., Chem. Commun. (16), 578–580 (1977)
147. Simonet, J., Rault-Berthelot, J.: Electrochemistry: a technique to form, to modify and to characterize organic conducting polymers. Prog. Solid State Chem. **21**(1), 1–48 (1991)
148. Skotheim, T., Inganäs, O., Prejza, J., Lundström, I.: Polypyrrole-semiconductor photovoltaic devices. Mol. Cryst. Liq. Cryst. **83**(1), 329–339 (1982)
149. Somanathan, N., Wegner, G.: Mechanical properties of conducting poly (3-cyclohexyl thiophene) films. Polymer **37**(10), 1891–1895 (1996)
150. Stefanescu, E., Daranga, C., Stefanescu, C.: Insight into the broad field of polymer nanocomposites: from carbon nanotubes to clay nanoplatelets, via metal nanoparticles. Materials **2**(4), 2095–2153 (2009)
151. Stejskal, J., Sapurina, I., Trchová, M.: Polyaniline nanostructures and the role of aniline oligomers in their formation. Prog. Polym. Sci. **35**(12), 1420–1481 (2010)
152. Street, G., Clarke, T.: Conducting polymers: a review of recent work. IBM J. Res. Dev. **25**(1), 51–57 (1981)
153. Su, S.J., Kuramoto, N.: Processable polyaniline–titanium dioxide nanocomposites: effect of titanium dioxide on the conductivity. Synth. Met. **114**(2), 147–153 (2000)
154. Sulong, A.B., Ramli, M.I., Hau, S.L., Sahari, J., Muhamad, N., Suherman, H.: Rheological and mechanical properties of carbon nanotube/Graphite/SS316L/polypropylene nanocomposite for a conductive polymer composite. Compos. B Eng. **50**, 54–61 (2013)
155. Talikowska, M., Fu, X., Lisak, G.: Application of conducting polymers to wound care and skin tissue engineering: a review. Biosens. Bioelectron. (2019)

156. Tanaka, H., Danno, T.: Effects of impurities and thermal isomerization on the electrical properties of undoped polyacetylene. Synth. Met. **17**(1–3), 545–550 (1987)
157. Tanner, D., Doll, G., Rao, A., Eklund, P., Arbuckle, G., MacDiarmid, A.: Optical properties of potassium-doped polyacetylene. Synth. Met. **141**(1–2), 75–79 (2004)
158. Teh, K.S., Takahashi, Y., Yao, Z., Lu, Y.W.: Influence of redox-induced restructuring of polypyrrole on its surface morphology and wettability. Sens. Actuators, a **155**(1), 113–119 (2009)
159. Tibbetts, G.G., Lake, M.L., Strong, K.L., Rice, B.P.: A review of the fabrication and properties of vapor-grown carbon nanofiber/polymer composites. Compos. Sci. Technol. **67**(7), 1709–1718 (2007)
160. Ting, T., Jau, Y., Yu, R.: Microwave absorbing properties of polyaniline/multi-walled carbon nanotube composites with various polyaniline contents. Appl. Surf. Sci. **258**(7), 3184–3190 (2012)
161. Trivedi, D.: Observation of ferromagnetism in polyaniline. Synth. Met. **121**(1), 1780–1781 (2001)
162. Tsukamoto, J., Takahashi, A.: Synthesis and electrical properties of polyacetylene yielding conductivity of 105 S/cm. Synth. Met. **41**(1–2), 7–12 (1991)
163. Virji, S., Huang, J., Kaner, R.B., Weiller, B.H.: Polyaniline nanofiber gas sensors: examination of response mechanisms. Nano Lett. **4**(3), 491–496 (2004)
164. Vishnuvardhan, T., Kulkarni, V., Basavaraja, C., Raghavendra, S.: Synthesis, characterization and ac conductivity of polypyrrole/Y 2 O 3 composites. Bull. Mater. Sci. **29**(1), 77–83 (2006)
165. Wan, M.: A template-free method towards conducting polymer nanostructures. Adv. Mater. **20**(15), 2926–2932 (2008)
166. Wang, J., Langhe, D., Ponting, M., Wnek, G.E., Korley, L.T., Baer, E.: Manufacturing of polymer continuous nanofibers using a novel co-extrusion and multiplication technique. Polymer **55**(2), 673–685 (2014)
167. Wang, Y., Zhu, C., Pfattner, R., Yan, H., Jin, L., Chen, S., Molina-Lopez, F., Lissel, F., Liu, J., Rabiah, N.I., Chen, Z., Chung, J.W., Linder, C., Toney, M.F., Murmann, B., Bao, Z.: A highly stretchable, transparent, and conductive polymer. Sci. Adv. **3**(3), e1602076 (2017)
168. Wang, Y.G., Li, H.Q., Xia, Y.Y.: Ordered whiskerlike polyaniline grown on the surface of mesoporous carbon and its electrochemical capacitance performance. Adv. Mater. **18**(19), 2619–2623 (2006)
169. Wei, X., Li, D., Jiang, W., Gu, Z., Wang, X., Zhang, Z., Sun, Z.: 3D printable graphene composite. Sci. Rep. **5**, 11181 (2015)
170. Weisman, J.A., Nicholson, J.C., Tappa, K., Jammalamadaka, U., Wilson, C.G., Mills, D.K.: Antibiotic and chemotherapeutic enhanced three-dimensional printer filaments and constructs for biomedical applications. Int. J. Nanomed. **10**, 357 (2015)
171. Wise, D.L.: Electrical and Optical Polymer Systems: Fundamentals: Methods, and Applications. CRC Press (1998)
172. Wu, C.G., Bein, T.: Conducting polyaniline filaments in a mesoporous channel host. Science **264**(5166), 1757–1759 (1994)
173. Wu, D., Zhang, Y., Zhang, M., Yu, W.: Selective localization of multiwalled carbon nanotubes in poly (ε-caprolactone)/polylactide blend. Biomacromol **10**(2), 417–424 (2009)
174. Wynne, K.J., Street, G.B.: Conducting polymers. A short review. Ind. Eng. Chem. Prod. Res. Dev. **21**(1), 23–28 (1982)
175. Xu, C., Fang, L., Huang, Q., Yin, B., Ruan, H., Li, D.: Preparation and surface wettability of TiO2 nanorod films modified with triethoxyoctylsilane. Thin Solid Films **531**, 255–260 (2013)
176. Yakhmi, J.V., Saxena, V., Aswal, D.K.: 2—Conducting polymer sensors, actuators and field-effect transistors. In: Banerjee, S., Tyagi, A.K. (eds.) Functional Materials, pp. 61–110. London, Elsevier (2012)
177. Yamamoto, T., Sanechika, K., Yamamoto, A.: Preparation of thermostable and electric-conducting poly (2, 5-thienylene). J. Polym. Sci.: Polym. Lett. Edition **18**(1), 9–12 (1980)

178. Yan, F., Xue, G., Chen, J., Lu, Y.: Preparation of a conducting polymer/ferromagnet composite film by anodic-oxidation method. Synth. Met. **123**(1), 17–20 (2001)
179. Yang, X., Dai, T., Zhu, Z., Lu, Y.: Electrochemical synthesis of functional polypyrrole nanotubes via a self-assembly process. Polymer **48**(14), 4021–4027 (2007)
180. Yeganeh, J.K., Goharpey, F., Moghimi, E., Petekidis, G., Foudazi, R.: Manipulating the kinetics and mechanism of phase separation in dynamically asymmetric LCST blends by nanoparticles. Phys. Chem. Chem. Phys. **17**(41), 27446–27461 (2015)
181. Yoshino, K., Tada, K., Yoshimoto, K., Yoshida, M., Kawai, T., Araki, H., Hamaguchi, M., Zakhidov, A.: Electrical and optical properties of molecularly doped conducting polymers. Synth. Met. **78**(3), 301–312 (1996)
182. Yu, J.H., Fridrikh, S.V., Rutledge, G.C.: Production of submicrometer diameter fibers by two-fluid electrospinning. Adv. Mater. **16**(17), 1562–1566 (2004)
183. Yu, Z., Niu, X., Liu, Z., Pei, Q.: Intrinsically stretchable polymer light-emitting devices using carbon nanotube-polymer composite electrodes. Adv. Mater. **23**(34), 3989–3994 (2011)
184. Yu, Z., Zhang, Q., Li, L., Chen, Q., Niu, X., Liu, J., Pei, Q.: Highly flexible silver nanowire electrodes for shape-memory polymer light-emitting diodes. Adv. Mater. **23**(5), 664–668 (2011)
185. Zaidi, N.A., Giblin, S., Terry, I., Monkman, A.: Room temperature magnetic order in an organic magnet derived from polyaniline. Polymer **45**(16), 5683–5689 (2004)
186. Zhang, B., Fu, R., Zhang, M., Dong, X., Wang, L., Pittman, C.U.: Gas sensitive vapor grown carbon nanofiber/polystyrene sensors. Mater. Res. Bull. **41**(3), 553–562 (2006)
187. Zhang, D., Zhang, L., Yang, K., Wang, H., Yu, C., Xu, D., Xu, B., Wang, L.M.: Superior blends solid polymer electrolyte with integrated hierarchical architectures for all-solid-state lithium-ion batteries. ACS Appl. Mater. Interfaces **9**(42), 36886–36896 (2017)
188. Zhang, F., Nyberg, T., Inganäs, O.: Conducting polymer nanowires and nanodots made with soft lithography. Nano Lett. **2**(12), 1373–1377 (2002)
189. Zhang, W., Blackburn, R.S., Dehghani-Sanij, A.A.: Effect of silica concentration on electrical conductivity of epoxy resin–carbon black–silica nanocomposites. Scripta Mater. **56**(7), 581–584 (2007)
190. Zhang, W., Dehghani-Sanij, A.A., Blackburn, R.S.: Carbon based conductive polymer composites. J. Mater. Sci. **42**(10), 3408–3418 (2007)
191. Zhang, X., MacDiarmid, A.G., Manohar, S.K.: Chemical synthesis of PEDOT nanofibers. Chem. Commun. (42), 5328–5330 (2005)
192. Zhang, Y., Zhu, C., Kan, J.: Synthesis and characterization of ferromagnetic polyaniline with conductivity in an applied magnetic field. J. Appl. Polym. Sci. **109**(5), 3024–3029 (2008)
193. Zhou, M., Qian, J., Ai, X., Yang, H.: Redox-active Fe(CN)64−-doped conducting polymers with greatly enhanced capacity as cathode materials for Li-ion batteries. Adv. Mater. **23**(42), 4913–4917 (2011)
194. Zhu, J., Wei, S., Zhang, L., Mao, Y., Ryu, J., Karki, A.B., Young, D.P., Guo, Z.: Polyaniline-tungsten oxide metacomposites with tunable electronic properties. J. Mater. Chem. **21**(2), 342–348 (2011)

Two-Dimensional Transition Metal Carbides and Nitrides (MXenes): Synthesis to Applications

Muhammad Zahir Iqbal and Saman Siddique

Abstract Recently, a novel family of two-dimensional materials, called MXenes, comprising of early transition metal nitrides and carbides was discovered with intriguing characteristics and potential applications. MXenes are synthesized by adopting various top down and bottom up approaches such as selective etching of "A" element from MAX phases result in a new MXene element, for instance, Ti_3C_2, V_2C, Ti_3CN, MoC_2, Ta_4C_3 etc. MXenes exhibit high metallic conductivity in which solid layers are bonded together with strong ionic, covalent and metallic bonds. The hydrophilic nature of MXene enhance its practical applications such as electrocatalyst, energy storage devices and in biomedical applications. Here, the chapter reviews the basic structure of newly discovered MXene materials, different synthesis techniques, structural, electrical and optical properties. Some potential applications in the field of biomedical, energy conversion and electrochemical energy storage systems and electrocatalyst are also presented in this chapter.

Keywords MXenes · Synthesis · Properties · Applications · Biomedical · Energy storage

1 Introduction

Two-dimensional materials (2DMs) have achieved a significant importance over the last decade due to exceptional structural, electrical and optical properties. Graphene was the first 2DM discovered in 2004 by Novoselov, comprising of atomically thin carbon sheets bonded covalently in a honeycomb structure [1]. Additionally, various two-dimensional materials have been synthesized such as hexagonal boron nitride (hBN) [2], silicene [3], arsenene [4], germanene, bismuthene [5, 6] and transition

M. Z. Iqbal (✉) · S. Siddique
Nanotechnology Research Laboratory, Faculty of Engineering Sciences, GIK Institute of Engineering Sciences and Technology, Topi, Khyber Pakhtunkhwa 23640, Pakistan
e-mail: zahir.upc@gmail.com

N. M. Mubarak et al. (eds.), *Contemporary Nanomaterials in Material Engineering Applications*, Engineering Materials,
https://doi.org/10.1007/978-3-030-62761-4_7

metal dichalcogenides [7]. Among various two-dimensional materials some are insulators such as hBN while some are semiconductors as transition metal dichalcogenides, silicene, arsenene and bismuthene. Two-dimensional materials are utilized in various scientific applications ranging from electronics to electrochemical energy storage and conversion devices such as field effect transistors, supercapacitors and lithium ion batteries [8–10].

Recently, a novel class of 2DMs entitled as "MXene" has been discovered with similar properties as that of other 2DMs. MXenes are categorized as early transition metal carbides and nitrides. MXenes are obtained by etching and chemical exfoliation of layered carbides in MAX phases, where M is the early transition metal, A is the IIIA and IVA group elements such as aluminum (Al) or silicon (Si) and X represents carbon or nitrogen atom and the suffix "ene" represents their similarity with that of graphene. The general formula used for MXene is $M_{n+1}AX_n$ where n ranges from 1 to 3 [11–13]. The first ever MXene material "named as $Ti_3C_2T_z$" was synthesized in 2011 by heating Ti_3AlC_2 in hydrofluoric acid at room temperature [14]. During this process, aluminum atoms are substituted by oxygen atoms, hydroxyl ions or fluorine atoms. The removal of aluminum layer effects the bonding strength of $M_{n+1}X_n$ layers and allow them to be separated easily. Due to the removal of Al layer from the MAX phases, the obtained product was termed as MXene due to 2D nature and similar properties as that of graphene such as high electrical conductivity and stability. Figure 1 represent the schematic illustration of MXene elements having single M layer or a mixture of two different M layers.

Here, M can be any element such as Ti, Ta, V, Nb etc. The MXenes studied so far includes V_2CT_x, $Ti_3C_2T_x$, $Ta_4C_3T_x$, $Nb_4C_3T_x$, Ti_2CT_x and Nb_2CT_x, here T_x denote the surface groups such as oxygen, fluorine and hydroxyl ions [16–18]. Some other MXene solutions have also been reported such as Ti_3CNT_x, $(V_{0.5}Cr_{0.5})_3C_2T_x$ and $(Ti_{0.5}Nb_{0.5})_2CT_x$ having thickness less than 1 nm [19, 20].

MXenes have distinctive properties of metallic conductivity and hydrophilic nature which exposed many promising applications in energy conversion and storage devices such as hydrogen evolution reactions, supercapacitors, sodium capacitors, potassium, lithium and sodium ion batteries [21–25]. The discovery and extensive utilization of MXene, extends their potential applications in the development of future nanoelectronics and energy device applications.

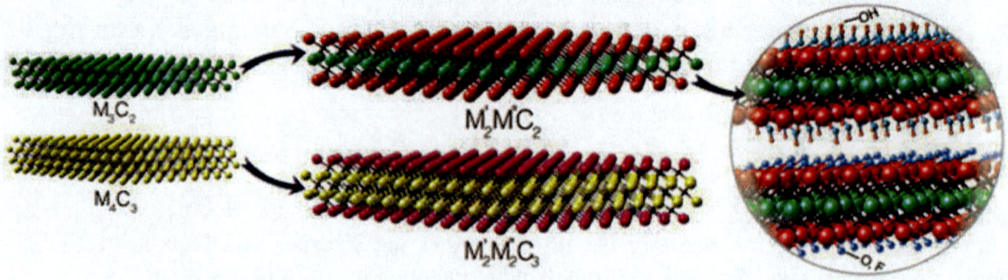

Fig. 1 Schematic of MXenes having single and bilayers of M element [15]

2 Synthesis of MXenes

Generally, two routes are employed for the synthesis of MXene including bottom-up approach and top-down approach. The bottom-up approach utilizes chemical vapor deposition (CVD) technique which is used to synthesize high quality film on a substrate. In most of the cases, the films grown by this method contains stacked multilayers of the material. On the other hand, the second method for the synthesis of MXene is the top-down approach including exfoliation of layered materials. This method is further subdivided into chemical and physical exfoliation. The synthesis technique is clearly understandable by the flowsheet diagram as shown in Fig. 2. Following section summarizes different techniques for the synthesis of MXene.

2.1 *Synthesis of MXene by Bottom-Up Approach*

In recent years, many bottom-up techniques have been exploited for the synthesis of transition metal nitrides and carbides such as CVD technique [26, 27], plasma enhanced pulsed laser deposition (PELPD) technique and template method [28]. The materials produced by using these techniques exhibits high crystallinity. However, the thin films produced by such methods are not single layer but consists of few thin layers having the similar electronic characteristics as that of other two-dimensional materials.

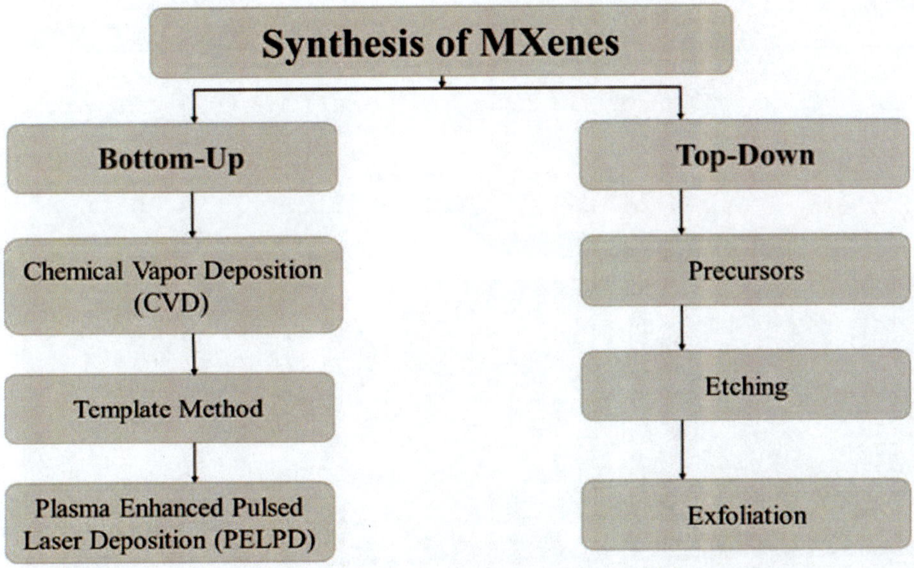

Fig. 2 Flowsheet diagram of synthesis of MXenes dividing in two main categories

Xu et al. was the first to report the synthesis of transition metal carbides crystals by using chemical vapor deposition technique in which copper and transition metal was used as a substrate [29]. The stack of copper and molybdenum foils was preheated at a high temperature of 1085 °C to produce thin films of α-Mo_2C crystals. The crystals were grown on liquid copper surface in the presence of methane and hydrogen. The liquid copper layer exhibits many advantages as it improves the decomposition process of methane by acting as a catalyst and it also provide a channel for controlled diffusion of molybdenum atoms to the liquid copper surface which leads to the formation of Mo_2C crystals on copper surface. Figure 3a is the schematic illustration of Mo_2C crystals that are grown on a Cu and Mo substrate. The as-grown ultrathin crystals of Mo_2C is based on various definite shapes such as rectangles, triangles and hexagons having 3–20 nm thickness. Moreover, after etching of copper by using $(NH_4)_2S_2O_8$, the crystals can easily be transferred on any desired target substrate. It was also observed that Mo atoms consisted of a symmetrical hexagon structure having rotational symmetry with three or six domains except octagon and rectangular crystals [30]. Figure 3b–d illustrate three different domain boundaries which have a strong impact on the conductivity of 2D materials. Further studies showed that the crystal shape of Mo_2C can be transformed from triangular to other polygonal structure by varying the amount of methane flow [31].

Similar to the formation of Mo_2C crystals, Wang et al. reported the growth of tantalum carbide (TaC) crystals by using CVD technique [34]. Ammonia as a source of nitrogen can be used for the synthesis of two-dimensional TaN crystals. In this case,

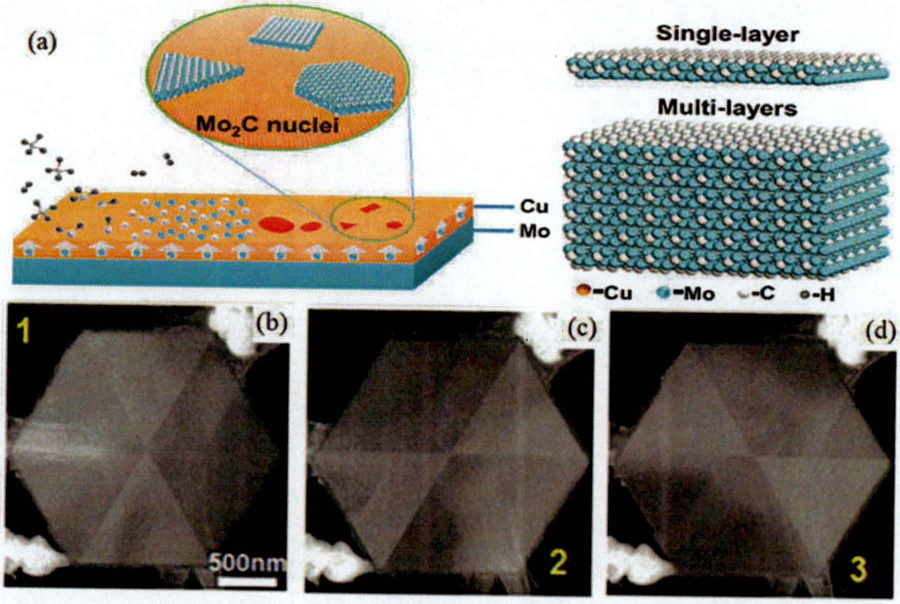

Fig. 3 **a** Schematic illustration of growth of two-dimensional Mo_2C crystals. **b–d** Transmission electron microscopy image indicating well-defined domain structure of Mo_2C crystal [30, 32, 33]

Cu foil at 1077 °C is kept in solid form during the whole procedure. Moreover, DFT calculations prove that the thickness of two-dimensional crystals can be controlled easily at lower temperature. The temperature below the melting point of Cu is sufficient to supply energy to Ta atoms to fill the vacant sites of Cu lattice. Just similar to transition metal carbides and nitrides, CVD method is also utilized for the synthesis of transition metal chalcogenides [35]. Previous reports demonstrate two-different methods to grow two-dimensional vertical heterostructures of Mo_2C and graphene. One method is the direct-step CVD in which methane was used in higher concentration for the formation of graphene [35]. Second method is two-step CVD in which graphene was initially grown on Cu substrate at low temperature (below melting point of copper) and then temperature was gradually raised to grow Mo_2C crystals under the graphene layer [36]. Figure 4a shows the schematic illustration of two-step method of using CVD to develop vertical heterostructures of Mo_2C/graphene. The crystals obtained by using two-step method exhibits high crystallinity and non-uniform strain domains which results in super conductive behavior.

Another technique used in the bottom-up approach is the template method in which two-dimensional transition metal oxides are utilized as a template which is further carbonized to form transition metal carbides [37, 38]. Xiao et al. discussed the formation of MoN nanosheets by using MoO_3 as a template forming a hexagonal structure as shown in Fig. 4b. The resulted nanosheets of MoN are uniform having a thickness of 0.71 nm as observed by using HR-TEM image in Fig. 4c. In another studies, MoN nanosheets are synthesized by using template method in which Mo filament treated with oxygen is used as a template nanosheets of MoO_3 are vertically

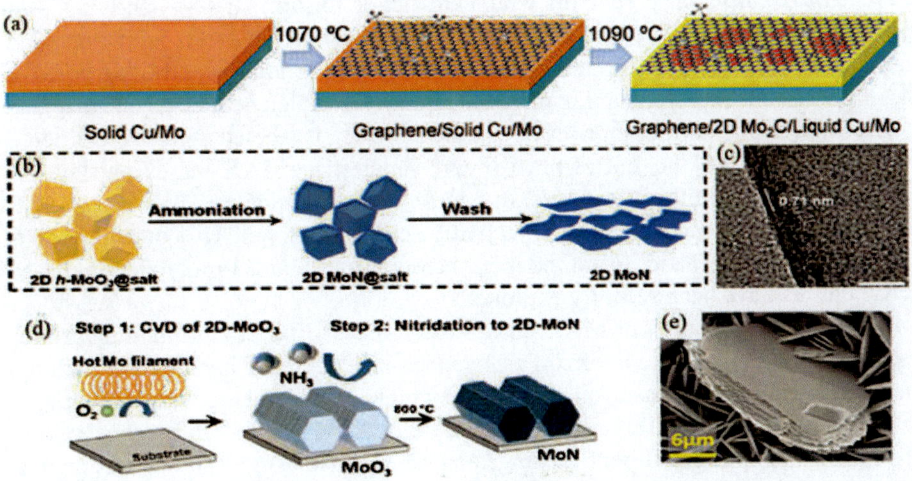

Fig. 4 **a** Schematic demonstration of two step CVD of two-dimensional Mo_2C/graphene heterostructure [36], **b** Schematic view of template method for synthesizing MoN nanosheets [37].**c** HRTEM image representing edge view of MoN nanosheets. **d** Schematic demonstration of template assisted process of ultrathin MoN nanosheets. **e** SEM image of vertically aligned MoN nanosheets grown on a FTO substrate [38]

grown over it [38]. Two-dimensional MoO_3 nanosheets was then heated at 800 °C in the presence of ammonia which transformed MoO_3 into MoN nanosheets resulting in a layered structure having 20–30 μm lateral size and thickness of about 5–40 nm as illustrated in Fig. 4d, e.

Another bottom-up approach is the PELPD, which is recently reported by Zhang et al. for synthesis of ultrathin Mo_2C FCC films on sapphire substrate [39, 40]. To deposit high quality films of Mo_2C, the substrate was heated at 700 °C by using methane plasma as carbon source followed by the generation of Mo vapor by pulsed laser. The resulting films possessed a thickness of 2–25 nm with low crystallinity as compared to CVD-grown nanosheets.

2.2 *Synthesis by Top-Down Approach*

Since the exploration of first MXene, $Ti_3C_2T_z$ family grew so fast that up till now more than 30 different arrangements have been designed by using precursor and etching processes. Following section will discuss different top-down approaches to synthesize MXene.

The MAX phases comprised of more than 130 different compositions having octahedral structure [41]. Three types of MAX phases have been reported such as 211, 312 and 413 having 1, 2 and 3 number of M layers between A layers, respectively. Over 30 different MXenes are obtained from aluminum containing MAX phases by etching aluminum. Some new layered solids such as $(MC)_n(AlSi)_4C_3$ and $(MC)_n(Al_3C_2)$ are recently reported as precursors which are etched to form $Hf_3C_2T_z$ and $Zr_3C_2T_z$, respectively [42, 43]. The structural or chemical order in MXenes is of profound significance and is directly related to the parent's phase order. Recently, two types of MAX phases are explored having phases 312 and 413 which is titled as out-of-plane ordered and another one comprises of 211 phase which is labeled as in-plane ordered MAX phases [44, 45]. For out-of-plane ordered quaternary MAX phases, one layer of M element which is either monolayer or double layer is sandwiched between another layer of M element such as $(Mo_{2/3}Ti_{1/3})_3AlC_2$ and $(Mo_{2/3}Sc_{1/3})_3AlC_2$ in which Mo is the front layer and Ti and Sc are the inner layers. More than 11 molybdenum based MAX phases have been recently exploited.

For the transformation of MAX phases to MXenes, different termination groups such as fluorine, oxygen or hydroxyl are used in place of etched layers. After etching, the obtained material is based on multilayers which are bonded together by hydrogen or van der Waal bonds. Ternary MAX phase element Ti_3AlC_2 can be converted to $Ti_3C_2T_z$ by adopting different etching techniques [14]. Among those techniques, one method is comprised of mixing Ti_3AlC_2 powder in hydrofluoric acid where bonds between M-A and M-X reacts with hydrofluoric acid resulting in the selective etching of aluminum layers. Etching condition depends on the structural order of MAX phases of parent element. Due to the corrosive and hazardous behavior of hydrofluoric acid (HF) a different route has been exploited to avoid the use of HF by replacing it with the mixture of hydrochloric acid (HCl) and fluoride salt [46]. Some other fluoride salts are sodium fluoride (NaF) and potassium fluoride (KF) which can be mixed with HCl and can be utilized instead of HF [47, 48]. Ammonium hydrogen bifluoride was also reported as a good contender to replace HF for the etching of Al from Ti_3AlC_2. Mostly, the element etched from MAX phases was aluminum but recently silicon (Si) was also etched from its MAX phase Ti_3SiC_2 by using HF as an oxidant [49].

After etching process, next step is to exfoliate the etched layers for obtaining few or multilayers of MXenes. But before exfoliation, the etched layers are frequently washed with distilled water to remove the residues of etching solution. Prewashing with sulphuric acid (H_2SO_4) and HCl is also helpful in dissolving the residual salts such as AlF_3 and LiF [50]. After washing the etched layers, different exfoliation methods such as direct sonication or liquid exfoliation can be adopted to obtain MXene layers. Direct sonication results in low yield of MXenes as compared to liquid exfoliation which gives higher yield up to 20 mg per mL solution. Different organic molecules such as dimethyl sulfoxide (DMSO) [51], iso-propylamine [52], tetra butyl ammonium hydroxide (TBAOH) [53], choline hydroxide and n-butylamine are used to carried out exfoliation of $Ti_3C_2T_z$, $(Mo_{2/3}Ti_{1/3})_3C_2T_z$, Ti_3CNT_z and V_2CT_z, respectively [54]. Figure 5a is the schematic representation of the preparation of MXene sheets by using top-down approach. MXenes obtained from the top-down approach exhibits some significant characteristics such as the chemical order in i- and o-MAX phases which does not deteriorate by the etching of Al. Some of the MXene elements that are obtained from o-MAX phase parent elements having Ti in the inner layers are $(Cr_{2/3}Ti_{1/3})_3C_2T_z$, $(Mo_{1/2}Ti_{1/2})_4C_3T_z$ and $(Mo_{2/3}Ti_{1/3})_3C_2T_z$ [12,

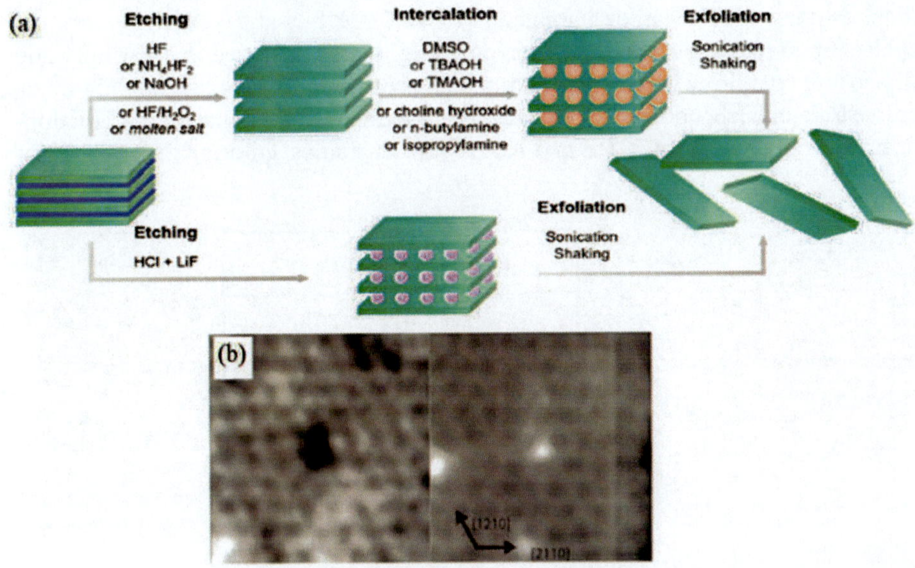

Fig. 5 **a** Schematic illustration of synthesizing MXene by top-down approach. **b** HRTEM image of $Ti_3C_2T_x$ showing atomic defects in single layer MXene [32]

55]. The inner layers play an important role towards the stability of the molecule during etching process. The etching and exfoliation of $Ti_3C_2T_z$ may arise some defects at atomic scale which are observed by transmission electron microscopy (TEM) as shown in Fig. 5b. It was observed that by enhancing the concentration of HF during etching ultimately increases the amount of defects. At microscopic level, some defects are observed in single layer of MXene which also depends on etching environments. One major drawback of conducting exfoliation by sonication is the excessive increase in the number of pores in Mo_2CT_z.

Recently, more than 30 different compositions of MXenes have been synthesized by using top-down approach and DFT calculations predict the existence of many more. New elements of MXene family and new etching processes are being discovered with time.

3 Properties of MXenes

3.1 Structural Properties

The novel 2D material, MXene, exhibits unique structural properties such as hydrophilic nature and active edge sites which enhance its significance in many practical applications. In a recent study, the geometry of Ti_2C and Ti_3C_2 MXene was optimized by adding methoxy functional group [56]. The lattice parameters are found to be 3.3338 and 3.2921 A° which is in good agreement with the theoretical estimation. The lattice parameters of $Ti_2C(OH)_2$ and $Ti_3C_2(OH)_2$ strongly depends upon the type of hydroxyl ion covering. It was observed that partially and fully covered methoxy or hydroxyl group increases the lattice parameters. Figure 6a demonstrate the dependence of Ti_2C and Ti_3C_2 lattice parameters with the functionalization groups such as AA, BB and AB. In further studies, tailoring the structure of

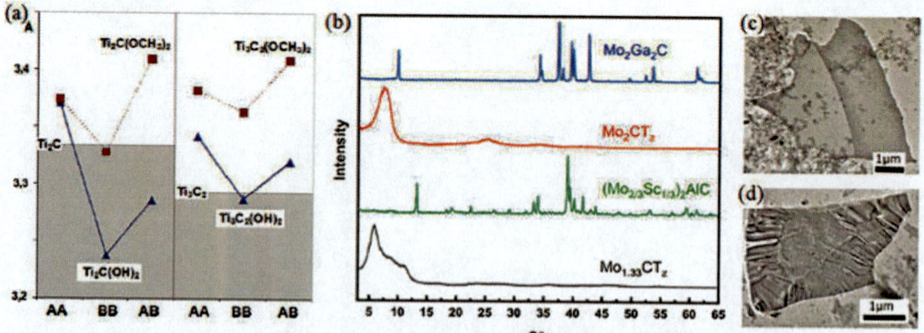

Fig. 6 **a** Optimation of lattice parameters for Ti_2C and Ti_3C_2 by varying surface covering of functionalized groups. **b** X-Ray diffraction pattern of various molybdenum based MXenes. **c, d** TEM images of Mo_2CT_x and $Mo_{1.33}CT_z$, respectively [56, 58]

MXene was investigated by adopting different etching protocols such as the transformation of i-MAX phase of $(Mo_{2/3}Y_{1/3})_2AlC$ into in-plane order $(Mo_{2/3}Y_{1/3})_2C$ and in-plane vacancy order (Mo1.33C) [57]. Electrochemical investigations reveal different results for both in-plane order element and in-plane vacancy order and it was observed that vacancy ordered MXene show relatively good results.

The atomic surface structure of MXene plays an important role in the electrocatalytic activity towards hydrogen evolution reaction (HER). Intikhab et al. synthesized $Mo_{1.33}CT_z$ and Mo_2CT_z to study the electrochemical behavior of these MXenes towards HER [58]. Results indicate that the in-plane ordered vacancy in $Mo_{1.33}CT_z$ causes to decrease the HER comparative to Mo_2CT_z. Figure 6b represents the XRD pattern of the parent's MAX phases and their corresponding MXenes. No peak was found at 61° which confirms that the large number of flakes parallel to the substrate belongs to the 110 family of planes. TEM images (Fig. 6c, d) show that Mo_2CT_z and $Mo_{1.33}CT_z$ includes single and multilayers. The catalytic activity is attributed to the basal planes which are confirmed by the orientation with respect to the substrate. It can be concluded that the structural arrangement of MXene is directly linked with the catalytic performance. Furthermore, Shuck et al. investigate the structural properties of $Ti_3C_2T_z$ MXene by their synthesis from graphite, carbon lampblack and titanium carbide and discuss the electrical conductivity and stability of these MXenes [59]. The obtained products contain the flakes of different sizes, composition, stability and conductivity. MXenes derived from graphite exhibits the highest stability and electrical conductivity as compared to titanium carbide and carbon lampblack derived MXenes.

3.2 Electrical and Magnetic Properties

Khazaei et al. investigated fabricated and investigated the electrical properties of various MXene types such as M_2C and M_2N where M is Sc, Ti, Zr, Cr, V along with some functional group such as F, O and OH [60]. It was evidenced that attachment of suitable functional group e.g. Ti_2C, Hf_2C, Sc_2C and Zr_2C MXenes show semiconducting behavior. Some theoretical studies ensure that Cr_2C and Cr_2N show magnetic behavior. Results reveal that MXenes show metallic as well as superconducting behavior. It is inferred theoretically that the $Sc_2C(OH)_2$ exhibits a direct band gap whereas other semiconductors possess an indirect band gap. The electronic properties of MXenes are affected by the type of functional group attached with it. For instance, F and OH functional groups can alter the electrical properties by receiving one electron from the surface whereas O group need two electrons to get stable. Pure MXenes show metallic behavior as their Fermi energy is in d band of the transition metals whereas some MXenes have p band below that of the d band having a small band gap between both bands. It was also found that transition metals show low electronegativity due to which these transition metals donate electrons to other atoms and itself becomes positively charged. Due to the magnetic nature of chromium compounds, spin polarized PBE method reveals that ground states of chromium carbide and chromium nitride show ferromagnetic behavior whereas other functionalized MXenes show nonmagnetic behavior. The magnetic properties of Cr_2C and Cr_2N can be tailored by applying strain. Figure 7a represents the resistivity versus temperature curves for Ti based films. Ti based films demonstrate the metallic behavior with the linear enhancement in resistivity with respect to temperature. It was observed that below 50 K temperature the decreasing resistivity attribute to the weak localization in this metallic behavior. The results were matched with the previous studies where $Ti_3C_2T_x$ was synthesized from sputter deposited epitaxial films in which the as-synthesized MXene show metallic response at high temperature and weak localization at low temperature. Magnetoresistance is also plotted in Fig. 7b at 10 K temperature. The negative magnetoresistance for Ti based films indicate that as-synthesized MXene is etched in hydrofluoric acid or ammonium hydrofluoride and also confirms the weak localization. Moreover, Miranda et al. investigated the electrical properties of monolayer $Ti_3C_2T_x$ MXene [61]. Results reveal the metallic nature of $Ti_3C_2T_x$ MXene in which carrier density can be altered by varying gate voltages. In the presence of magnetic field, electrical transport measurements show a quadratic increase in the conductance. It can be observed from field effect measurements that at backgate voltage of 18 V there occurs an abrupt change in the conductance slope. At region below 18 V charge carrier mobility is found to be 0.2 cm^2/Vs whereas in region where gate voltage is greater than 18 V the mobility is 0.8 cm^2/Vs. The study indicates that $Ti_3C_2T_x$ MXene exhibits potential applications in transparent conductive films. The metallic nature of $Ti_3C_2T_x$ is very favorable for superconductor applications.

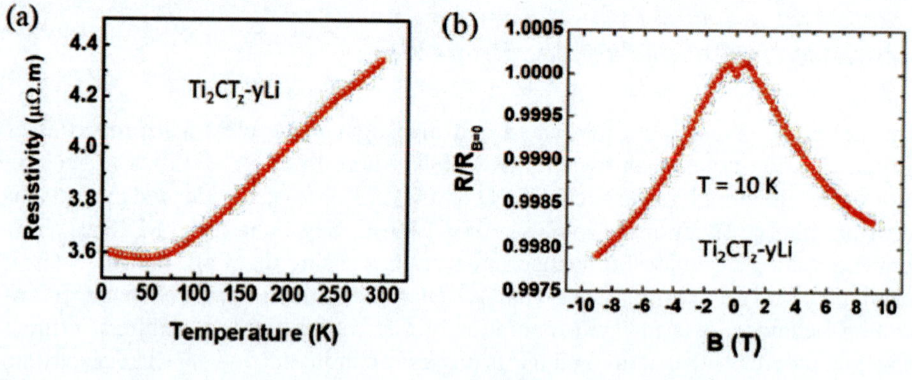

Fig. 7 **a** Temperature dependent resistivity measurement of Ti2CTz-yLi nanosheets. **b** Magnetoresistive curve of Ti2CTz-yLi film [62]

3.3 Optical and Charge Transport Properties

The optoelectronic properties of early discovered MXene ($Ti_3C_2T_z$) was investigated by spray coating its colloidal solution to fabricate conductive thin films [17]. The evaluation of optical properties of $Ti_3C_2T_z$ was studied by using UV-VIS spectrophotometry. Light is absorbed by the films in ultraviolet region and it was evidenced that absorption peak at 270 nm corresponds to glassy substrate whereas band in the range of 700–800 nm represents pale greenish color. In the visible region these films show a flat absorption response due to which these films can be used for display applications. It was also observed that by increasing the film thickness the transmittance lowers down while at about 70 nm the transmittance is retained about 43.8%. About 40–90% transmittance is most suitable for transparent conductive applications Along with good optoelectronic properties $Ti_3C_2T_z$ possesses good mechanical properties as well. The sheets were folded on a flexible substrate and sheet resistance was measured as function of bending radii. During measurements the sheets were held folded for about 5 min and after removing stress the sheets were recovered immediately showing an excellent mechanical stability. As various MXene types exhibit an indirect band gap which limits their utilization in optoelectronic devices. However, the issue can be resolved by transforming an indirect band gap to the direct band gap by applying tensile strength. The transition from indirect to direct band gap can be performed by different structures and functional groups. Further studies referred that semiconducting MXenes show better stability towards biaxial strain and have the capability to maintain their stability for excessive strain [63]. The unique structural, electromagnetic, optoelectronic and mechanical properties of MXenes makes it a promising material for many practical applications.

Owing to high mobility and suitable band gap of two-dimensional materials, Lai et al. investigate the electrical transport properties of Ti_2CT_x MXene field-effect transistors by varying the surface functional groups such as F, O and OH through different synthesis techniques [64]. Results indicate that Ti_2CT_x demonstrate a high mobility of 10^4 cm^2/Vs at room temperature and temperature dependent study reveals the superconducting behavior of Ti_2CT_x. Results confirm better electric charge transport property of MXene by optimizing the treatment methods.

4 Applications of MXenes

MXenes show potential applications in various field of research such as energy conversion and storage systems, electrocatalytic reactions such as hydrogen evolution reaction (HER) and oxygen reduction reaction (ORR) and in the field of biomedical research. Different applications of MXenes based on their structural and optoelectronic properties are discussed as follows.

4.1 Biomedical Applications of MXenes

The attractive physiochemical properties of MXenes makes them a promising candidate in different biomedical applications such as biosensors, bio-imaging, antibacterial activity and drug delivery systems [65–69]. Xue et al. synthesized Ti_3C_2 quantum dots (MQDs) by using a hydrothermal technique and investigate their bio-imaging applications [70]. The prepared MQDs can be attributed to quantum confinement due to hydrothermal process and the size of MQDs can be changed by tailoring the reaction temperature. The uniformity in the nanoscale size of MQDs allow them to penetrate through the cells and have potential applications in bio-imaging. Similar work was done by Zhou and his coworkers, who synthesized $Ti_3C_2T_x$ Mxene QDs in dimethyl-formamide (DMF) and explored their potential in cell-imaging [71]. Besides cell-imaging, MXenes can also be used in computed tomography (CT), magnetic resonance (MR) and photo-acoustic (PA) imaging. Ta_4C_3 MXene having tantalum element makes it more suitable for CT imaging due to high atomic number [66, 72].

In the current age, cancer is the most lethal disease and different therapeutic strategies have been adopted to destroy cancer cells. Among those strategies, chemotherapy, radiotherapy and photothermal therapy is mostly common [73]. In photothermal therapy, light is transformed to the cancer sites and restrict the growth of tumor cells [74]. MXenes have high absorption and facilitate in the photo thermal-conversion efficiency (PCE). Lin et al. utilized Nb2C MXene as a highly efficient photo thermal agent in photo thermal extirpation of mouse tumor. In contrast to PTT, chemotherapy not only deteriorates the cancer cells but also cause damage to normal tissues. Therefore, an effective and controllable drug delivery system should be designed to reduce the harm to normal tissues. MXenes have prolific advantages in providing drug delivery channels due to its large surface area. Recently discovered MXene (Ta_4C_3) are used as X-Ray CT contrast agent [75]. Another potential application of MXenes is found in antibacterial activity such as photodynamic therapy (PDT) for killing cells [68]. A conducting bridge is placed over an insulating lipid bilayer which facilitates the transfer of electrons from intercellular components to the environment and cause the death of cells.

4.2 Energy Storage Applications

Electrochemical energy storage devices are a clean source to store renewable energy. Variety of two-dimensional materials (carbon, fullerene, graphene, carbon nanotubes etc.) are used in energy storage applications due to their admirable characteristics such as high mechanical strength, large surface area and high electrical conductivity [76–78]. Similarly, MXenes also show promising applications in energy conversion and storage devices such as photoelectrochemical cell (PEC), supercapacitors, lithium ion batteries, electro-catalysts and sodium ion batteries [70, 79, 80]. The hydrophilic nature of MXene suggests it a powerful candidate for the photo-anodes in photoelectrochemical water splitting in which incident photons strike the surface of photo-anode to liberate electrons and holes which individually react with OH^- and H^+ to form water and hydrogen. Yu et al. synthesized Ti_3C_2 nanoparticles to synthesize a metal oxide/MXene heterostructure and use the heterostructure for photo-anode in photoelectrochemical water splitting [81]. A significant increase in the photoconductivity and stability is observed for the as-synthesized heterostructure. Furthermore, Chen et al. fabricated inverse opal photonic crystals of titania and Ti_3C_2 quantum dots heterostructure [82]. Titania enables efficient charge separation and thus enhances the charge transportation resulting in an increase in photoconductivity of PEC based sensor. Bismuth vanadate was also utilized as photo-anode in PEC water oxidation. Due to lack of charge transport and slow kinetics in bismuth vanadate, Yan et al. synthesized MXene flakes on the surface of $BiVO_4$ grown on a fluorine doped tin oxide substrate [83]. The as-prepared composite of Ti_3C_2/Mxene reveals a remarkable increase in the photocurrent density up to approximately 3 $mAcm^{-2}$ and photo-conversion efficiency up to 0.78%. The coating of Ti_3C_2 on the surface of $BiVO_4$ facilitates in the efficient charge separation and reduce the recombination of photo-generated electrons and holes due to unique anisotropic structure. The photoelectrochemical properties of as-prepared composite was investigated in three electrode cell configuration with 1 M solution of potassium borate. All results of photoelectrochemical measurements are demonstrated in Fig. 6. J-V curves show a higher photocurrent density of 3.45 $mAcm^{-2}$ for Ti_3C_2/MXene composite and a higher efficiency of 0.78% as shown in Fig. 8a, b. The electrochemical stability of the as-prepared composite was also studied by using chronoamperometry measurements as illustrated in Fig. 8c. Results reveal that $Ti_3C_2/BiVO_4$ composite is more stable as compared to pristine $BiVO_4$. Ti_3C_2 (MXene) causes to enhance the photoconductivity and improves the photocatalytic activity of $BiVO_4$ as well.

Furthermore, MXenes play an important role in energy storage devices such as supercapacitors and batteries. Supercapacitors are divided into three types which includes electric double layer capacitor, pseudocapacitor and hybrid supercapacitors. Supercapacitors have more power density and quick charge/discharge rate while in contrast the batteries have large energy density and high specific capacities [84]. Carbon based materials are found very effective as electrode materials for energy storage systems. Carbonaceous materials such as activated carbons, graphene, fullerene, carbon nanotubes etc. are employed for electrode materials in supercapacitors [85–87]. Similarly, the emerging class of two-dimensional material MXene is also frequently used in supercapacitor applications due to its layered MAX phases and remarkable electrical properties. In supercapacitors charge stores at electrode/electrolyte interface through physical adsorption. The fast charging discharging capability of supercapacitors and long cyclic durability enhance its utilization in vehicles and power stations. In one studies, the etching of aluminum from T_3AlC_3 results in the synthesis of tantalum carbide Ta_4C_3 which was electrochemically tested for supercapacitor electrode in 0.1 M H_2SO_4 electrolyte [88]. From structural analysis it was found that Ta_4C_3 exhibits a hexagonal lattice structure and electrochemical analysis reveals better performance of as-synthesized electrode material in energy

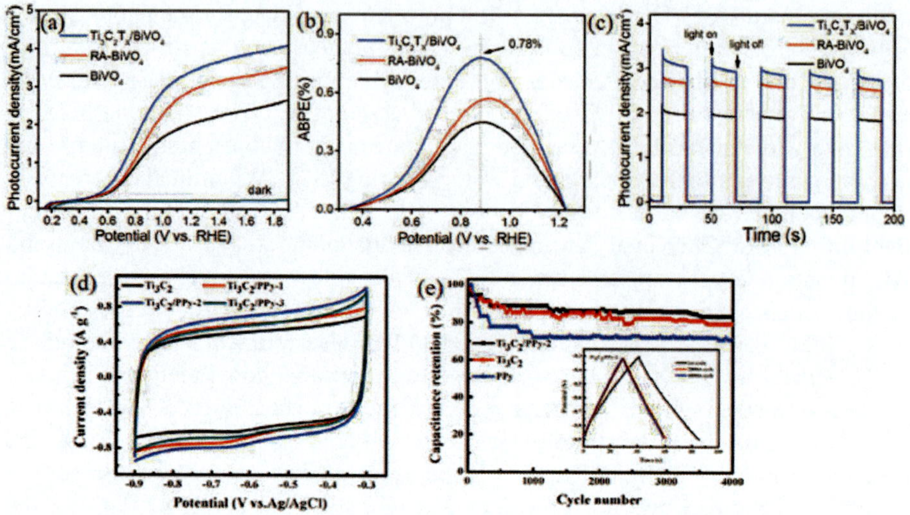

Fig. 8 **a** Linear sweep voltammetry measurements of $BiVO_4$, RA-$BiVO_4$ and $Ti_3C_2T_x/BiVO_4$ photoanodes. **b** Applied bias photon to current conversion efficiency of $BiVO_4$, RA-$BiVO_4$ and $Ti_3C_2T_x/BiVO_4$ photoanodes for PEC water splitting. **c** Chronoamperometry measurements of $BiVO_4$, RA-$BiVO_4$ and $Ti_3C_2T_x/BiVO_4$ photoanodes under1.23 v versus RHE [83]. **d** Cyclic voltammetry measurements of Ti_3C_2 and Ti_3C_2/ppy composites at 10 mV/s scan rate. **e** Cyclic stability test of Ti_3C_2/ppy-2, Ti_3C_2 and ppy electrodes for 4000 cycles at 1 A/g [89]

storage applications. The estimated specific capacitance was found to be 481 F/g with a cyclic retention of about 89% after 2000 cycles of continuous charge/discharge. Furthermore, the selective etching of Al from V_4AlC_3 results in multilayered V_4C_3 MXene which show a gravimetric capacitance of 209 F/g with a cyclic retention of about 97% after 10,000 cycles. The factors which contribute in high capacitance is wide interlayer spacing approximately 0.466 nm, pore volume of about 0.047 cm^3/g, good hydrophilicity and large surface area of about 31.35 m^2/g. Similarly, further studies suggest that dispersed nanoparticles of polypyrrole (ppy) on Ti_3C_2 nanosheets may act as a promising material for supercapacitor electrodes [89]. Electrochemical tests were performed in three electrode assembly by using platinum plate as a counter electrode, Ag/AgCl as reference electrode and Ti_3C_2/ppy nanocomposite as working electrode and Na_2SO_4 as electrolyte. The cyclic voltammetry curves show a specific capacitance of approximately 184 F/g at scan rate of 2 mV/s with a cyclic stability of 83.33% after 4000 cycles as shown in Fig. 6d, e. The addition of polypyrrole expand the interlayer spacing of Ti_3C_2 MXene which facilitates in the ion and charge transportation. The electrochemical behavior of nanocomposite is improved due to the synergistic effect of both MXene and polypyrrole.

Similar to supercapacitors, MXenes are also employed for anode material in rechargeable batteries such as (Li, K, Na) ion batteries due to their dominant characteristics [90]. In case of Li ion batteries, Li act as electron donor (anode) and lithium cobalt oxide as electron acceptor (cathode) separated by an ionic electrolyte. Initially, reduction reaction occurs at cathode liberating lithium ions during charging whereas in discharging process, lithium ions releases from anode and move towards cathode. Intercalation process took place during charging whereas deintercalation process occur during discharging of lithium ion batteries. Due to continuous shuttling of lithium ions from anode to cathode such types of cells are also known as rocking chair cells. MXenes play a crucial role in secondary batteries due to the dominant structural and electrical properties. Variety of MXenes such as V_2C, Ti_2C, Sc_2C and Nb_2C are theoretically estimated to have high specific capacities as compared to M_3X_2 and M_4X_3. Sun et al. has synthesized Ti_3C_2 through exfoliation of Ti_3AlC_2 and use Ti_3C_2 as anode for lithium ion batteries [91]. Electrochemical analysis reveals a high capacity of 123.6 mAh/g having a columbic efficiency of about 47%. The volumetric capacitance of Li ion battery is a major issue which restrict its practical applications such as electric vehicles. To overcome this limitation, Wang et al. fabricated a MXene hybrid of $Fe_3O_4@Ti_3C_2$ by using ultrasonication process [92]. Electrochemical performance of as-prepared MXene hybrid was evaluated by using a coin cell type geometry. The CV curves of MXene hybrid was obtained at a scan rate of 0.2 mV/s. Broad reduction peaks were found for Ti_3C_2 at a potential of 0.67 V and 0.28 V whereas a strong reduction peak at a potential of 0.69 V was found for pristine Fe_3O_4 and $Fe_3O_4@Ti_3C_2$ MXene hybrid. The CV results show a high reversible specific capacity of about 747.4 mAh/g for $Fe_3O_4@Ti_3C_2$-2.5 sample. The cyclic durability of the abovementioned sample was electrochemically studied. Results reveal a good cyclic performance of $Fe_3O_4@Ti_3C_2$ MXene. In this case, the obtained volumetric capacitance was 2038 mAh/g. Therefore, it can be concluded that such type of hybrid MXenes are promising anode materials for energy storage devices.

Along with exceptional characteristics of electrode and anode in energy storage devices, MXenes also exhibit dominant features as a catalyst in hydrogen evolution reaction (HER). Among various two-dimensional materials MoS_2 has been widely explored for different electrocatalytic applications and theoretical studies confirm that edge sites are responsible for all catalytic activities. Therefore, to improve the efficiency of HER it is necessary to enhance the exposure of edge sites. Due to similar characteristics of MXenes as other two-dimensional materials, it is also considered to possess good catalytic performance. In this regard, Seh et al. reported the application of MXenes as an effective electrocatalyst for the very first time [93]. Both theoretical and experimental studies were analyzed to predict MXene as an active catalyst for HER. Initially, Mo_2CT_x and Ti_2CT_x was theoretically studied and it was found that Mo_2CT_x exhibit higher catalytic activity in HER as compared to Ti_2CT_x. Basal planes of Mo_2CT_x are electrocatalytically more active towards HER as compared to MoS_2. Density functional theory (DFT) calculations estimates that Mo_2CT_x have less Gibbs energy which confirms its higher HER performance. Along with theoretical studies, experimental analysis was also performed to investigate the catalytic performance and stability of MXenes. For experimental studies, Mo_2CT_x and Ti_2CT_x was etched from Mo_2Ga_2C and Ti_2AlC by using hydrofluoric acid. The synthesis results in the attachment of functional group (T_x) on basal planes of MXenes. Linear sweep voltammetry measurements were conducted as shown in

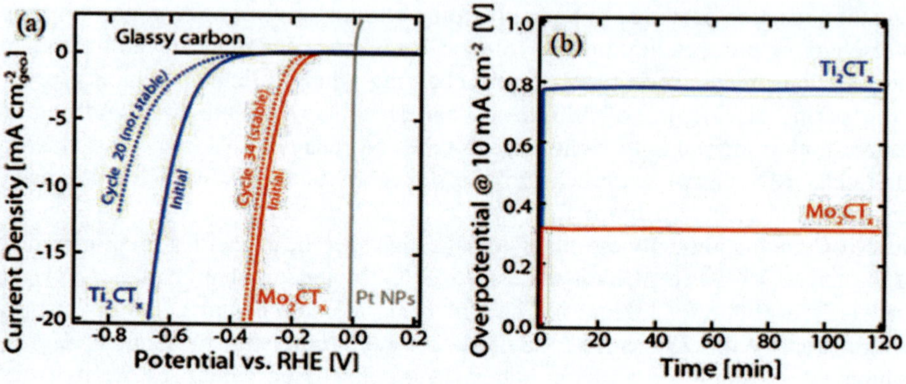

Fig. 9 **a** Linear sweep voltammograms of Mo_2CT_x and Ti_2CT_x compared with bare glassy carbon electrode. **b** Chronopotentiometry measurement of Mo_2CT_x and Ti_2CT_x at a constant current density of 10 mA/cm^2 [93]

Fig. 9a to investigate the catalytic activity of as-synthesized MXenes towards HER. It was observed that catalytic activity of Ti_2CT_x is continuously decreasing showing the unstable nature of Ti_2CT_x whereas Mo_2CT_x demonstrate comparatively a higher HER activity with a slight decrease upon continuous cycling indicating more stability of Mo_2CT_x. The stability of both MXenes were evaluated at a constant current density of 10 mA/cm2 as plotted in Fig. 9b and results reveal no substantial change in the overpotential of Mo_2CT_x and Ti_2CT_x. However, after conducting stability measurement, both MXenes were again tested by LSV which shows that HER activity of Ti_2CT_x was further decreased. In short, both theoretical and experimental studies verify a remarkable HER performance of Mo_2CT_x which opens the doorway for the development of other two-dimensional materials with excessive active basal planes.

Following table illustrate the comparison of different electrode materials and their corresponding properties in supercapacitors and lithium ion batteries (Table 1).

5 Conclusion

In summary, a new class of two-dimensional materials based on early transition metal carbides and nitrides are explored which can be synthesized by etching A element from MAX phases along with the attachment of surface functional groups such as OH, F and O. The term MXene emphasize on their two-dimensional structure and highlights the loss of Al (A element) from its MAX phase. Up to date, numerous MXene elements have been exploited such as Ta_4C_3, Ti_2C, V_2C, Nb_2C, NB_4C_3 etc. Many new elements have been theoretically predicted. Various synthesis process results in the formation of multilayered flakes such as sonication process. Therefore, to obtain a single thin layer of MXene some new synthesis techniques are required. The 2D morphology and high electronic conductivity of MXenes makes

Table 1 Comparative analysis of various two-dimensional materials based electrodes for supercapacitor and lithium ion batteries

Electrode materials	Specific capacitance (F/g)	Current density (A/g)/scan rate	Electrolytes	References
Reduced graphene oxide/carbon black	79	1	H_2SO_4	[94]
rGO/MWCNTs/polypyrrole	82	0.5	KCl	[95]
CNT/Mn_3O_4/Graphene	73	0.5	KCl	[96]
Graphene hydrogel	186	1	H_2SO_4	[97]
Graphene/PANI	210	0.3	H_2SO_4	[98]
$Ti_3C_2T_x$/carbon nano fibers	70 mF/cm^2	50 mV/s	H_2SO_4	[99]
Two dimensional titanium carbide	520 mF/cm^2	2 mV/s	KOH	[100]
Ti_2CT_x	180 mAh/g	20 mA/g	Na+aqueous electrolyte	[101]
Ti_2CT_x	51	1	KOH	[79]
V_4C_3	209	50 mV/s	H_2SO_4	[102]

them promising materials for practical applications such as in supercapacitors and lithium ion batteries. The most frequently used MXene for supercapacitor electrode is Ti_3C_2 due to its electrochemically stable nature. The excessive edge sites marked MXenes as useful candidate for electrocatalyst in HER activity. Many experimental and theoretical studies such as DFT calculations are currently employing to explore new MXene elements with extraordinary properties and potential applications.

References

1. Novoselov, K.S., et al.: Electric field effect in atomically thin carbon films. Science **306**(5696), 666–669 (2004)
2. Watanabe, K., Taniguchi, T., Kanda, H.: Direct-bandgap properties and evidence for ultraviolet lasing of hexagonal boron nitride single crystal. Nat. Mater. **3**(6), 404 (2004)
3. Lalmi, B., et al.: Epitaxial growth of a silicene sheet. Appl. Phys. Lett. **97**(22), 223109 (2010)
4. Kamal, C., Ezawa, M.: Arsenene: two-dimensional buckled and puckered honeycomb arsenic systems. Phys. Rev. B **91**(8), 085423 (2015)
5. Ni, Z., et al.: Tunable bandgap in silicene and germanene. Nano Lett. **12**(1), 113–118 (2011)
6. Aktürk, E., Aktürk, O.Ü., Ciraci, S.: Single and bilayer bismuthene: stability at high temperature and mechanical and electronic properties. Phys. Rev. B **94**(1), 014115 (2016)
7. Wang, Q.H., et al.: Electronics and optoelectronics of two-dimensional transition metal dichalcogenides. Nat. Nanotechnol. **7**(11), 699 (2012)
8. Fiori, G., et al.: Electronics based on two-dimensional materials. Nat. Nanotechnol. **9**(10), 768 (2014)

9. Rao, C., Gopalakrishnan, K., Maitra, U.: Comparative study of potential applications of graphene, MoS2, and other two-dimensional materials in energy devices, sensors, and related areas. ACS Appl. Mater. Interfaces **7**(15), 7809–7832 (2015)
10. Wang, X., et al.: Quantum dots derived from two-dimensional materials and their applications for catalysis and energy. Chem. Soc. Rev. **45**(8), 2239–2262 (2016)
11. Naguib, M., et al.: 25th anniversary article: MXenes: a new family of two-dimensional materials. Adv. Mater. **26**(7), 992–1005 (2014)
12. Anasori, B., et al.: Two-dimensional, ordered, double transition metals carbides (MXenes). ACS Nano **9**(10), 9507–9516 (2015)
13. Ng, V.M.H., et al.: Recent progress in layered transition metal carbides and/or nitrides (MXenes) and their composites: synthesis and applications. J. Mater. Chem. A **5**(7), 3039–3068 (2017)
14. Naguib, M., et al.: Two-dimensional nanocrystals produced by exfoliation of Ti3AlC2. Adv. Mater. **23**(37), 4248–4253 (2011)
15. Zhang, C., et al.: Two-dimensional MXenes for lithium-sulfur batteries. InfoMat **2**(4), 613–638 (2020)
16. Ying, G., et al.: Conductive transparent V2CTx (MXene) films. FlatChem **8**, 25–30 (2018)
17. Hantanasirisakul, K., et al.: Fabrication of Ti3C2Tx MXene transparent thin films with tunable optoelectronic properties. Adv. Electron. Mater. **2**(6), 1600050 (2016)
18. Eklund, P., Rosen, J., Persson, P.O.Å.: Layered ternary M n + 1AX n phases and their 2D derivative MXene: an overview from a thin-film perspective. J. Phys. D Appl. Phys. **50**(11), 113001 (2017)
19. Naguib, M., et al.: Two-dimensional transition metal carbides. ACS Nano **6**(2), 1322–1331 (2012)
20. Naguib, M., et al.: New two-dimensional niobium and vanadium carbides as promising materials for Li-ion batteries. J. Am. Chem. Soc. **135**(43), 15966–15969 (2013)
21. Yan, J., et al.: Flexible MXene/graphene films for ultrafast supercapacitors with outstanding volumetric capacitance. Adv. Func. Mater. **27**(30), 1701264 (2017)
22. Zhu, M., et al.: Highly flexible, freestanding supercapacitor electrode with enhanced performance obtained by hybridizing polypyrrole chains with MXene. Adv. Energy Mater. **6**(21), 1600969 (2016)
23. Dall'Agnese, Y., et al.: Two-dimensional vanadium carbide (MXene) as positive electrode for sodium-ion capacitors. J. Phys. Chem. Lett. **6**(12), 2305–2309 (2015)
24. Naguib, M., et al.: MXene: a promising transition metal carbide anode for lithium-ion batteries. Electrochem. Commun. **16**(1), 61–64 (2012)
25. Gao, G., O'Mullane, A.P., Du, A.: 2D MXenes: a new family of promising catalysts for the hydrogen evolution reaction. ACS Catal. **7**(1), 494–500 (2016)
26. Lee, Y.H., et al.: Synthesis of large-area MoS2 atomic layers with chemical vapor deposition. Adv. Mater. **24**(17), 2320–2325 (2012)
27. Gogotsi, Y.: Chemical vapour deposition: transition metal carbides go 2D. Nat. Mater. **14**(11) (2015)
28. Halim, J., et al.: Synthesis and characterization of 2D molybdenum carbide (MXene). Adv. Func. Mater. **26**(18), 3118–3127 (2016)
29. Xu, C., et al.: Large-area high-quality 2D ultrathin Mo 2 C superconducting crystals. Nat. Mater. **14**(11), 1135 (2015)
30. Liu, Z., et al.: Unique domain structure of two-dimensional α-Mo2C superconducting crystals. Nano Lett. **16**(7), 4243–4250 (2016)
31. Geng, D., et al.: Controlled growth of ultrathin Mo2C superconducting crystals on liquid Cu surface. 2D Mater. **4**, 011012 (2017)
32. Verger, L., et al.: Overview of the synthesis of MXenes and other ultrathin 2D transition metal carbides and nitrides. Curr. Opin. Solid State Mater. Sci. (2019)
33. Qiao, J.-B., et al.: One-step synthesis of van der Waals heterostructures of graphene and two-dimensional superconducting $\alpha - M o 2 C$. Phys. Rev. B **95**(20), 201403 (2017)

34. Wang, Z., et al.: Metal immiscibility route to synthesis of ultrathin carbides, borides, and nitrides. Adv. Mater. **29**(29), 1700364 (2017)
35. Deng, R., et al.: Graphene/Mo2C heterostructure directly grown by chemical vapor deposition. Chin. Phys. B **26**(6), 067901 (2017)
36. Xu, C., et al.: Strongly coupled high-quality graphene/2D superconducting Mo2C vertical heterostructures with aligned orientation. ACS Nano **11**(6), 5906–5914 (2017)
37. Xiao, X., et al.: Salt-templated synthesis of 2D metallic MoN and other nitrides. ACS Nano **11**(2), 2180–2186 (2017)
38. Joshi, S., et al.: Facile synthesis of large area two-dimensional layers of transition-metal nitride and their use as insertion electrodes. ACS Energy Lett. **2**(6), 1257–1262 (2017)
39. Zhang, F., et al.: Plasma-enhanced pulsed-laser deposition of single-crystalline M o 2 C ultrathin superconducting films. Phys. Rev. Mater. **1**(3), 034002 (2017)
40. Zhang, Z., et al.: Substrate orientation-induced epitaxial growth of face centered cubic Mo 2 C superconductive thin film. J. Mater. Chem. C **5**(41), 10822–10827 (2017)
41. Barsoum, M.W.: MAX Phases: Properties of Machinable Ternary Carbides and Nitrides. Wiley (2013)
42. Zhou, J., et al.: A two-dimensional zirconium carbide by selective etching of Al3C3 from nanolaminated Zr3Al3C5. Angew. Chem. Int. Ed. **55**(16), 5008–5013 (2016)
43. Zhou, J., et al.: Synthesis and electrochemical properties of two-dimensional hafnium carbide. ACS Nano **11**(4), 3841–3850 (2017)
44. Liu, Z., et al.: (Cr 2/3 Ti 1/3) 3 AlC 2 and (Cr 5/8 Ti 3/8) 4 AlC 3: new MAX-phase Compounds in Ti–Cr–Al–C System. J. Am. Ceram. Soc. **97**(1), 67–69 (2014)
45. Anasori, B., et al.: Mo2TiAlC2: a new ordered layered ternary carbide. Scripta Mater. **101**, 5–7 (2015)
46. Lakhe, P., et al.: Process safety analysis for Ti3C2T x MXene synthesis and processing. Ind. Eng. Chem. Res. **58**(4), 1570–1579 (2019)
47. Liu, F., et al.: Preparation of Ti3C2 and Ti2C MXenes by fluoride salts etching and methane adsorptive properties. Appl. Surf. Sci. **416**, 781–789 (2017)
48. Liu, F., et al.: Preparation of high-purity V2C MXene and electrochemical properties as Li-ion batteries. J. Electrochem. Soc. **164**(4), A709–A713 (2017)
49. Halim, J., et al.: Transparent conductive two-dimensional titanium carbide epitaxial thin films. Chem. Mater. **26**(7), 2374–2381 (2014)
50. Ghidiu, M., et al.: Ion-exchange and cation solvation reactions in Ti3C2 MXene. Chem. Mater. **28**(10), 3507–3514 (2016)
51. Mashtalir, O., et al.: Intercalation and delamination of layered carbides and carbonitrides. Nat. Commun. **4**, 1716 (2013)
52. Mashtalir, O., et al.: Amine-assisted delamination of Nb2C MXene for Li-ion energy storage devices. Adv. Mater. **27**(23), 3501–3506 (2015)
53. Omomo, Y., et al.: Redoxable nanosheet crystallites of MnO2 derived via delamination of a layered manganese oxide. J. Am. Chem. Soc. **125**(12), 3568–3575 (2003)
54. Naguib, M., et al.: Large-scale delamination of multi-layers transition metal carbides and carbonitrides "MXenes". Dalton Trans. **44**(20), 9353–9358 (2015)
55. Meshkian, R., et al.: Theoretical stability and materials synthesis of a chemically ordered MAX phase, Mo2ScAlC2, and its two-dimensional derivate Mo2ScC2 MXene. Acta Mater. **125**, 476–480 (2017)
56. Enyashin, A.N., Ivanovskii, A.L.: Structural and electronic properties and stability of MX enes Ti2C and Ti3C2 functionalized by methoxy groups. J. Phys. Chem. C **117**(26), 13637–13643 (2013)
57. Persson, I., et al.: Tailoring structure, composition, and energy storage properties of MXenes from selective etching of in-plane, chemically ordered MAX phases. Small **14**(17), 1703676 (2018)
58. Intikhab, S., et al.: Stoichiometry and surface structure dependence of hydrogen evolution reaction activity and stability of MoxC MXenes. J. Catal. **371**, 325–332 (2019)

59. Shuck, C.E., et al.: Effect of Ti3AlC2 MAX phase on structure and properties of resultant Ti3C2Tx MXene. ACS Appl. Nano Mater. (2019)
60. Khazaei, M., et al.: Novel electronic and magnetic properties of two-dimensional transition metal carbides and nitrides. Adv. Func. Mater. **23**(17), 2185–2192 (2013)
61. Miranda, A., et al.: Electronic properties of freestanding Ti3C2Tx MXene monolayers. Appl. Phys. Lett. **108**(3), 033102 (2016)
62. Halim, J., et al.: Electronic and optical characterization of 2D Ti2C and Nb2C (MXene) thin films. J. Phys. Condens. Matter **31**(16), 165301 (2019)
63. Cui, J., et al.: Strain-tunable electronic structures and optical properties of semiconducting MXenes. Nanotechnology **30**(34), 345205 (2019)
64. Lai, S., et al.: Surface group modification and carrier transport properties of layered transition metal carbides (Ti2CTx, T:–OH,–F and–O). Nanoscale **7**(46), 19390–19396 (2015)
65. Liu, H., et al.: A novel nitrite biosensor based on the direct electrochemistry of hemoglobin immobilized on MXene-Ti3C2. Sens. Actuators B Chem. **218**, 60–66 (2015)
66. Dai, C., et al.: Two-dimensional tantalum carbide (MXenes) composite nanosheets for multiple imaging-guided photothermal tumor ablation. ACS Nano **11**(12), 12696–12712 (2017)
67. Chen, X., et al.: Ratiometric photoluminescence sensing based on Ti 3 C 2 MXene quantum dots as an intracellular pH sensor. Nanoscale **10**(3), 1111–1118 (2018)
68. Rasool, K., et al.: Antibacterial Activity of Ti3C2T x MXene. ACS Nano **10**(3), 3674–3684 (2016)
69. Huang, K., et al.: Two-dimensional transition metal carbides and nitrides (MXenes) for biomedical applications. Chem. Soc. Rev. **47**(14), 5109–5124 (2018)
70. Xue, Q., et al.: Photoluminescent Ti3C2 MXene quantum dots for multicolor cellular imaging. Adv. Mater. **29**(15), 1604847 (2017)
71. Zhou, L., et al.: Titanium carbide (Ti3C2Tx) MXene: a novel precursor to amphiphilic carbide-derived graphene quantum dots for fluorescent ink, light-emitting composite and bioimaging. Carbon **118**, 50–57 (2017)
72. Cai, Y., et al.: Diketopyrrolopyrrole–triphenylamine organic nanoparticles as multifunctional reagents for photoacoustic imaging-guided photodynamic/photothermal synergistic tumor therapy. ACS Nano **11**(1), 1054–1063 (2017)
73. Naunheim, K.S., et al.: Preoperative chemotherapy and radiotherapy for esophageal carcinoma. J. Thorac. Cardiovasc. Surg. **103**(5), 887–895 (1992)
74. Robinson, J.T., et al.: Ultrasmall reduced graphene oxide with high near-infrared absorbance for photothermal therapy. J. Am. Chem. Soc. **133**(17), 6825–6831 (2011)
75. Bashkatov, A., et al.: Optical properties of human skin, subcutaneous and mucous tissues in the wavelength range from 400 to 2000 nm. J. Phys. D Appl. Phys. **38**(15), 2543 (2005)
76. Li, X., et al.: Carbon and graphene quantum dots for optoelectronic and energy devices: a review. Adv. Func. Mater. **25**(31), 4929–4947 (2015)
77. Ross, R.B., et al.: Endohedral fullerenes for organic photovoltaic devices. Nat. Mater. **8**(3), 208 (2009)
78. Arico, A.S., et al.: Nanostructured materials for advanced energy conversion and storage devices. In: Materials For Sustainable Energy: A Collection of Peer-Reviewed Research and Review Articles from Nature Publishing Group, pp. 148–159. World Scientific (2011)
79. Rakhi, R.B., et al.: Effect of postetch annealing gas composition on the structural and electrochemical properties of Ti2CT x MXene electrodes for supercapacitor applications. Chem. Mater. **27**(15), 5314–5323 (2015)
80. Xie, Y., et al.: Prediction and characterization of MXene nanosheet anodes for non-lithium-ion batteries. ACS Nano **8**(9), 9606–9615 (2014)
81. Yu, X., et al.: Ti3C2 MXene nanoparticles modified metal oxide composites for enhanced photoelectrochemical water splitting. Int. J. Hydrog. Energy **44**(5), 2704–2710 (2019)
82. Chen, X., et al.: Ti3C2 MXene quantum dots/TiO2 inverse opal heterojunction electrode platform for superior photoelectrochemical biosensing. Sens. Actuators B Chem. (2019)

83. Yan, D., et al.: A BiVO4 film photoanode with re-annealing treatment and 2D thin Ti3C2TX flakes decoration for enhanced photoelectrochemical water oxidation. Chem. Eng. J. **361**, 853–861 (2019)
84. Peng, X., et al.: Two dimensional nanomaterials for flexible supercapacitors. Chem. Soc. Rev. **43**(10), 3303–3323 (2014)
85. Portet, C., et al.: High power density electrodes for carbon supercapacitor applications. Electrochim. Acta **50**(20), 4174–4181 (2005)
86. Zhang, L.L., Zhou, R., Zhao, X.: Graphene-based materials as supercapacitor electrodes. J. Mater. Chem. **20**(29), 5983–5992 (2010)
87. Kaempgen, M., et al.: Printable thin film supercapacitors using single-walled carbon nanotubes. Nano Lett. **9**(5), 1872–1876 (2009)
88. Syamsai, R., Grace, A.N.: Ta4C3 MXene as supercapacitor electrodes. J. Alloy. Compd. **792**, 1230–1238 (2019)
89. Wu, W., et al.: Enhanced electrochemical performances of organ-like Ti3C2 MXenes/polypyrrole composites as supercapacitors electrode materials. Ceram. Int. **45**(6), 7328–7337 (2019)
90. Tang, Q., Zhou, Z., Shen, P.: Are MXenes promising anode materials for Li ion batteries? Computational studies on electronic properties and Li storage capability of Ti3C2 and Ti3C2X2 (X = F, OH) monolayer. J. Am. Chem. Soc. **134**(40), 16909–16916 (2012)
91. Sun, D., et al.: Two-dimensional Ti3C2 as anode material for Li-ion batteries. Electrochem. Commun. **47**, 80–83 (2014)
92. Wang, Y., et al.: Fe 3 O 4@ Ti 3 C 2 MXene hybrids with ultrahigh volumetric capacity as an anode material for lithium-ion batteries. J. Mater. Chem. A **6**(24), 11189–11197 (2018)
93. Seh, Z.W., et al.: Two-dimensional molybdenum carbide (MXene) as an efficient electrocatalyst for hydrogen evolution. ACS Energy Lett. **1**(3), 589–594 (2016)
94. Wang, Y., et al.: Graphene/carbon black hybrid film for flexible and high rate performance supercapacitor. J. Power Sources **271**, 269–277 (2014)
95. Yang, C., et al.: All-solid-state asymmetric supercapacitor based on reduced graphene oxide/carbon nanotube and carbon fiber paper/polypyrrole electrodes. J. Mater. Chem. A **2**(5), 1458–1464 (2014)
96. Gao, H., et al.: Flexible all-solid-state asymmetric supercapacitors based on free-standing carbon nanotube/graphene and Mn3O4 nanoparticle/graphene paper electrodes. ACS Appl. Mater. Interfaces. **4**(12), 7020–7026 (2012)
97. Xu, Y., et al.: Flexible solid-state supercapacitors based on three-dimensional graphene hydrogel films. ACS Nano **7**(5), 4042–4049 (2013)
98. Wu, Q., et al.: Supercapacitors based on flexible graphene/polyaniline nanofiber composite films. ACS Nano **4**(4), 1963–1970 (2010)
99. Levitt, A.S., et al.: Electrospun MXene/carbon nanofibers as supercapacitor electrodes. J. Mater. Chem. A **7**(1), 269–277 (2019)
100. Zhan, C., et al.: Computational discovery and design of MXenes for energy applications: status, successes, and opportunities. ACS Appl. Mater. Interfaces (2019)
101. Velpula, G., et al.: Graphene meets ionic liquids: fermi level engineering via electrostatic forces. ACS Nano (2019)
102. Wang, X., et al.: Two-dimensional V4C3 MXene as high performance electrode materials for supercapacitors. Electrochim. Acta **307**, 414–421 (2019)

Chalcogenides Nanocrystals and Its Applications

Arunachalam Arulraj, U. Mehana Usmaniya, Govindan Senguttuvan, Vadivel Sivakumar, and Mohammad Khalid

Abstract Chalcogenides nanocrystals, one of the emerging hot topics in the field of research offers the characteristics that are unparalleled by conventional materials in their respective field of applications. In recent years, chalcogenides nanocrystals are holding significant impact in various applications such as lighting, displays, photovoltaics, sensors, imaging devices, etc. owing to their shape, size and tunability of composition. Moreover, most of the chalcogenides nanocrystals are deployed in large-scale devices due to its sustainability, inexpensive and environmentally friendly. This chapter emphasizes the concise overview of chalcogenides nanocrystals, sketching the state of the art of synthesizing materials along with strategies to circumvent the limitations arise during ternary and quaternary compositions. A discussion highlighting on shape, size, composition controlled growth of chalcogenides are provided. Followed by the synthesis, functional properties (structural, optical, electronic and magnetic) and its ubiquitous applications are addressed in a detailed manner. Finally, as a summary, track of the rapidly advancing field and their key challenges have been discussed.

Keywords Chalcogenides · Nanocrystals · Synthesis route · Functional properties · Engineering applications

A. Arulraj (✉) · M. Khalid
Graphene & Advanced 2D Materials Research Group (GAMRG), School of Science and Technology, Sunway University, No. 5, Jalan Universiti, Bandar Sunway, 47500 Petaling Jaya, Selangor, Malaysia
e-mail: arulnanotech@gmail.com

U. Mehana Usmaniya
Electropyro Metallurgy, CSIR-Central Electrochemical Research Institute, Karaikudi, India

G. Senguttuvan · V. Sivakumar
Department of Physics, University College of Engineering-BIT Campus, Tiruchirappalli, India

N. M. Mubarak et al. (eds.), *Contemporary Nanomaterials in Material Engineering Applications*, Engineering Materials,
https://doi.org/10.1007/978-3-030-62761-4_8

1 Introduction

Nanotechnology was one of the tremendous technologies of the twenty-first century which rapidly expands its branches in the wide area of science. It can be simply described with its dimension ranging in the order of few (1–100 nm). Production of materials in the nanometer-scale drastically changes the properties compared with the bulk one. Owing to its smaller size and properties, it placed its benchmark in diverse fields of industrial applications. More innovative concepts, products, and tools were developed and reported in the short spam. In this regard, nanocrystals (NCs), one of the key elements in the branch of nanotechnology gains rising interest starting from the synthesis of the crystals to end-user products across diverse disciplines. The fundamental concepts involving the processing and development of NCs are advancing at an astounding rate. NCs can exert control over their size and shape as well as able to tune their physical and/or chemical properties depending on their end-user applications for target applications [1, 2]. The most common features exciting in the NCs are its size, narrow bandgap, large surface to volume ratio, and quantum confinement which intensifies the development of NCs in recent years. Based on its properties, numerous novel NCs materials, including semiconductor, plasmonics, metals, carbon-based materials, nitrides, chalcogenides, etc., were developed in the past few decades and even some of the NCs were commercialized towards targeted applications [3, 4].

Among the different kinds of NCs materials, chalcogenides NCs was one of the promising candidates with huge interest owing to its exciting properties like high carrier mobility, thermal conductivity, morphology, etc. The properties of the chalcogenides NCs can also be engineered by varying its composition or by combining with different materials like semiconductors, metals, etc., [2]. The exciting prospects and challenges in the growth of chalcogenides NCs motivated research community to extensively investigate its growth methodology and properties of the different chalcogenides materials like copper sulfide (CuS), molybdenum selenides (MoSe), bismuth telluride (BiTe), copper indium gallium sulfide (CIGS) and so on [5–7]. The remarkable properties possessed by the chalcogenides NCs find its applications in different fields such as sensors, photovoltaics, bio-imaging, lasers, optical switches, amplifiers, etc. In this chapter, a brief description of chalcogenides NCs starting from its synthesis approaches followed by functional properties (crystal structure, electronic structure, optical, electrical and magnetic), composites, device engineering with final wrap up on its future scope.

2 Synthesis Methods

Chalcogenides NCs can be produced by approaching various synthesis routes. Almost the successful protocol of synthesizing chalcogenides NCs have been derived by understanding the fundamentals of science permits in optimizing the products with its desired properties. The quality of NCs depends on the precursors, choice of solvents, temperatures, etc. In the way of providing a stereotyped recipe for synthesizing chalcogenides NCs, the five most prevalent methods are chosen from the literature and described in this section. The prevailing method of synthesis includes (i) colloidal synthesis, (ii) solvothermal, (iii) template-assisted, (iv) Kirkendall effect-induced, and (v) cation exchange method. Brief elucidations of each approach are as follows.

2.1 Colloidal Synthesis

The exquisite control of chalcogenides NCs growth is undoubtedly produced by approaching one the most common method called "colloidal-synthesis". It holds precedence owing to its monodispersity of NCs along with its high crystallinity and high dispersion of precursors in organic solvents. In this approach, it typically involves thermal decomposition of precursors like organometallic and/or inorganic salts at its high boiling point in the oxygen-free conditions. Upon heating the chalcogenides NCs precursors it chemically transforms into active molecular or atomic species (monomers) and it turns results in the formation of colloidal chalcogenides NCs. The synthesis process includes of several consecutive stages starting from nucleation-growth-isolation to the end of products (chalcogenides NCs). A typical colloidal setup consists of a three-necked round bottom flask connected to a condenser and a Schlenk line to provide an oxygen-free environment for the synthesis. To avoid the presence of water and oxygen molecules in the reaction flask which induces pyrophoric precursors in reactions, the flask is subjected to evacuate at 100 °C for an hour before starting the reaction.

Colloidal synthesis encompasses both the heat-up and hot-injection protocols. In the "heat-up" process, it involves mixing metal and chalcogen precursors in organic solvents at low temperatures followed by heating up of reaction at the desired temperature to initiate nucleation and growth of chalcogenides NCs. Large scale production of NCs can be achieved by this approach. After attaining the nucleation stage of chalcogenides NCs, the supersaturation temperature is maintained for a prolonged time, which may lead to the broadened size distribution of NCs. Bawendi and co-workers pioneered the hot-injection technique to synthesize the monodisperse cadmium (Cd) chalcogenides NCs in the year of 1993 [8]. Using this approach, controlled growth of NCs with its desired size and shape can be achieved through inducing a high degree of supersaturation upon injection [9, 10]. Even though both heat-up and hot-injection processes involve the growth of crystals upon the burst of nucleation, better uniformity can be done by the process of injection resulting in a single nucleation event [8, 11–13]. Moreover, a hot-injection process not only limited in the production of chalcogenides NCs it can also be extended for different materials like metal NCs [14], metal oxides NCs [15] and so on by choosing appropriate combinations of precursors, solvents and other agents. Since the protocol for synthesizing variety of NCs are same, mostly it results in uniform and rapid crytsal formtion.

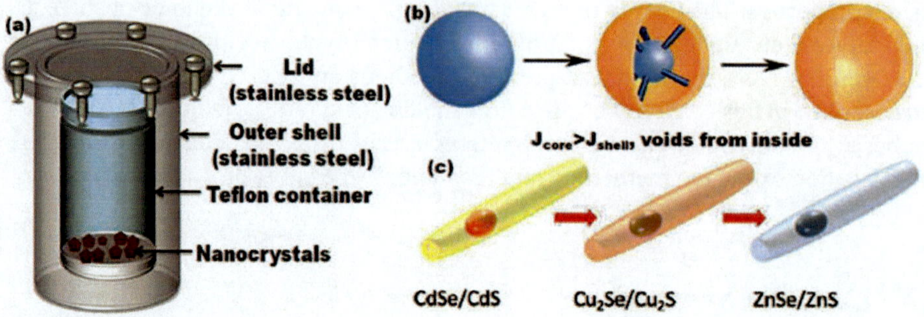

Fig. 1 **a** Schematic illustration of the solvothermal reactor; **b** core/shell NCs synthesized by Kirkendall effect. Reproduced with permission of American Chemical Society [23]; **c** schematic representation of nanohetrostructures of CdSe/CdS, Cu_2Se/Cu_2S, ZnSe/ZnS (**b** and **c**). Reproduced with permission of American Chemical Society [46]

2.2 Solvothermal Synthesis

Solvothermal synthesis alike the hydrothermal synthesis method to produce the monodisperse NCs at low temperatures with the main difference of source of solvents used [8]. In hydrothermal water will be used as a source whereas in solvothermal route, solvents (organic and/or inorganic agents) will be employed as a source. Due to its ease of access and reactivity of solvents under elevated pressure and temperature, it draws considerable attention in the research community [16, 17]. To withstand the temperature and pressure over a prolonged period during the reaction, specialized autoclave made up of stainless steel covering the Teflon liner inside the vessel has been used for the synthesis of NCs (Fig. 1a) [17, 18]. During the reactions, the reactant and solvents undergo high pressure with certain temperature, ultimately results in the subsequent formation of crystals from the dissolved materials in the solvents. Since it doesn't involve complicated process, researchers find it to be a flexible method for producing high-quality NCs.

2.3 Template Assisted Synthesis

For the production of one-dimensional nanostructures (nanowires, nanotubes, etc.) in a well-distributed manner, template-assisted synthesis was one of the best proven method. Using this approach, different nanostructure specifically of hollow shapes can be obtained by designing the templates for the desired structural frameworks [19–21]. The uniqueness of this approach was its large versatility concerning pore's length and diameter [22]. The designing of templates can be done either by (1) hard templates (physical scaffolds for deposition of the desired material) and/or (2) soft templates (contain ligands, surfactants, and polymers) [16]. Besides its versatility growth of nanostructures, it also holds some drawbacks such as the issue in template removal (selective etching acids required to remove hard templates) as well as the high production cost of template materials, which impede the up-scaling of products (NCs) for practical applications.

2.4 *Kirkendall Effect*

Kirkendall effect is a diffusion-related phenomenon that occurs between the reactive phases sharing an interface with the merit of avoiding templates removal [23, 24]. The first report by Yin et al. for the synthesis of plethora materials using this approach turns the research community to focus on this effect for the production of a variety of NCs which includes metal oxides and chalcogenides [25–27]. For the production of hollow nanostructures using the Kirkendall effect, it involves either as one-pot and/or two-step reaction process as illustrated in Fig. 1b. Initially, it involves the reaction of solid particles at the outer shell (consider as A) followed by the solution or gas phase (consider as B) reacts with A in turns results in the formation of nanostructured composites (AB) [23]. Moreover, at the reaction, the core element A diffused through AB composites and creates a cavity inside, thereby removes the interior core [23, 25, 26, 28]. Owing to its ease of accessibility, functional nanomaterials, including chalcogenides NCs with controlled hollow interiors and shell thickness have been obtained and used for potential applications such as catalysis, sensing, drug delivery, and optical devices [29–31]. Till dated of research in the field of chalcogenides NCs growth the Kirkendall effect was the most established approach with its compositions. There exists numerous report on the production of chalcogenides NCs with different nanostructures like nanocages [32], nanoboxes [33, 34], polyhedrons [35] and nanoplates [36] using the Kirkendall effect.

2.5 *Cation Exchange Method*

Cation exchange method was one of the interesting synthesize approaches to grow chalcogenides NCs with novel structure, phases, and composition which are intricate by the conventional synthesis route. Even though it was not considered to be a direct synthesis approach, its growth mechanism relies on the cationic solution/solvents used in the reaction which permits the exchange of cations within the NCs lattice to form unique shaped NCs and its compositions [37–40]. As the name stands in this approach, the cation exchange takes place by exhibiting a higher diffusion rate than anions. Alivisatos and their research group reported the transformation of cadmium selenides (CdSe) chalcogenides NCs into silver selenide (Ag_2Se) chalcogenides NCs and their analogues by replacement of metal cations within its crystal lattice (i.e. Cd^{2+} with Ag^{+}) using cation exchange method [41]. The cationic exchange takes place at an ambient condition with a short time owing to the lower activation barrier of NCs which stipulates the rapid exchange of cationic within seconds. Using this approach spontaneous production of NCs can be achieved, for example, the transformation of cadmium sulfur (CdS) to copper sulfur (Cu_2S) has been carried out in a second at RT [42]. Also, using this approach reverse exchange reaction (i.e. Cu_2S to CdS) was made possible by inducing an excessive amount of Cd^{2+} ions with its preferential solvents [43]. The affinity of metal ions to the solvent plays a leading role in the cation exchange method which can be predicted using Pearson's HSAB theory [44]. For a better understanding of HSAB theory, an example of Cd^{2+} ions was taken since Cd^{2+} was treated as hard acid and so for the type of solvent chosen for reactions should also be hard base like methanol which helps in preferential formation of resultant products. Wide variety of nanostructured NCs can be produced by adopting this method, for example, the copper indium sulfide (CIS) nanoplates have been obtained from the conversion of copper sulfide ($Cu_{2-x}S$) NCs using cations exchange method [36], octopod and sphalerite shaped cadmium selenides (CdSe) NCs are obtained by using copper selenide ($Cu_{2-x}Se$) NCs as parent material in the conversion process [45]. Li and co-workers developed Cu_2Se/Cu_2S and ZnSe/ZnS rods like nanoheterostructures by adopting the procedure of the cation exchange process (Fig. 1c) [46]. All the synthesized materials (CdSe/Cds, Cu_2Se/Cu_2S and ZnSe/ZnS) are observed to depict similar morphologies of nanorods like structure (Fig. 2a, b, c) but the PL spectra of each material differ due to its compositions (Fig. 2d, e, f) [46]. Later the same research group reported the possibility of selective exchange of cations in the Cu_2Se/Cu_2S core/shell with nanorods like morphology [47]. Stam et al. demonstrated the production of luminescent NCs with rod-like morphology for the CISe/CIS heterojunction materials by implementing a sequential topotactic cation exchange reaction [48]. As well as complex NCs materials like CZTSSe NCs, Cu_2SeS NCs, etc. can be produced using this cation exchange method by controlling the volume and concentration of the precursors [49–51]. One of the attractive features of this approach was by adjusting the concentration and relative volume of cations substitution researchers can play in partial and complete conversion of NCs depending on the individual needs for their applications [42, 52, 53].

3 Functional Properties

In recent decades, the transition from micro to nanoscale dimension provides a tremendous path to meet out the modern needs and of NCs with a diameter smaller than double Bohr exciton radius reveals pronounced effect depending on its size and electronic structure. The functional properties of chalcogenides NCs were determined from its size, bandgap, energy band structure, composition, defects and so on. Nanocrystals, particularly of chalcogenides based, possess a broad spectrum of functional properties due to its quantum sized confinement, plasmonics, structural and compositional versatility which have been exploited in diverse applications. Researchers have developed a variety of chalcogenides NCs starting from binary

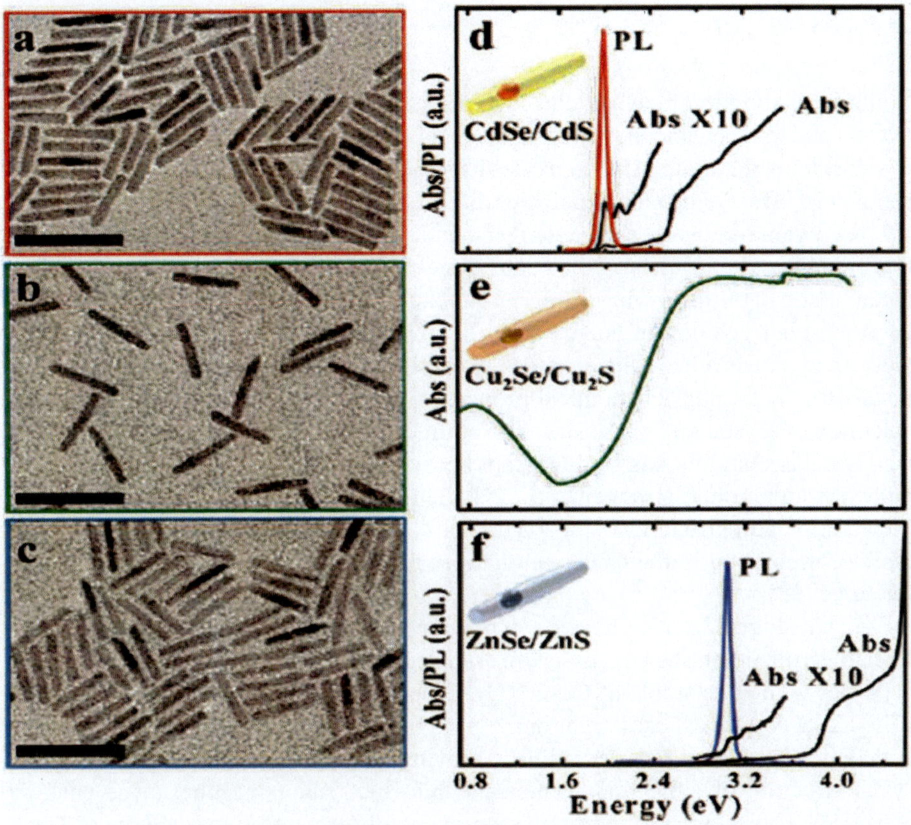

Fig. 2 **a, b, c**) Observed TEM micrographs of CdSe/CS, Cu_2Se/Cu_2S, ZnSe/ZnS and **d, e, f** optical spectra of CdSe/CS, Cu_2Se/Cu_2S, ZnSe/ZnS. Reproduced with the permission of the American Chemical Society [46]

(ZnSe NCs) to quinary (CZTSSe NCs) compounds, providing extraordinary flexibility for materials designing and studied their properties and thereby tuned its properties depending on its field of applications. For example, the synthesis of ternary CIS NCs with direct bandgap finds its applications as an absorber in photovoltaic devices for the replacement of toxic cadmium telluride (CdTe) compound. Even though varieties of chalcogenides NCs are synthesized and reported, still various compounds remain unexplored in this field [54]. In this section, an overview on the important properties (crystal structure, bandgap, plasmonics, absorption coefficient, conductivity, etc.) of chalcogenides NCs is described.

3.1 Crystal Structure

The prime distinction between chalcogenides and other materials like oxides is its covalent and ionic character, in transition metal chalcogenides (TMCs) it shows more covalent character while in metal oxides its ionic resembling will be of fluorides. The pronounced effect of covalent character in TMCs was due to its low electronegativity difference value between the chalcogens and metals which tends to lead a strong bonding of s and p orbitals of the chalcogen with the transition metal. The bonding orbitals result in the narrowing of energy gaps since the valence level of chalcogenides lies at a higher position of energy than the oxides [55, 56]. Apart from the energy levels, there exists a huge difference between the chalcogenides and oxides in their crystal structures, physicochemical properties, and their functional properties. The evaluation of crystal and phase structure of the chalcogenides NCs are more complicated and longstanding unsolvable issues because of its polymorphs, stoichiometric, deficiency and atomic arrangements [57]. Generally, the chalcogenides highly rely on the temperature treatment and at a higher degree, there exists more possibility of disorder orientation in the chalcogen lattices resulting in the complexity of crystal structures [58].

Each chalcogenides NCs have its own phases and crystal structures that are distinguishable from others, here let us have a brief description of the crystal phase structure properties with an example of Cu_2S NCs. Based on the composition of copper (Cu) in the Cu_2S NCs it exposes eight dominant phase structures (low-Chalcocite; high Chalcocite; djurleite; digenite; anilite; roxbyite; and covellite) with a characterized crystal structure of either face-centered cubic (FCC) or hexagonal close packing (hcp) [59]. The schematic representation of the different phase structures of Cu_2S was shown in Fig. 3 [60–62].

Since there exist different phases of Cu_2S an important question arises in the research community regarding the phase structure matching with the standard X-ray diffraction (XRD) database files (JCPDS). Since most of its phases exhibit the

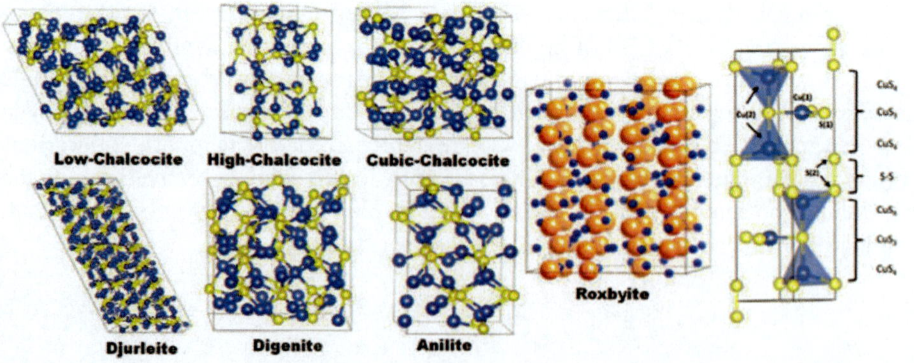

Fig. 3 Schematic illustration of the different phase structure of Cu_2S NCs. Reproduced with the permission of Applied Physics Letters and American Chemical Society [60–62]

same kinds of patterns like chalcocite (Cu_2S) and djurleite ($Cu_{1.94}S$) in the XRD with distinguishable crystal and electronics structures. XRD patterns of Cu_2S NCs reported by Alivisatos and co-workers described that Cu_2S NCs exhibits chalcocite phase under inert atmosphere and upon exposure to air it transforms immediately into djurleite phase [63]. The same research group also observed the dynamic structural transformation of Cu_2S NCs using high-resolution transmission electron microscopy (HRTEM). The irradiation from the electron beam source used in HRTEM helps in determining the transformation of phase structure from low to high chalcocite structure with respect to a temperature ranging between ~100–440 °C, beyond the temperature of ~440 °C the transformation of high into cubic chalcocite has been observed [64]. The phase transformation temperature also depends on the particle size which should be greater than 6 nm [65]. Similarly, Zhao et al. reported the insights in assigning the correct phase from the XRD patterns [66]. At elevated temperature, the Cu atoms in the Cu_2S NCs are unusually mobile, thereby making it partially an ionic conductor [67, 68]. Moreover, transformation involves in the arrangement of S atoms in its lattices leads to several metastable phases of $Cu_{2\text{-}x}S$ [69] resultant in the vacancies of Cu. These vacancies may due to loss of Cu to either oxygen or carbon dioxide in its grain boundary or free surface [43, 70]. The experimental and theoretical investigations of Cu_2S conclude that it was intrinsically unstable at ambient conditions and degrades rapidly [57, 71, 72]. Hence from the application point, it was more critical in fine-tuning the phase structures of the chalcogenides NCs due to its high sensitivity towards its stoichiometric and defects.

3.2 *Optical Properties*

The most attractive features of the NCs are its enhanced optical properties due to its particle size and quantum confinement effect. Tuning the optical properties of chalcogenides NCs was one of the hot topics in the field of research for the past few decades. Some varieties of NCs like plasmonics NCs, quantum dots (QDs) NCs delivers higher light absorption coefficient by covering a wide range of electromagnetic spectrum thereby potentially finds its applications in the field of laser therapy, bio-imaging [73–76], photonics [77], photovoltaics [78–80], sensors [81–84], etc. In plasmonics, NCs the optical property (plasmons) originated from the collections of free charge carriers and tends to be localized by satisfying the conditions of surface plasmon resonances thereby it also referred to be localized surface plasmon resonance (LSPR) [77]. The LSPR properties are not only achieved in plasmonics NCs (Ag, Au based NCs) it can also be achieved in semiconducting NCs too with optimizing free carrier concentrations [43]. The discovery of LSPR property in chalcogenides based NCs created a new path in the manipulation of light (nearly to NIR region) through tuning the intrinsic doping concentration and tailoring its stoichiometric ratio [85]. The research community made their efforts in manipulating the LSPR frequency in chalcogenides NCs by altering its various parameters such as composition, shape, and size. In the year of 2009, a report on the synthesis of $Cu_{2\text{-}x}S$ NCs with absorption in the regime of NIR was proposed by Zhao and co-workers. In this report, the $Cu_{2-x}S$ NCs were synthesized by three different approaches viz.: hydrothermal, Solventless thermolysis and sonoelectrochemical methods. In these different methods, parameters like pH, reduction potential, the concentration of precursors are been varied and are results in the production of different phase structures with composition ranging from djurleite ($Cu_{1.97}S$) to chalcocite (CuS). The obtained product with different compositions correspondingly exhibits its plasmonics absorption (Fig. 4a) and the author also stated that the djurleite phase was existed to be more stable than chalcocite at ambient conditions [66]. Similarly, the morphology (shape) of the materials will also have prolonged effect in its properties and the same theme was applicable for the chalcogenides based NCs too. The shape of the chalcogenides NCs was consistently found to have an effect in its optical properties. Hsu et al. demonstrated the shape dependent LSPR properties for $Cu_{2-x}S$ nanodisks shaped NCs [86, 87] the synthesized nanodisks shaped NCs splits resonances in two modes: (i) longitudinal (i.e. in-plane with low energy) and (ii) transverse (i.e. out-plane with high energy). Two asymmetric plasmonics band at NIR regime were observed at 1600–1900 nm for out-plane mode and 3100 nm for in-plane mode (Fig. 4b). From the spectra clear blue shift was observed with increasing in aspect ratio of nanodisks attributing to higher free densities and also these LSPR modes can also be modulated by tuning the concentration as well as aspect ratio of nanodisks [87].

Besides the composition and shape, the size of the NCs will also play a drastic role in fine-tuning (altering) the properties. The size of the NCs majorly influences the bandgap of the synthesized NCs materials [43]. In transition metal chalcogenides NCs, the cationic metals like Cd, Cu, Mo, etc. will always induce the vacancy defects results in the widening of bandgap [86, 87]. These widening of bandgap may be attributed to free carriers leading to lower the energy tail in the absorption spectrum. Pan et al. demonstrated a noticeable shift in LSPR towards lower energy upon the incorporation of metal cations like Ce^{4+}, In^{3+} [88]. Some kind of chalcogenides NCs

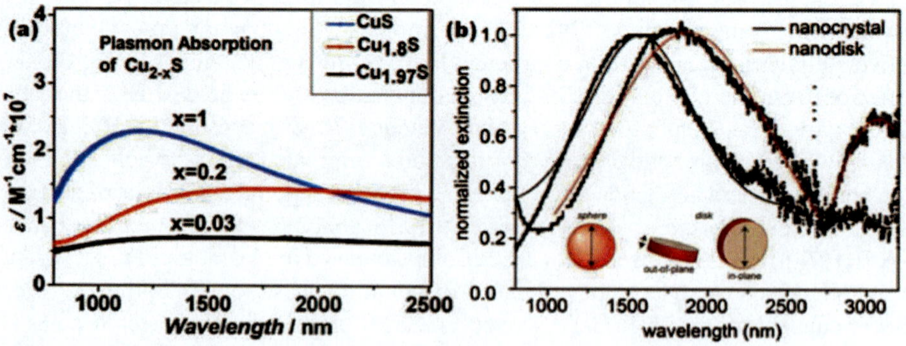

Fig. 4 **a** Extinction spectra of $Cu_{2-x}S$ NCs with various compositions (x = 0, 0.2, 0.03). Reproduced with permission of American Chemical Society [66] and **b** Shape dependent LSPRs property of spherical and nanodisks shaped NCs. Reproduced with permission of American Chemical Society [87]

like CIS, CISe, etc. are of great interests owing to its high quantum yield, NIR spectral absorption, large Stroke's shift, extended PL lifetimes (up to ns) [89–91].

3.2.1 Nonlinear Optics

Transition metal chalcogenides NCs express promising nonlinear optical properties, owing to its large bandgap energy, wide emission over the electromagnetic spectrum, strong quantum confinement. The nonlinear optics of the NCs can be described with a function of dielectric changes induced from its lattice temperatures and an increase in carriers. For example, copper chalcogenides NCs deliberates strong interband transitions in the visible regime and Cu chalcogenides NCs exhibit mighty interband transitions in the visible region and weak absorption in the mid-IR and NIR regime. The plasmonics effect in the chalcogenides NCs induces in the enhancement of inherent optical nonlinear response arises from photon-photon interaction finds its applications as nonlinear optical materials [92–94]. The enhanced nonlinearity property of NCs helps in reduced power consumption in photonic applications. Moreover, the fastening of optical signals at a response time of femtosecond can be achievable by speeding the response time of plasmonics excitation. Thus based on the phase structure and surface properties of the NCs the optical properties can be modulated.

3.3 Electronic Properties

The electronic properties of the chalcogenides NCs can be determined from its band structures. The prime differentiation existing between the chalcogenides and oxides was its higher covalence energy which results in narrowing of energy gaps. Apart from its covalence energy the size of the NCs as well as the electronegativity difference value existing between the transition metals and chalcogens induces the electronically defective sites. In general, the highly delocalized p orbitals in the chalcogens help in exhibiting strong electronic hybridization. Hybridization strength between two electronic states determined from the energy separation and wave function [95]. Upon hybridization of p and d orbitals, it results in a significant contribution towards near-edge valence bands. Considering the example of $CuGaS_2$ NCs, the 3d electrons of the Cu strongly hybridize with 3p states of S subsequently the valence band energy level will get push upward and resulting in smaller bandgap energy (2.34 eV) [95].

3.4 Magnetic Properties

The size and defects of the chalcogenides NCs play a role in the spinning of orbitals which tends to explore the effect of magnetism. The smaller the size of NCs higher the number of spins on the surface of NCs and the defective sites at cation and/or anion causes uncompensated spinning of orbitals resulting in the reduction of net magnetization [96]. The possibilities of employing chalcogenides NCs as magnetic materials involves either any one of the following approaches: (i) doping and/or (ii) alloying or (iii) heterostructures or (iv) conjugation of magnetic molecules/ions (such as Ni, Fe, Cr, Mn, Co, etc.) with the chalcogenides NCs [97–100]. From the family of chalcogenides NCs iron and chromium-based compounds are the most investigated NCs in the field of magnetism. The iron-based chalcogenides NCs ($CuFeS_2$) possess antiferromagnetic behaviour with an average magnetic moment of Fe atoms reported being 1.75 μB and a Neel temperature around 823 K [101]. Similarly, the report on chromium-based chalcogenides NCs ($CuCr_2Se_4$) had a pronounced effect of magneto-optic at room temperature and possessed ferromagnetic behaviour [102, 103]. Some other reports on the chromium-based chalcogenides NCs stated that the property of magnetism relies on the size as well as the synthesis approach of the NCs too. For example, $CuCr_2Se_4$ NCs synthesized by microwave method the saturation magnetization value was observed at 15 emu g^{-1} and coercive of 80 Oe [104, 105]. The same compound synthesized by the Solvothermal approach the particles were observed to agglomerate with a size of 25–200 nm (inhomogeneous) express the magnetism of 2.3 μB per Cr atom [106]. Thus the magnetism recorded for the chalcogenides NCs are better than that of oxides and the magnetism value highly depends on its size and defects.

4 Composites

The formation of composites using chalcogenides NCs with oxides, metals, ligands, polymers, etc. was one of the most interesting field which involves interdisciplinary research. The understanding of its mechanism and application behavior is not yet clear creating a scope for the researcher to focus. In this section, glimpses of the nanocomposites with chalcogenides NCs (ternary, quaternary) were described by considering the point of view on its properties and applications. Initially let's start with having a glance at metallic chalcogenides NCs composites which are reported to be remarkable once based on its composition and morphologies. The multinary metallic chalcogenides NCs based on indium (In) or gallium (Ga) metallic composites tolerates stoichiometric deviation, doping and/or alloying. Considering example of ternary composites synthesis of silver indium sulfur ($AgIn_2S_4$) NCs, the composition and stoichiometric of metals as well as chalcogens can be varied depends on the need of their applications. Sometimes there was also the possibility of replacement of metals Ag with zinc (Zn) without compromising much change in the phase and crystal structure[107]. Similarly, the quaternary composites have also been reported and the most interesting NCs of this type were Cu_2ZnSnS_4 which exhibits different compounds within its lattice symmetry. Such type of chalcogenides NCs is found to have distinguishable features than others. Apart from its stoichiometric and composition of metallic chalcogenides NCs the stable and metastable phases of the chalcogenides NCs were also made possible [108–110]. For example in the synthesis of

CuAuS NCs, the lattices in Cu and Au tends to form stable kesterite and stannate phases making the possibility of substitution of molecules/ions in its lattice without losing its crystallinity [108, 109].

Bailey and co-workers reported a range of cadmium based chalcogenides NCs, including CdSe, CdTe, CdSeTe by a hot-injection method using cadmium oxide as metal precursor and selenium, tellurium metals as chalcogen precursor. The ratio of selenium and telluride was weighed to relatively compensate the amount of Cd [111]. Following this report, Gurusinghe et al. mentioned that the bowing constant (i.e., bandgap energy vs composition relation) should be large enough to yield bandgaps lower than binary compositions (such as Cdse or CdTe) particularly for the alloy NCs [112]. These binary compositions with alloy or core/shell had energy emission in the range of 800 nm (near IR) as shown in Fig. 5a with its corresponding morphology observed using TEM images was given in Fig. 5b [111]. Weller's research group demonstrated core/shell and multilayered chalcogenides NCs (CdS-HgS-CdS) by

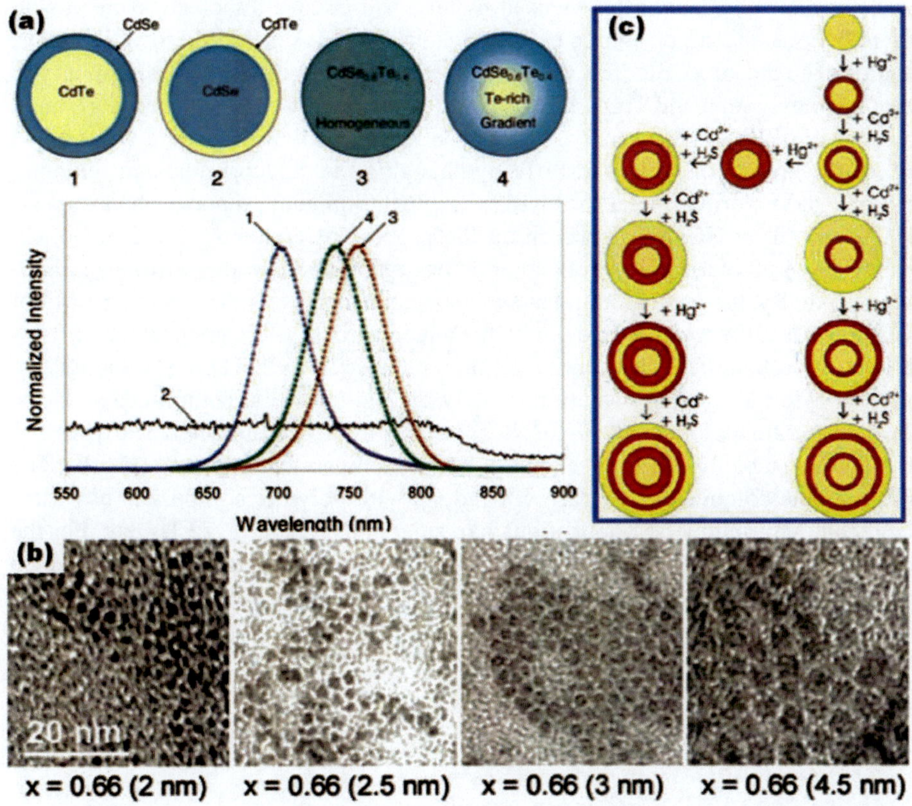

Fig. 5 **a** Schematic illustrations of QDs (core–shell CdTe- CdSe dots, reversed core–shell dots, homogeneous alloyed dots, gradient alloyed dots) with its corresponding fluorescence emission spectra; **b** TEM micrographs with increased NCs size and **c** CdS–HgS structures. Reproduced with permission of American Chemical Society [111, 116]

approaching an aqueous synthetic approach. For the synthesis of CdS aqueous solution of cadmium ions (precursor), H_2S (chalcogen source) and hexametaphosphate (stabilizer) were taken. Maintaining a higher pH rate starts the nucleation of CdS and by lowering the pH NCs growth takes place respectively. Then by approaching the surface ion-exchange method, the mercury ions were added to cover the CdS core upon pH neutralization. As a final step, the Cd ions were further deposited over the HgS shell by retuning to low pH value [113]. From this synthesis approach report, it was to be noted that during synthesis, the pH of the solution has been taken into account and varied at each stage to attain the balanced growth of nucleation towards core/shell NCs [114, 115]. Followed by this report, Dorfs et al. and Braun et al. approached the same synthesis route to develop quantum dot and wells in core/shell formation with variation in the thickness of the shell. The schematic representation of the core/shell growth approach was illustrated in Fig. 5c [116–118].

Zamkov and co-workers demonstrated the direct growth of PbS, PbSe, and PbTe QDs on TiO_2 substrates using a hetero-epitaxial hot injection approach. For the synthesis of PbS QDs, PbO was used as lead source and dissolved in oleic acid (OA)/ODE, as chalcogen source elemental sulfur was taken and dissolved in ODE solution. The prepared solutions were injected into the round-necked flask containing degassed oleylamine and TiO_2 substrates. The hot-injection allows direct formation and growth of PbS NCs over the TiO_2 substrates. A similar procedure was carried out for the growth of PbSe and PbTe by changing the chalcogen source selenium for PbSe and tellurium metal for PbTe [119, 120]. Sun et al., reported the synthesis of CdHgTe alloy NCs by approaching the process of aqueous growth treatment followed by polymerization to obtain good transparent NCs with narrow size distribution. Initially, the precursor and chalcogen source corresponding to the growth of CdHgTe NCs alloy were carried out by the aqueous route and then the aqueous NCs product was transferred into organic solution (octadecyl-4-vinyl benzyl dimethylammonium chloride, ligands) where thermally driven copolymerization takes places. The obtained final product (CdHgTe NCs) exhibits high quantum yield nearly to 45% and delivers stable luminescence with respect to temperature and time (Fig. 6). The NCs with the combination of narrow bandgap with polymer composites provide a significant path in diverse applications like solar cells, displays, LEDs, etc. For the benefits of readers, few NCs possessing narrow bandgap are listed in Table1 which helps new researchers to point out the materials for their desired applications [121, 122].

5 Application of Chalcogenides in Engineering Application

Chalcogenides NCs are gaining rising interest and provides their access to the researchers in the diverse field of disciplines, including electronic devices, solar panels, displays, sensors, amplifiers, biological engineering, medical systems, etc. [121, 123, 124]. The chalcogenides NCs are not only limited to its engineering applications, but it also contributes to developing interdisciplinary research. As the

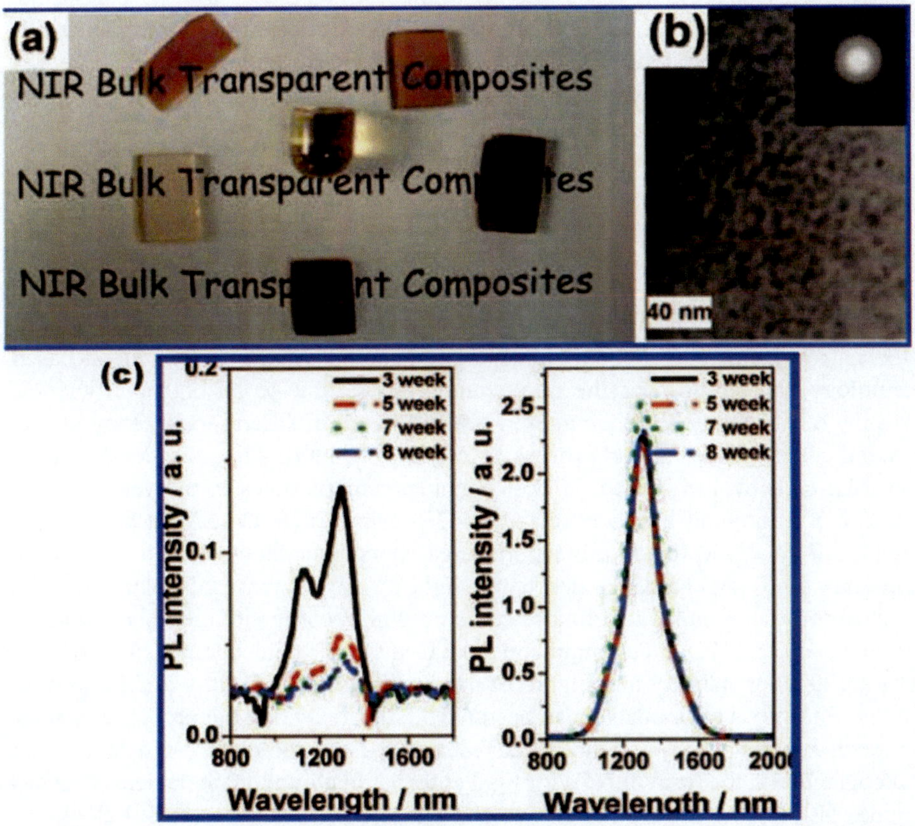

Fig. 6 **a** Optical image of different shaped CdHgTe NCs with polymer bulk composites; **b** TEM micrograph with SAED pattern of $Cd_{0.23}Hg_{0.77}Te$ NCs with polymer composite and **c** PL spectra of CdHgTe NCs in aqueous solution and CdHgTe NCs with polymer composites. Reproduced with permission of American Chemical Society [121]

Table 1 Different chalcogenides NCs with narrow bandgap

Chalcogenides NCs	Bandgap (eV)
PbS	0.37
PbTe	0.28
CdTe	1.5
HgSe	−0.06
$Hg_xCd_{1-x}Te$	−0.3 to 1.5
$HgS_{1-x}Se_x$	−0.06 to 0.5
Ag_2Se	0.15
Bi_2Te_3	0.16
$CuInSe_2$	1.0
$CuInS_2$	1.2
$AgInS_2$	1.2

glimpses mentioned above, there are numerous methods of synthesis approach available to produce the chalcogenides NCs as of its desired size, shape and properties towards the applications. In this section, some of the interesting applications of chalcogenides NCs are discussed.

5.1 Photovoltaics

The two most essential qualifications to survive as photovoltaic materials are its direct narrow bandgap and high light absorption coefficient. Currently, silicon-based technology was ruling over the photovoltaic market due to its higher efficiency. Still, the most commonly encountered issues existed in silicon technology are its elevated cost and purification process [125]. To sort out this issue, second generation solar cells evolved based on chalcogenides nanoparticles in the year of 1954 with 6% of photo-conversion efficiency (PCE) using Cu_xS, Cu_xS/CdS heterojunction [126, 127]. Followed by this report, researchers made their effort in developing numerous kinds of chalcogenides based light-absorbing materials with different compositions/compounds and finally come up with recorded efficiency of 22% for CIGSe solar cells [128]. The main concerns faced in the chalcogenides based solar cells are its poor stability and high cost due to the usage of indium and gallium in CIGSe. Enormous efforts have also been devoted to analyzing the problems associated with this technology and motivated research to make their interest in developing chalcogenides in the form of NCs for light absorber in photovoltaic devices. Chalcogenides NCs owing to its smaller size and quantum confinement find it as one of the potential candidates to use as photovoltaic materials. There are several kinds of chalcogenides NCs with different compositions such as binary (CdSe), ternary (CIS), quaternary (CZTS) compounds of NCs were effectively used as a light absorber in the photovoltaic devices.

The chalcogenides NCs processed by a solution-based approach can be used as a planar configuration type of solar cells (photovoltaics). The absorber (chalcogenides NCs) finds to be a low-cost approach and provides higher PCE have great significance towards NCs. Futuristic strategies on limiting the process of sintering chalcogenides NCs, compatibility of NCs solvents for flexible substrates, thinning the absorbing layer, designing/pattering the solar cells configurations helps in boosting the performance of the devices. Moreover, by choosing proper solvents, ligands for composites, heat treatment progress the life span time of chalcogenides NCs by reducing the surface traps, size of NCs, the introduction of doping, etc. resultants in boosting the performance of the solar cells [129, 130]. Beyond their usage, as a light absorber in solar cells, it also been used as luminescent materials for solar concentrators, counter electrodes in third-generation solar cells (dye-sensitized solar cells) and so on.

Koreal et al. demonstrated CISe QDs NCs with reasonable PCE, but the key point to be noted from the report is the performance of the device relies on the size of NCs. A decrease in the size of CISe NCs from ~7 to 2 nm resulted in the enhancement of open-circuit voltage (Voc) of the solar cells (Fig. 7a) [131]. The research group also

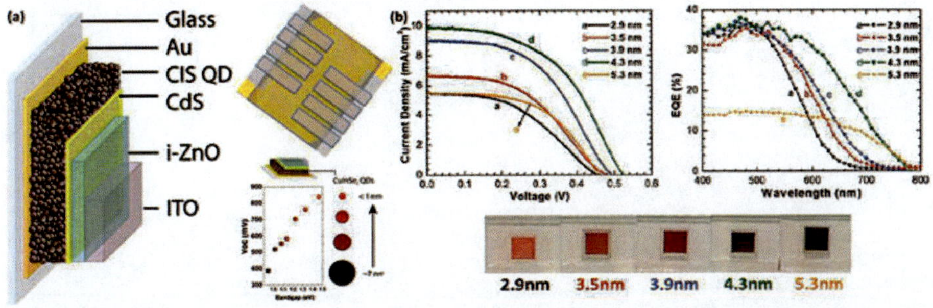

Fig. 7 **a** Schematic illustration of CISe QDs NCs in heterojunction solar cells. Reproduced with permission of American Chemical Society [131]; **b** photovoltaic (JV) performances of TiO_2/CIS QDs NCs ranging from 2–5 nm with its EQE and optical images. Reproduced with permission of American Chemical Society [133]

stated that the decrease in size of NCs leads to the band to band process as well as trap assisted recombination increases the Voc of the solar cells. Similar to the frame of work, Luther and co-workers reported the synthesis of PbS QDs NCs using Hines method and used in the PbS-ZnO heterojunction solar cells. PbS NCs synthesized with tightly defines size distribution plays a role in enhancement in the performance of the solar cells [132]. Generally, the size of the materials will be directly related to its bandgap in inverse relation. The same thematic has also applied in NCs, the decrease in bandgap leads to an increase in the size of the NCs. But it differs in the case of QDs based chalcogenides NCs such as Pb, Cd, etc. since there is no size-dependent electron injection takes place in QDs NCs [133]. Moreover, reasonable EQEs have been obtained despite its dominance of defects in the charge separation process. The PCE was observed to increase with increasing size of CIS QDs which surprisingly is the reverse effect of the trend (Fig. 7b). From the observed PCE of QDs NCs, it was concluded that the QDs ranging in the size of 4–5 nm would produce good PCE [133]. Chalcogenides NCs also used along with a combination of other absorbing materials like perovskite ($CH_3NH_3PbI_3$) for heterojunction solar cells. For example, in the heterojunction solar cells, CIS NCs replaced the role of TiO_2 in architecture ITO/CIS/Al_2O_3/$CH_3NH_3PbI_3$/Ag resultant in the better enhancement in the absorption of photons and recorded PCE of~5% [134]. Similar to report on CIS, Das et al. synthesized CIS/ZnS core/shell QDs NCs and used as an inorganic hole transporting layer in the perovskite solar cells with a PCE of 8% [135].

For further enhancement in the performance of the solar cells as well as towards commercialization researchers find the availability of suitable chalcogenides materials in the form of NCs based inks. Various kinds of chalcogenides NCs inks like CIS, CZTS, CuSbS, CIZS, CZTGeSSe, etc. were developed and used as a light absorber in solar cells applications. Chalcogenides NCs inks facilitate in large scale fabrication of solar cells (modules), particularly in the roll-to-roll deposition process. The light absorption coefficient can be improved by introducing suitable elements such as Zn, Al, Au, Ag, etc. Brik and co-workers reported the synthesis and properties

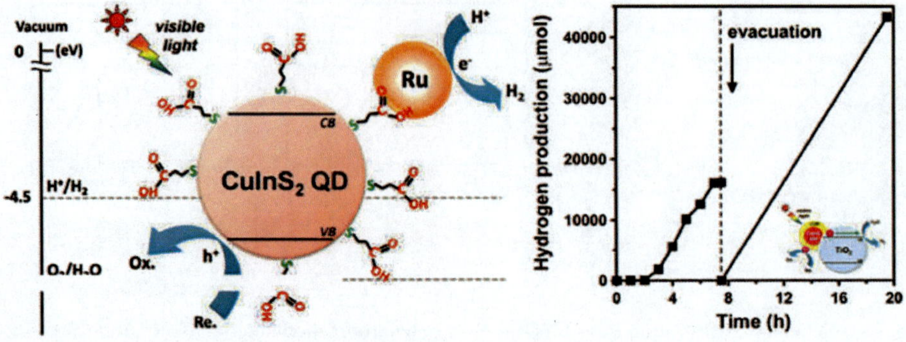

Fig. 8 Schematic representation of MPA-coated CIS QDs NCs loaded with Ru as a hydrogen evolution catalyst (left) and its hydrogen production activity (right). Reproduced with permission of Elsevier [154]

of Al-doped CGS; Vahidshad et al. reported the production of Fe and Zn doped CIS NCs [136, 137].

5.2 *Sensors*

Sensors are analytical devices that detect and provide some information by interacting with the environment and it can be of different types depending on its interaction. NCs owing to its large surface to volume ratio, strong phonon quenching and high quantum confinement it have been used as active sensing elements, transducer component, and direct specific sensors. The surface modification of chalcogenides NCs also been achieved for designing the specific kinds of nanosensors. Chalcogenides QDs NCs are used in the fluorescent/luminescent sensors because of their high quantum yield. QDs NCs emits changes in the PL upon adsorption of molecules on the surface it [138]. For instance, CIS QDs NCs the interaction of Cu^{2+} molecules with physical adsorption of environment induces reduction of Cu^{2+} to Cu^{+} generates electron–hole recombination via electron transfer process results in the quenching of PL [139]. In general, PL based sensors depends on the fluorescence quenching/enhancement where Cu based chalcogenides NCs are also employed as acceptors in the fluorescence resonance energy transfer (FRET) [140].

Chemiluminescence (CL)-based sensors the compound to interact with the physical environment will be in the form of a label. In this scenario, the chalcogenides NCs (CuS NCs) will be employed in the form of nanotag as a sensing element in the CL-based sensor [141]. A huge volume of reports existed on chalcogenides NCs for CL based sensors with high selectivity/sensitivity [98]. Similarly, for the electrochemical sensors, electrochemical signal arises due to electron interaction with the molecules/species [142]. The determination of decomposition activity of $_2O_2$ using chalcogenides NCs composites with ligands is well recognized for its activity

[143, 144]. Cu_2S NCs decorated with multiwalled carbon nanotubes (MWCNTs) for amperometric glucose sensors with high sensitivity range have been reported. The enrichment in the performance of the hybrid nanocomposites NCs can be explained from the synergetic catalytic activity of Cu_2S NC and the MWCNTs helps in the formation of electrical networking which permits to deliver excellent performance of the biosensor [145]. Followed by this report, several kinds of sensors were developed using compositing chalcogenides NCs with biological agents (amphiphilic polymer, bioconjugates, polymers, etc.) [146–148].

5.3 Catalytic Applications

Chalcogenides NCs are not only limited its application in the field of solar energy conversion (photovoltaics) it was also used in the production of solar fuels mainly of hydrogen energy as well as degradation of pollutants using photocatalytic process. The compositions, size and variable valences of the chalcogenides NCs play a key role in the mechanism of the photocatalytic process [149, 150]. Chalcogenides NCs hold their uniqueness in catalytic activity by providing multiple reactions at particular reaction sites, ease of separation from products, light absorbance, plasmonics enhancement, etc. [151, 152]. The production of solar fuels (hydrogen activity) carried out using photoelectrochemical (PEC) cells similar setup like PV. In the PEC cells, the materials (chalcogenides NCs) either be dispersed in the solution or deposited over the conducting electrode for hydrogen production. In 1992, hydrogen (H_2) evolution activity of CIS was first demonstrated by Kobayakawa et al. using chemical precipitate synthesized CIS NCs and the hydrogen evolution was carried out under aqueous sulfur solution with the source of xenon arc lamp. From the report, it is conceptualized that the stoichiometric compositions of CIS play an important in the enhancement of hydrogen evolution [153]. Later on numerous chalcogenides, NCs materials were evolved and used as a catalytic material for the production of H_2. For example, 400 μmol/h of H_2 evolution rate was recorded for the ruthenium (Ru) photo deposited CIS QDs. The evolution rate was also further improved by making composites with semiconducting NCs. CIS QDs were attached with TiO_2 NCs using MPA linkers and over the surface of the CIS QDs CdS layer was photo deposited. The schematic illustration of the nanocomposites with its hydrogen production activity was illustrated in Fig. 8. In the heterostructure, electrons will be photogenerated at the CIS QDs and then electrons are transferred to the conduction band of TiO_2 which generates H_2 by reacting with H_2O and achieved the H_2 evolution rate of 3310 μmol/h with a quantum efficiency of 41% [154]. Furthermore, the chalcogenides NCs also demonstrate excellent catalytic property in the other areas of catalytic applications, including oxygen evolution, oxygen reduction, photochemistry and as well as in synthetic organic chemistry.

5.4 Energy Storage

Electrochemical energy storage device (batteries, supercapacitors) composed of single/several electrochemical cells consisting of anode and cathode separated by the electrolyte solution. So far lithium (Li) based energy storage devices recorded its benchmark activity in the field of battery market due to its higher output efficiency [155]. Besides its higher output power there remain to be some unsolved challenges like cost, toxicity, complicated synthesis approach, etc. in the existing technology. Modern generation needs are focused keenly on the robust development of materials with portable applications. With the support of nanotechnology, the unsolved challenges are been handled by synthesizing materials with different nanostructures, making compositions and so on. Yamakawa and co-workers reported a detailed investigation of the reaction mechanism of Li with CuS NCs. The formation of Li_xCuS was achieved by a two-step reaction method. Initially, the intercalation formation between Li and CuS was observed followed by which the conversion reaction of the Li_xCuS insertion phase (Fig. 9) [156].

Due to the higher electronic conductivity of the sulfide phases, there may be a higher probability for the cause of intercalations of ions at the Li_xCuS/Cu interfaces. Moreover, during the intercalations at the interfaces, the Li^+ ions gets diffuses rapidly results in the induced formation of Li_2S contributing towards the formation of composition Li_2S/Li_xCuS. Hence in the growth mechanism of Li_xCuS, the Cu particles might have less impact and allow the higher possibilities for the growth of Li_2S [157].

Looking from the supercapacitor's perspective of the chalcogenides, its high theoretical value and valence states of metal constituents (specifically of sulfur groups) had its impact in the field of applications. Various sulfide-based materials and its compounds have been used as active materials for supercapacitor. An interesting report proposed by Qian and co-workers about the synthesis of CuS nanotubes as active materials for supercapacitor. The CuS NCs with nanotube like morphology delivers the output performance of 2175 F/g at the current density of 15 A/g. The higher performance of the devices was briefed in the report relating to its large surface area, high electron transfer passage and porosity of CuS nanotubes (made of NCs), faster redox rate and charge transfer resistance. The obtained output power was observed to 17 folds increase in value to that of the existing report on CuS-CNT composites [158]. Even though wide varieties of chalcogenides based materials are reported in the field of energy storage devices still some of the challenges like boosting the capacity retention and stability of the devices are need to be addressed [159].

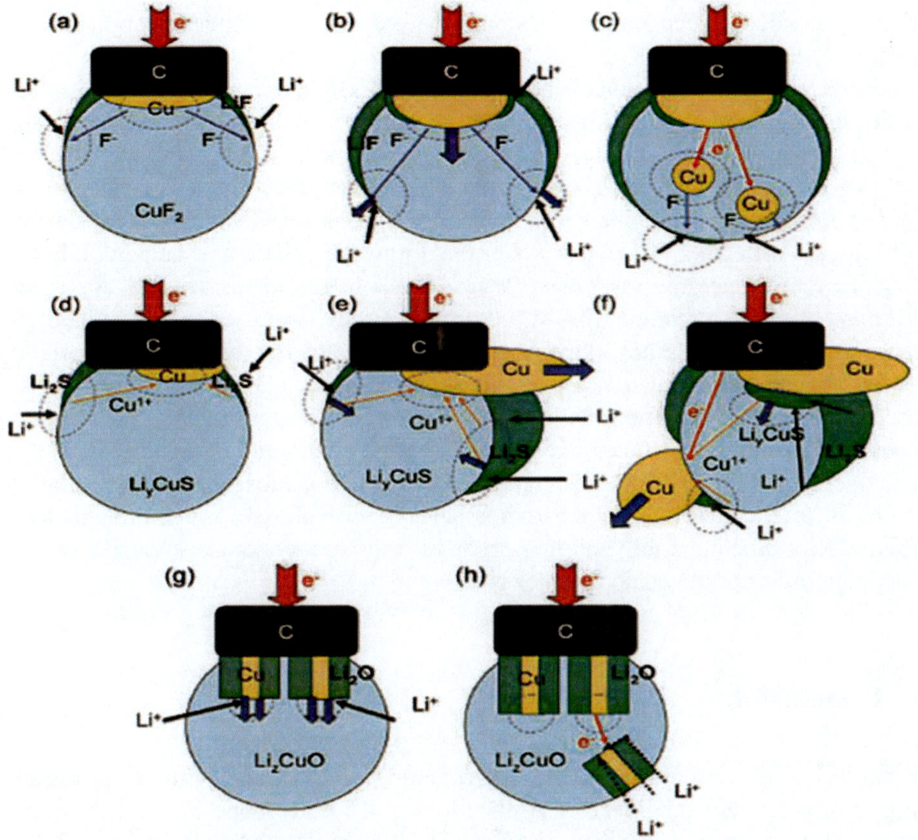

Fig. 9 Schematics illustrating possible reaction pathways for the three systems, assuming a single electrical contact to a LizCuXn **a–c** CuF_2, **d–f**, CuS, and **g, h** CuO. Reproduced with the permission of the American Chemical Society [156]

5.5 *Bio-Application*

Apart from the energy sectors, the chalcogenides NCs also place its path in the field of biological sciences. The tuning ability of size, shape, and compositions of chalcogenides NCs finds its usage for targeted drug delivery, therapeutic devices, bio labelling, radiations therapy and so on. The most complicated problem existing in the biological fields is its compatibility. Proper chalcogenides NCs have poor compatibility in integrating with the biological environment. To make the biocompatible chalcogenides NCs the surface modifications treatment will be carried out which in turn helps in tuning its properties. In the surface treatment of NCs ligands and/or polymeric agents were most commonly used. For example, the polyethylene glycol (PEG) was most often used with the chalcogenides NCs to make to be biocompatible. Since the PEG was highly hydrophilic in nature as well as non-toxic it helps in binding with proteins, molecules or cells. Holding the advantage of its

PEG composited chalcogenides NCs can be used in the therapeutics applications [160].

Likewise, in the application of chemotherapy, NCs hold their uniqueness of higher carrier ability of anti-cancer drugs. The loading of drugs can be maximized by playing with the surface modification of NCs and there is plenty of literature available on this subject. Among the various treatments, the simplest approach was surface functionalization with respect to its biomolecules. For example, Gelatin gel will be used to conjugate with CuS NCs to carry over the drugs, the gelatin gels help in building enzymes that induce the responsive drugs [161]. In the intensive field of cancer treatment, the NIR radiation (700–1200 nm) source will be used owing to its high transparency towards tissues, skins, blood fluid, and cells. For these kinds of specific needs, the materials like CdS QDs NCs will be used which absorbs a wide range of electromagnetic spectrum particularly in the near and IR region [162, 163]. Sometimes the plasmonics particles like Ag, Au will be anchored on the surface of the chalcogenides NCs for the usage in targeted drug delivery, biosensors, etc. Similarly, several different functionalization routes have been explored to make the chalcogenides NCs biocompatible and there remain futuristic works to be carried out in biocompatibility of the chalcogenides NCs.

6 Conclusion

In summary, an outline about the chalcogenides NCs along with its synthesis approach, properties, engineering applications and future scope in this field were discussed. A comprehensive discussion was made concerning the effects of stoichiometric compositions and the size of chalcogenides NCs with different compositions (binary, ternary, quaternary). The fundamentals of materials properties, including phase structure, optical properties, electrical properties, and magnetic properties, were explained in detail. A focus on the engineering applications (photovoltaics, sensors, biological, catalytic) of the chalcogenides NCs were also detailed in the chapter. Notwithstanding the importance of the chalcogenides NCs it finds its path in the diverse field of science and technology. Even though it holds its benchmarking effects in various fields, still there remains much complexity to solve. Many interesting riddles still exist in the development of NCs, for instance: no reports available on exploring sizes/shapes of NCs to enhance its properties in the end-user applications. Likewise, some of the challenges in the field of chalcogenides NCs growth are its versatility, compositions, stability which provides an open path to beginners to involve in the field of research.

References

1. Shi, W., Hughes, R.W., Denholme, S.J., Gregory, D.H.: CrystEngComm **12**, 641 (2010)
2. Van Der Stam, W., Berends, A.C., De Mello Donega, C.: ChemPhysChem **17**, 559 (2016)
3. Rao, C.N., Deepak, F., Gundiah, G., Govindaraj, A.: Prog. Solid State Chem. **31**, 5 (2003)
4. Yarema, O., Bozyigit, D., Rousseau, I., Nowack, L., Yarema, M., Heiss, W., Wood, V.: Chem. Mater. **25**, 3753 (2013)
5. Vikraman, D., Hussain, S., Akbar, K., Adaikalam, K., Lee, S.H., Chun, S.H., Jung, J., Kim, H.S., Park, H.J.: ACS Omega **3**, 5799 (2018)
6. Jiang, G., Yi, J., Miao, L., Tang, P., Huang, H., Zhao, C., Wen, S.: Sci. Rep. **8** (2018)
7. Akhavan, V.A., Goodfellow, B.W., Panthani, M.G., Steinhagen, C., Harvey, T.B., Stolle, C.J., Korgel, B.A.: J. Solid State Chem. **2** (2012)
8. Murray, C.B., Norris, D.J., Bawendi, M.G.: J. Am. Chem. Soc. **115**, 8706 (1993)
9. Park, J., Joo, J., Soon, G.K., Jang, Y., Hyeon, T.: Angew. Chemie Int. Ed. **46**, 4630 (2007)
10. De Mello Donegá, C., Liljeroth, P., Vanmaekelbergh, D.: Small **1**, 1152 (2005)
11. Peng, X., Wickham, J., Alivisatos, A.P.: J. Am. Chem. Soc. **120**, 5343 (1998)
12. Peng, Z.A., Peng, X.: J. Am. Chem. Soc. **124**, 3343 (2002)
13. Yin, Y., Alivisatos, A.P.: Nature **437**, 664 (2005)
14. Tao, A.R., Habas, S., Yang, P.: Small **4**, 310 (2008)
15. Jun, Y.W., Choi, J.S., Cheon, J.: Angew. Chemie Int. Ed. **45**, 3414 (2006)
16. Gao, M.R., Xu, Y.F., Jiang, J., Yu, S.H.: Chem. Soc. Rev. **42**, 2986 (2013)
17. Walton, R.I.: Chem. Soc. Rev. **31**, 230 (2002)
18. Cushing, B.L., Kolesnichenko, V.L., O'Connor, C.J.: Chem. Rev. **104**, 3893 (2004)
19. Yu, X., Cao, C., Zhu, H., Li, Q., Liu, C., Gong, Q.: Adv. Funct. Mater. **17**, 1397 (2007)
20. Kuo, C.H., Chu, Y.T., Song, Y.F., Huang, M.H.: Adv. Funct. Mater. **21**, 792 (2011)
21. Jiao, S., Xu, L., Jiang, K., Xu, D.: Adv. Mater. **18**, 1174 (2006)
22. Martín, J., Maiz, J., Sacristan, J., Mijangos, C.: Polymer (Guildf). **53**, 1149 (2012)
23. Wang, W., Dahl, M., Yin, Y.: Chem. Mater. **25**, 1179 (2013)
24. Gao, J., Li, Q., Zhao, H., Li, A., Liu, C., Gong, Q., Qi, L.: Chem. Mater. **20**, 6263 (2008)
25. Cabot, A., Smith, R.K., Yin, Y., Zheng, H., Reinhard, B.M., Liu, H., Alivisatos, A.P.: ACS Nano **2**, 1452 (2008)
26. Yin, Y.: Science **304**(80), 711 (2004)
27. Fan, H.J., Gösele, U., Zacharias, M.: Small **3**, 1660 (2007)
28. Ibáñez, M., Fan, J., Li, W., Cadavid, D., Nafria, R., Carrete, A., Cabot, A.: Chem. Mater. **23**, 3095 (2011)
29. Lou, X.W., Archer, L.A., Yang, Z.: Adv. Mater. **20**, 3987 (2008)
30. Zhang, Q., Wang, W., Goebl, J., Yin, Y.: Nano Today **4**, 494 (2009)
31. Shan, Z.W., Adesso, G., Cabot, A., Sherburne, M.P., Syed Asif, S.A., Warren, O.L., Chrzan, D.C., Minor, A.M., Alivisatos, A.P.: Nat. Mater. **7**, 947 (2008)
32. Cao, H., Qian, X., Wang, C., Ma, X., Yin, J., Zhu, Z.: J. Am. Chem. Soc. **127**, 16024 (2005)
33. Chen, C.C., Herhold, A.B., Johnson, C.S., Alivisatos, A.P.: Science **276**(80), 398 (1997)
34. Wang, Z., Peng, F., Wu, Y., Yang, L., Zhang, F., Huang, J.: CrystEngComm **14**, 3528 (2012)
35. Xiao, G., Zeng, Y., Jiang, Y., Ning, J., Zheng, W., Liu, B., Chen, X., Zou, G., Zou, B.: Small **9**, 793 (2013)
36. Mu, L., Wang, F., Sadtler, B., Loomis, R.A., Buhro, W.E.: ACS Nano **9**, 7419 (2015)
37. Kriegel, I., Rodríguez-Fernández, J., Wisnet, A., Zhang, H., Waurisch, C., Eychmüller, A., Dubavik, A., Govorov, A.O., Feldmann, J.: ACS Nano **7**, 4367 (2013)
38. Akkerman, Q.A., Genovese, A., George, C., Prato, M., Moreels, I., Casu, A., Marras, S., Curcio, A., Scarpellini, A., Pellegrino, T., Manna, L., Lesnyak, V.: ACS Nano **9**, 521 (2015)
39. Rivest, J.B., Jain, P.K.: Chem. Soc. Rev. **42**, 89 (2013)
40. Gupta, S., Kershaw, S.V., Rogach, A.L.: Adv. Mater. **25**, 6923 (2013)
41. Dong, H.S., Hughes, S.M., Yin, Y., Alivisatos, A.P.: Science **306**(80), 1009 (2004)
42. Kelly, D., Singh, A., Barrett, C.A., O'Sullivan, C., Coughlan, C., Laffir, F.R., O'Dwyer, C., Ryan, K.M.: Nanoscale **3**, 4580 (2011)

43. Luther, J.M., Jain, P.K., Ewers, T., Alivisatos, A.P.: Nat. Mater. **10**, 361 (2011)
44. Pearson, R.G.: J. Am. Chem. Soc. **85**, 3533 (1963)
45. Deka, S., Miszta, K., Dorfs, D., Genovese, A., Bertoni, G., Manna, L.: Nano Lett. **10**, 3770 (2010)
46. Li, H., Brescia, R., Krahne, R., Bertoni, G., Alcocer, M.J.P., D'Andrea, C., Scotognella, F., Tassone, F., Zanella, M., De Giorgi, M., Manna, L.: ACS Nano **6**, 1637 (2012)
47. Miszta, K., Gariano, G., Brescia, R., Marras, S., De Donato, F., Ghosh, S., De Trizio, L., Manna, L.: J. Am. Chem. Soc. **137**, 12195 (2015)
48. Van Der Stam, W., Bladt, E., Rabouw, F.T., Bals, S., De Mello Donega, C.: ACS Nano **9**, 11430 (2015)
49. De Trizio, L., Prato, M., Genovese, A., Casu, A., Povia, M., Simonutti, R., Alcocer, M.J.P., D'Andrea, C., Tassone, F., Manna, L.: Chem. Mater. **24**, 2400 (2012)
50. Park, J., Kim, S.W.: J. Mater. Chem. **21**, 3745 (2011)
51. Lesnyak, V., George, C., Genovese, A., Prato, M., Casu, A., Ayyappan, S., Scarpellini, A., Manna, L.: ACS Nano **8**, 8407 (2014)
52. Ha, D.H., Caldwell, A.H., Ward, M.J., Honrao, S., Mathew, K., Hovden, R., Koker, M.K.A., Muller, D.A., Hennig, R.G., Robinson, R.D.: Nano Lett. **14**, 7090 (2014)
53. Robinson, R.D., Sadtler, B., Demchenko, D.O., Erdonmez, C.K., Wang, L.W., Alivisatos, A.P.: Science **317**(80), 355 (2007)
54. Zhang, X., Stevanović, V., D'Avezac, M., Lany, S., Zunger, A.: Phys. Rev. B Condens. Matter Mater. Phys. **86**, (2012)
55. Genin, H.S., Ibers, J.A.: Transit. Met. Sulphides, pp. 1–35. Springer Netherlands, Dordrecht (1998)
56. Bouroushian, M.: pp. 1–56 (2010)
57. Lukashev, P., Lambrecht, W.R.L., Kotani, T., Van Schilfgaarde, M.: Phys. Rev. B Condens. Matter Mater. Phys. **76**, (2007)
58. Yu, L., Luo, K., Chen, S., Duan, C.G.: CrystEngComm **17**, 2878 (2015)
59. Goble, J.R.: The Relationship between Crystal Structure, Bonding and Cell Dimensions in the Copper Sulfides. The Canadian Mineralogist (1985)
60. Xu, Q., Huang, B., Zhao, Y., Yan, Y., Noufi, R., Wei, S.H.: Appl. Phys. Lett. **100** (2012)
61. Powell, A.E., Hodges, J.M., Schaak, R.E.: J. Am. Chem. Soc. **138**, 471 (2016)
62. Morales-García, A., Soares, A.L., Dos Santos, E.C., De Abreu, H.A., Duarte, H.A.: J. Phys. Chem. A **118**, 5823 (2014)
63. Wu, Y., Wadia, C., Ma, W., Sadtler, B., Alivisatos, A.P.: Nano Lett. **8**, 2551 (2008)
64. Zheng, H., Rivest, J.B., Miller, T.A., Sadtler, B., Lindenberg, A., Toney, M.F., Wang, L., Kisielowski, C., Alivisatos, A.P.: Science **333**(80), 206 (2011)
65. Rivest, J.B., Fong, L.K., Jain, P.K., Toney, M.F., Alivisatos, A.P.: J. Phys. Chem. Lett. **2**, 2402 (2011)
66. Zhao, Y., Pan, H., Lou, Y., Qiu, X., Zhu, J., Burda, C.: J. Am. Chem. Soc. **131**, 4253 (2009)
67. Hirahara, E.: J. Phys. Soc. Jpn. **6**, 428 (1951a)
68. Hirahara, E.: J. Phys. Soc. Jpn. **6**, 422 (1951b)
69. Chakrabarti, D.J., Laughlin, D.E.: Bull. Alloy Phase Diagrams **4**, 254 (1983)
70. Sands, T.D., Washburn, J., Gronsky, R.: Phys. Status Solidi **72**, 551 (1982)
71. Zhao, Y., Burda, C.: Energy Environ. Sci. **5**, 5564 (2012)
72. Partain, L.D., McLeod, P.S., Duisman, J.A., Peterson, T.M., Sawyer, D.E., Dean, C.S.: J. Appl. Phys. **54**, 6708 (1983)
73. Cheng, L., Wang, C., Feng, L., Yang, K., Liu, Z.: Chem. Rev. **114**, 10869 (2014)
74. Jain, P.K., Huang, X., El-Sayed, I.H., El-Sayed, M.A.: Acc. Chem. Res. **41**, 1578 (2008)
75. Hu, M., Chen, J., Li, Z.Y., Au, L., Hartland, G.V., Li, X., Marquez, M., Xia, Y.: Chem. Soc. Rev. **35**, 1084 (2006)
76. Huang, X., Jain, P.K., El-Sayed, I.H., El-Sayed, M.A.: Lasers Med. Sci. **23**, 217 (2008)
77. Ozbay, E.: Science **311**(80), 189 (2006)
78. Ferry, V.E., Munday, J.N., Atwater, H.A.: Adv. Mater. **22**, 4794 (2010)
79. Pillai, S., Green, M.A.: Sol. Energy Mater. Sol. Cells **1** (2010)

80. Catchpole, K.R., Polman, A.: Opt. Express **16**, 21793 (2008)
81. Mayer, K.M., Hafner, J.H.: Chem. Soc. Rev. **8**, 435 (2009)
82. Liu, N., Tang, M.L., Hentschel, M., Giessen, H., Alivisatos, A.P.: Nat. Mater. **10**, 631 (2011)
83. Giannini, V., Fernández-Domínguez, A.I., Heck, S.C., Maier, S.A.: Chem. Rev. **111**, 3888 (2011)
84. Stewart, M.E., Anderton, C.R., Thompson, L.B., Maria, J., Gray, S.K., Rogers, J.A., Nuzzo, R.G.: Chem. Rev. **108**, 494 (2008)
85. Comin, A., Manna, L.: Chem. Soc. Rev. **43**, 3957 (2014)
86. Hsu, S.W., Bryks, W., Tao, A.R.: Chem. Mater. **24**, 3765 (2012)
87. Hsu, S.W., On, K., Tao, A.R.: J. Am. Chem. Soc. **133**, 19072 (2011)
88. Pan, Z., Mora-Seró, I., Shen, Q., Zhang, H., Li, Y., Zhao, K., Wang, J., Zhong, X., Bisquert, J.: J. Am. Chem. Soc. **136**, 9203 (2014)
89. Kolny-Olesiak, J., Weller, H.: ACS Appl. Mater. Interfaces **5**, 12221 (2013)
90. Li, L., Pandey, A., Werder, D.J., Khanal, B.P., Pietryga, J.M., Klimov, V.I.: J. Am. Chem. Soc. **133**, 1176 (2011)
91. Xie, R., Rutherford, M., Peng, X.: J. Am. Chem. Soc. **131**, 5691 (2009)
92. Li, Z.G., Qian, X.M., Xiao, Z.G., Wei, T.H., Song, Y.L.: Chem. Phys. Lett. **612**, 219 (2014)
93. Mary, K.A.A., Unnikrishnan, N.V., Philip, R.: APL Mater. **2** (2014)
94. Della Valle, G., Scotognella, F., Kandada, A.R.S., Zavelani-Rossi, M., Li, H., Conforti, M., Longhi, S., Manna, L., Lanzani, G., Tassone, F.: J. Phys. Chem. Lett. **4**, 3337 (2013)
95. Zhang, Y., Xi, L., Wang, Y., Zhang, J., Zhang, P., Zhang, W.: Comput. Mater. Sci. **108**, 239 (2015)
96. Demortière, A., Panissod, P., Pichon, B.P., Pourroy, G., Guillon, D., Donnio, B., Bégin-Colin, S.: Nanoscale **3**, 225 (2011)
97. Cheng, C.Y., Ou, K.L., Huang, W.T., Chen, J.K., Chang, J.Y., Yang, C.H.: ACS Appl. Mater. Interfaces **5**, 4389 (2013)
98. Ding, C., Zhong, H., Zhang, S.: Biosens. Bioelectron. **23**, 1314 (2008)
99. Sitbon, G., Bouccara, S., Tasso, M., Francois, A., Bezdetnaya, L., Marchal, F., Beaumont, M., Pons, T.: Nanoscale **6**, 9264 (2014)
100. Demillo, V.G., Liao, M., Zhu, X., Redelman, D., Publicover, N.G., Hunter, K.W.: Colloids Surfaces A Physicochem. Eng. Asp. **464**, 134 (2015)
101. Teranishi, T.: J. Phys. Soc. Jpn. **16**, 1881 (1961)
102. Fedorov, V.A., Kesler, Y.A., Zhukov, E.G.: Inorg. Mater. **39** (2003)
103. Brändle, H., Schoenes, J., Wachter, P., Hulliger, F., Reim, W.: Appl. Phys. Lett. **56**, 2602 (1990)
104. Kim, D., Gedanken, A., Tver'yanovich, Y.S., Lee, D.W., Kim, B.K.: Mater. Lett. **60**, 2807 (2006)
105. Rusnak, A.N., Kim, D., Parameswaran, S., Patra, C.R., Trofimov, V.B., Harpness, R., Gedanken, A., Tver'yanovich, Y.S.: J. Non. Cryst. Solids **352**, 2885 (2006)
106. Ramesha, K., Seshadri, R.: Solid State Sci. **6**, 841 (2004)
107. Stroyuk, O., Raevskaya, A., Gaponik, N.: Chem. Soc. Rev. **47**, 5354 (2018)
108. Fan, F.J., Wu, L., Yu, S.H.: Energy Environ. Sci. **7**, 190 (2014)
109. Aldakov, D., Lefrançois, A., Reiss, P.: J. Mater. Chem. C **1**, 3756 (2013)
110. Coughlan, C., Ibáñez, M., Dobrozhan, O., Singh, A., Cabot, A., Ryan, K.M.: Chem. Rev. **117**, 5865 (2017)
111. Bailey, R.E., Nie, S.: J. Am. Chem. Soc. **125**, 7100 (2003)
112. Gurusinghe, N.P., Hewa-Kasakarage, N.N., Zamkov, M.: J. Phys. Chem. C **112**, 12795 (2008)
113. Mews, A., Banin, U., Kadavanich, A.V., Alivisatos, A.P.: Berichte Der Bunsengesellschaft Für Phys. Chemie **101**, 1621 (1997)
114. Lifshitz, E., Porteanu, H., Glozman, A., Weller, H., Pflughoefft, M., Echymüller, A.: J. Phys. Chem. B **103**, 6870 (1999)
115. Fradkin, L., Langof, L., Lifshitz, E., Rogach, A., Gaponik, N., Weller, H., Eychmüller, A.: ChemPhysChem **4**, 1203 (2003)
116. Dorfs, D., Henschel, H., Kolny, J., Eychmüller, A.: J. Phys. Chem. B **108**, 1578 (2004)

117. Dorfs, D., Eychmüller, A.: Nano Lett. **1**, 663 (2001)
118. Braun, M., Burda, C., El-Sayed, M.A.: J. Phys. Chem. A **105**, 5548 (2001)
119. Acharya, K.P., Khon, E., Oconner, T., Nemitz, I., Klinkova, A., Khnayzer, R.S., Anzenbacher, P., Zamkov, M.: ACS Nano **5**, 4953 (2011)
120. Acharya, K.P., Hewa-Kasakarage, N.N., Alabi, T.R., Nemitz, I., Khon, E., Ullrich, B., Anzenbacher, P., Zamkov, M.: J. Phys. Chem. C **114**, 12496 (2010)
121. Sun, H., Zhang, H., Ju, J., Zhang, J., Qian, G., Wang, C., Yang, B., Wang, Z.Y.: Chem. Mater. **20**, 6764 (2008)
122. Kershaw, S.V., Susha, A.S., Rogach, A.L.: Chem. Soc. Rev. **42**, 3033 (2013)
123. Yang, Z., Chen, C.Y., Roy, P., Chang, H.T.: Chem. Commun. **47**, 9561 (2011)
124. Debnath, R., Bakr, O., Sargent, E.H.: Energy Environ. Sci. **4**, 4870 (2011)
125. Lu, L., Yang, H.X.: Appl. Energy **87**, 3625 (2010)
126. Bube, R.H.: **281** (1998)
127. Reynolds, D.C., Leies, G., Antes, L.L., Marburger, R.E.: Phys. Rev. **96**, 533 (1954)
128. Wagner, S., Shay, J.L., Migliorato, P., Kasper, H.M.: Appl. Phys. Lett. **25**, 434 (1974)
129. Lan, X., Voznyy, O., Kiani, A., García de Arquer, F.P., Abbas, A.S., Kim, G.H., Liu, M., Yang, Z., Walters, G., Xu, J., Yuan, M., Ning, Z., Fan, F., Kanjanaboos, P., Kramer, I., Zhitomirsky, D., Lee, P., Perelgut, A., Hoogland, S., Sargent, E.H.: Adv. Mater. **28**, 299 (2016)
130. Kramer, I.J., Minor, J.C., Moreno-Bautista, G., Rollny, L., Kanjanaboos, P., Kopilovic, D., Thon, S.M., Carey, G.H., Chou, K.W., Zhitomirsky, D., Amassian, A., Sargent, E.H.: Adv. Mater. **27**, 116 (2015)
131. Panthani, M.G., Stolle, C.J., Reid, D.K., Rhee, D.J., Harvey, T.B., Akhavan, V.A., Yu, Y., Korgel, B.A.: J. Phys. Chem. Lett. **4**, 2030 (2013)
132. Luther, J.M., Gao, J., Lloyd, M.T., Semonin, O.E., Beard, M.C., Nozik, A.J.: Adv. Mater. **22**, 3704 (2010)
133. Jara, D.H., Yoon, S.J., Stamplecoskie, K.G., Kamat, P.V.: Chem. Mater. **26**, 7221 (2014)
134. Chen, C., Li, C., Li, F., Wu, F., Tan, F., Zhai, Y., Zhang, W.: Nanoscale Res. Lett. **9** (2014)
135. Lv, M., Zhu, J., Huang, Y., Li, Y., Shao, Z., Xu, Y., Dai, S., Appl, A.C.S.: Mater. Interfaces **7**, 17482 (2015)
136. Brik, M.G., Ma, C.G.: J. Phys. D. Appl. Phys. **46**, (2013)
137. Vahidshad, Y., Nawaz Tahir, M., Iraji Zad, A., Mirkazemi, S.M., Ghasemzadeh, R., Huesmann, H., Tremel, W.: J. Phys. Chem. C **118**, 24670 (2014)
138. Wu, P., Zhao, T., Wang, S., Hou, X.: Nanoscale **6**, 43 (2014)
139. Isarov, A.V., Chrysochoos, J.: Langmuir **13**, 3142 (1997)
140. Sapsford, K.E., Berti, L., Medintz, I.L.: Angew. Chemie Int. Ed. **45**, 4562 (2006)
141. Zhang, S., Zhong, H., Ding, C.: Anal. Chem. **80**, 7206 (2008)
142. Chaubey, A., Malhotra, B.D.: Biosens. Bioelectron. **17**, 441 (2002)
143. Belle, C., Rammal, W., Pierre, J.L.: J. Inorg. Biochem. **99**, 1929 (2005)
144. Masarwa, M., Cohen, H., Meyerstein, D., Hickman, D.L., Bakac, A., Espenson, J.H.: J. Am. Chem. Soc. **110**, 4293 (1988)
145. Lee, H., Yoon, S.W., Kim, E.J., Park, J.: Nano Lett. **7**, 778 (2007)
146. Tolansky, S., Forester, G.O.: Proc. Phys. Soc. **50**, 826 (1938)
147. Liu, L., Hu, R., Law, W.C., Roy, I., Zhu, J., Ye, L., Hu, S., Zhang, X., Yong, K.T.: Analyst **138**, 6144 (2013)
148. Wang, J.J., Wang, Y.Q., Cao, F.F., Guo, Y.G., Wan, L.J.: J. Am. Chem. Soc. **132**, 12218 (2010)
149. Coronado, J., Fresno, F., Hernandez-Alonso, M.D., Portela, R.: Design of Advanced Photocatalytic Materials for Energy and Environmental Applications (n.d.)
150. Hernandez-Ramirez, A.: Photocatalytic Semiconductors. Springer International Pu (2016)
151. Roucoux, A., Schulz, J., Patin, H.: Chem. Rev. **102**, 3757 (2002)
152. Narayanan, R., El-Sayed, M.A.: J. Phys. Chem. B **109**, 12663 (2005)
153. Kobayakawa, K., Teranishi, A., Tsurumaki, T., Sato, Y., Fujishima, A.: Electrochim. Acta **37**, 465 (1992)
154. Li, T.L., Da Cai, C., Yeh, T.F., Teng, H.: J. Alloys Compd. **550**, 326 (2013)
155. Armand, M.: **414**, 359 (2001)

156. Yamakawa, N., Jiang, M., Grey, C.P.: Chem. Mater. **21**, 3162 (2009)
157. Etienne, A.: J. Electrochem. Soc. **117**, 870 (1970)
158. Qian, L., Tian, X., Yang, L., Mao, J., Yuan, H., Xiao, D.: RSC Adv. **3**, 1703 (2013)
159. Dampier, F.W., Dampier, F.W.: J. Electrochem. Soc. **128**, 2501 (1981)
160. Howard, M.D., Jay, M., Dziubla, T.D., Lu, X.: J. Biomed. Nanotechnol. **4**, 133 (2008)
161. Zha, Z., Zhang, S., Deng, Z., Li, Y., Li, C., Dai, Z.: Chem. Commun. **49**, 3455 (2013)
162. Vogel, A., Venugopalan, V.: Chem. Rev. **103**, 577 (2003)
163. Weissleder, R.: Nat. Biotechnol. **19**, 316 (2001)

Quantum Dots Synthesis and Application

Jaison Jeevanandam, Satheesh Kumar Balu, Swetha Andra, Michael K. Danquah, Manisha Vidyavathi, and Murugesan Muthalagu

Abstract Zero dimensional nanostructures that are electronically confined in all directions are called quantum dots. These nanosized dots are usually crystallized semiconductor with enhanced properties of fluorescence. Unlike the one- and two-dimensional nanoparticles such as thin films and rods, quantum dots can be fabricated from a widespread range of elements such as metals, metal complexes, carbon, and rare earth elements. Quantum dots can be used as molecular carriers without any loss of energy whilst enhancing the properties and functional characteristics of the foreign molecule. Thus, quantum dots are utilized in applications including electronics, delivery of drugs, solar panels, medical imaging and waste treatment. The present chapter discusses various approaches for synthesizing quantum dots and their associated physical and chemical properties. In addition, nanocomposites are discussed along with their recent applications in various fields as they are synthesized by blending different materials with quantum dots.

Keywords Quantum dots · Nanocomposites · Nanoparticles · Electron confinement · Zero-dimension

---

J. Jeevanandam
CQM - Centro de Química da Madeira, MMRG, Universidade da Madeira, Campus da Penteada, Funchal 9020-105, Portugal

S. K. Balu · M. Vidyavathi
Department of Ceramic Technology, Anna University, Chennai 600025, Tamil Nadu, India

S. Andra · M. Muthalagu
Department of Textile Technology, Anna University, Chennai 600025, Tamil Nadu, India

M. K. Danquah (✉)
Chemical Engineering Department, University of Tennessee, Chattanooga, TN 37403, USA
e-mail: michael-danquah@utc.edu

N. M. Mubarak et al. (eds.), *Contemporary Nanomaterials in Material Engineering Applications*, Engineering Materials,
https://doi.org/10.1007/978-3-030-62761-4_9

1 Introduction

Nanoparticles are generally classified into three types based on their dimensionality namely; one-dimensional (example: nanosized tubes of carbon), two-dimensional (example: graphene layers) and the three-dimensional (example: nanocrystals) particles. The dimensional classification of nanoparticles is based on the confinement of electrons in directional axis [74]. For instance, carbon nanotubes are formed by confining electrons in y, z-axis and allowing them to transport only in the x-axis which is nanometer in diameter [157]. Quantum dots are introduced as a novel type of nanoparticles, which can confine electrons in each of the three dimensions and are named as nanostructures with zero dimensions [209]. Quantum dots are generally below 10 nm in size that can exhibit quantum effects [159]. Quantum effects include inelastic electron tunneling [56], quantum tunneling [212], Fermi-Dirac distribution [136], quantum gravitational [230] and non-local effects [162]. It is noteworthy that the synthesis approaches of quantum dots play a crucial role in deciding their properties [76]. Quantum dots are widely synthesized via physical [4], chemical [181] and biogenic approaches [155]. Chemical approaches are extensively used for quantum dot fabrication to be beneficial in electrical and electronic purposes [134]. However, these approaches possess drawbacks such as ability to trigger toxic reactions in humans and the environment, which hinder their potential to be used in biological applications [132]. Thus, biogenic approaches to synthesize quantum dots which possess enhanced biocompatibility, bioavailability and less toxicity were introduced [217]. Mostly, these two approaches are used to fabricate quantum dots and physical approaches are used to coat them over nano/micro/bulk substrate [135, 191]. Certain quantum dots were also synthesized using physical approaches however, they lack stability, specific quantum effects and involves high cost during synthesis [184]. The selection of perfect synthesis approach [16] and strict characterization methods [156] is essential in the fabrication of quantum dots for desired applications. Quantum dots gained much attention towards research due to their widespread applications, ranging from electronics [85] to photo catalysts [182] and from biomedical [180], pharmaceutical [79] to carbon sequestration [57]. Thus, the present chapter discusses various approaches for synthesizing quantum dots and their associated physical and chemical properties. Subsequently, nanocomposites which are synthesized by blending different materials with quantum dots are also discussed along with their recent applications in material engineering and various other fields.

2 Synthesis Approaches

The nanosized dots in quantum size and semiconductor properties have gained incredible attention in photonics due to their inimitable opto-electronic properties that are significantly different from bulk substances [22]. By tuning their size, the quantum dots have enhanced effects in conventional solar cells, lights with solid-state characteristics, and labeling applications in biology [66]. This means by using various synthesis approaches, it is possible to obtain quantum dots with enhanced absorption and emission spectra [5]. Thus, the synthesis of quantum dots can be achieved by using three core strategies; biological, chemical and physical methods.

2.1 *Physical Approaches*

Various synthetic approaches have been used for the fabrication of nanosized dots based on the involvement of materials and size range requirements. The physical approach usually involves growth along with nucleation of the particles in the Vapor phase [22]. Synthesis through physical approaches can hold various advantages such as; simple method, maintenance of desired pressure, temperature, deposition rate, uniform deposition of the material on the substrate and materials growth in various forms such as powder and thin films [7].

2.1.1 Vapor Deposition

The development of quantum dots through the vapor phase methods usually involves the deposition of quantum dots in an atom-by-atom process [5, 88]. One of such methods is the molecular beam epitaxy (MBE), where the thin films of PbTe/CdTe are grown epitaxial at high or ultra-high vacuum with the rate of deposition, characterized below 3,000 nm/h [20]. The dot size is successfully reduced by altering the thickness of PbTe epilayers on the matrix of CdTe through MBE. The well-organized reduction of quantum PbTe/CdTe dot-size was determined by growing layers of PbTe with the thickness of 1–10 nm on the 50 nm thick CdTe layers and annealed at 340 °C for 10 min. The variation in the layer thickness of the PbTe results in distinct sizes and shapes, which is confirmed through across sectional, dark-field TEM analysis. Thus, the control over the size allows quantum dot luminescence tuning over a range of 2.2–3.7 μm spectra. This study demonstrates emission in the ultra-broadband region from quantum dots with multilayers, which can be used in super luminescent diodes [55]. Among vapor phase methods, physical vapor deposition is the most common technique for quantum dot fabrication which includes sputtering, laser ablation and thermal vapor deposition.

Sputtering is the technique in which the target surface is bombarded with a gaseous ion that is applied between target and substrate to eject the atoms or molecules from the target to be deposited on a substrate in the vacuum chamber under high voltage acceleration [14]. Recently, Dahi et al. [37] synthesized CdSe quantum dots through radio frequency magnetron sputtering. The size and shape of QDs are controlled and developed at target to substrate distance of 13.5 cm with 2.2×10^{-1} mbar of chamber pressure. The mean diameter of CdSe quantum dots is 7.2 nm which was obtained by the radio frequency power of 14 W and 7.5 min time of deposition. It is noteworthy that there is a quantum CdSe dot size increment with an increase in the RF power and deposition rate [37]. Likewise, the surface material is removed via laser irradiation in laser ablation technique. The high laser beams power hit the target surface, which rapidly increases the temperature of the absorbing material and converts them into plasma [90]. Moreover, Horoz et al. [65] synthesized quantum CdSe dots via laser ablation technique in aqueous medium and studied their applications as photovoltaics. During the synthesis, sufficient quantity of CdSe QDs was collected after running the laser ablation for a long duration. Later, QDs formation was indicated by the color change of the solution. Wurtzite CdSe quantum dot structure with an average of 5 nm size was obtained, which is confirmed by XRD and TEM analysis. The CdSe QDs were integrated with ZnO nanowires without using any ligands to fabricate the solar cells with quantum sensitization. The nanowires of ZnO were developed by chemical vapor deposition (CVD) approach using the FTO as a substrate and the formation of surface contaminations was avoided by annealing and altering growth conditions. The obtained CdSe QDs were free of ligands that reduce the blockade of transferring and extracting the electron between QDs and ZnO nanowire and potentially improved the solar cell efficiency with quantum sensitization. Additionally, the performance of the cell with 0.43 mA/cm^2 as current density and 0.48 V as voltage was obtained, which recommends that the material possess great potential in photovoltaic applications [65]. Another novel technique for quantum dot fabrication is the thermal evaporation deposition, which involves the heating of solid materials in a vacuum chamber. The material gets vaporized and hits the substrate. Niobium pentoxide (Nb_2O_5) quantum dots were already reported to be synthesized through thermal evaporation technique. It was described by Dhawan et al. [43] that Nb_2O_5 quantum dots of distinct sized were obtained via a thermal evaporation method by varying the deposition rate duration [43].

Another significant technique is the chemical vapor deposition, which contrasts with the other techniques. In this approach, the growth of semiconductor material is mediated by chemical reactions and not by physical deposition. It was demonstrated by Fan et al. [49] that quantum graphene dots are developed on copper substrate by atmospheric pressure CVD (CGQds). A uniform size distributed CGQDs of 5–15 nm was produced by altering parameters such as growth time, growth cycle and the methane flow rate [49]. In another report, Kumar et al. [100, 101] used chitosan for nitrogen doped dots of graphene fabrication (N-GQDs) via atmospheric pressure CVD. In this experiment, the chitosan serves as carbon material for the development of graphene dots and the nitrogen gets doped in the quantum nanosized dots. The fabricated N-GQDs were in the average diameter of 10–15 nm [101]. Even though various physical techniques have been successfully used for quantum dots fabrication with size tunability, there are some intrinsic disadvantages such as the usage of costly instruments and tediousness associated with the separation from the substrate [138].

2.2 Chemical Approaches

Chemical synthetic approaches have been extensively used as an alternate to physical approaches for quantum dots preparation with desirable size, colloidal nature and properties by tuning the precursor concentration, reaction time, use of various solvents and temperature [9, 88]. The formation of quantum dots through chemical approach follows two unique steps namely; nucleation and growth [176]. Li et al. [110, 113, 115] studied the influence of solvent reaction on temperature and polarity on the growth along with nucleation of the quantum dots using ether as solvent with different polarity. It is exciting to observe that the quantum dot size decreased with increase in solvent polarity, whereas, increasing the reaction temperature has led to an increment in the quantum size of the dots [110]. In this chemical approach, the colloidal and water-soluble quantum dots are synthesized using certain exclusive methods such as a hot-injection method, non-injection method, solvothermal and template assisted synthesis methods.

The injection with heat technique comprises of cold precursor in addition to hot metal-organic precursors, typically at the temperature of 300 °C, which primes the instantaneous CdSe nuclei formation and size reduction of CdSe nanocrystals by increasing the growth temperature [40]. In recent times, Osman et al. [133] developed a highly luminescent CdSe quantum dots through the hot injection method in a single stage. The synthesis yielded six samples with six distinct colors such as orange, green, yellow, blue, red and purple with the spectral range of 445–643 nm, by altering the molar ratios of cadmium to zinc and selenium to sulfur [133]. Likewise, Mahajan et al. [122] synthesized CdSe semiconductor crystal using injection with heat method, in which the gap of nanocrystal band is reduced by the steady integration of sulphur into CdSe [122]. However, this method possesses limitations such as the requirement of operating in harsh conditions, which hinders the large-scale production.

The non-injection method was introduced to overcome the limitation of hot injection method, where the reagents are loaded at optimum temperature and refluxed with heat for initiating growth along with nucleation of nanocrystals, which is a capable large scale production method [214]. Moreover, Chen et al. [29] has developed quantum perovskite cesium-lead halide dots through non-injection method. Size tunable quantum nanosized dots are attained via alteration of reaction temperature and also the emission spectra positions that are conveniently tuned in the ultraviolet-near infrared region (360–700 nm) [29]. Moreover, the solvothermal synthesis method is simple, cheap, nontoxic and environment friendly to produce quantum dots. This method involves heating of precursor with solvents in a impenetrable autoclave for maintaining an elevated temperature and pressure to attain enhanced nanocrystal [198]. Zaho et al. [219] developed amphiphilic, 2–5 nm sized carbon dots through solvothermal method. The viability of the prepared carbon dots was tested against HeLa cells up to 500 mg/ml and the viability of the cells was noted to be up to 85%, which indicates the biocompatibility of the material [219]. Similarly, quantum graphene dots (GQDs) with two different series such as GQDs with the same size and different oxidation state as well as GQDs with different size and similar surface chemistry for tuning the photoluminescence was reported by Qi et al. [139] via solvothermal method. The electrochemiluminescence (ECL) was adopted to study the photoluminescence (PL) mechanism and surface chemistry of the two different series of GQDs. This method provides exclusive materials and the opportunity to investigate the PL, with respect to the surface chemistry and the altering sizes of yielded quantum nanosized dots. The results of the study revealed that the larger size and higher oxidation in the surface led to the GQDs shift of PL towards red. The ECL technique also showed the evidences of conjugate GQDs structures, which has negligible surface defects and are responsible for the exclusive PL phenomenon [139].

Other than the above mentioned methods, template assisted synthesis method is one of such promising methods, which is convenient for controlling the morphology of quantum dots with ease [116]. This method comprises of nuclei growth that invariably nucleates at the holes and the defect of the electrode substrate. Templates such as porous polymer membranes and anodic aluminum oxides (AAO) have been widely utilized to prepare semiconductor nanocrystals. The AAO template has several advantages in producing quantum dots with uniform diameter [15, 72]. CdSe/ZnS quantum dots embedded polymer composites were developed by Weaver et al. [190] via template assisted synthesis. The surface quantum nanosized dots properties were tailored by encapsulating the quantum dots in various commercially available polymers. It was found that encapsulation of quantum dots in polymers resulted in either loss or gain in their optical emission, which is due to the presence of octadecyl amine in the solution and also resulted in gelation [190]. Another experimental work by Kim et al. [91] demonstrated the synthesis of nanoporous CdSe quantum dot masks to yield a regular nanostructured film on the graphene base plane. The thin silica film was incorporated on the graphene surface as the template of nanoporous mask. The hexagonal nanostructure array was obtained with the dot size of 10 nm. This indicates that the utilization of graphene with a nickel layer and nanoporous silica mask are required to obtain CdSe quantum dots in an array structure on the base graphene plane, which leads to the mechanical and electrical property enhancement [91]. In both physical and chemical approaches, concentration of precursor and reducing, stabilizing or physical energy is the crucial factor, which determines the size of quantum dots, and eventually decides their properties. Further, chemical methods also possess certain limitations, similar to physical synthesis approach, especially in terms of using toxic precursor and stabilizing agents to fabricate quantum dots, which are not advisable to be used in biomedical applications [77].

2.3 *Biological Synthesis*

Quantum dots have wide range of applications in photocatalysis, energy conversion, optoelectronic devices and chemical sensors. Moreover, they have great potential in biomedical field including; delivery of drugs, bioimaging, biomarkers and biosensors [154, 183]. It is expected that quantum dots should have properties such as high stability, less toxicity, biocompatibility, bioavailability and bioactivity, due to their enhanced biomedical applications [68]. However, traditional synthesis methods such as physical and chemical approaches often result in usage of costly equipment, hazardous chemicals which are not cost effective and environment friendly. Thus, biosynthesis methods for quantum dots fabrication are introduced to avoid the challenges of conventional synthesis methods [18]. The synthesis of quantum dots through biological methods involve; plants, bacteria, fungi, algae and virus. In recent times, these methods have proven to be highly beneficial and grabbed the attention of researchers for the scaled-up quantum nanosized dot production.

2.3.1 Plant Synthesis

Plants show a vital character in quantum nanodots production, compared to any other synthesis methods. Plants are widely used in quantum dot synthesis due to their high availability, safe to handle and the presence of phytochemicals including alkaloids, phenols, terpenoids, quinines, flavonoids and tannins which acts as stabilizing and reducing agent to facilitate the formation of zero dimensional particles [3]. A recent study testified the fabrication of green quantum carbon dots via sugarcane bagasse pulp. The results showed that the average particle size of the green carbon dots was 4.1 ± 0.17 nm [169]. In another study, fluorescent nitrogen doped dots of carbon (N-CDs) was fabricated via green chemistry using fruit extract of *Actinidia deliciosa* as a source of carbon and ammonium as a dopant of nitrogen. The obtained spherical N-CDs emitted fluorescence in blue color with 3.59 nm as an average particle size. These N-CDs were tested for its cytotoxicity against L-929 and activity of anticancer against breast cancer cell lines (MCF-7). The results demonstrate that the 'as prepared' green synthesized N-CDs shows high biocompatibility in L-929 cell lines with the cell viability of <90% and the cell viability of MCF-7 was <80%. The photocatalytic activity against Rhodamine B dye using N-CDs as a nanocatalyst was also studied in the presence of $NaBH_4$. The results without adding N-Cds demonstrate that in the presence of NaBH4, the RhB was decreased to a minimum level after 2 h, without any catalyst. In the case of N-CDs, the intensity of RhB gradually decreased and the reaction was completed in 10 min, which indicates the high catalytic activity of N-CDs [6]. In another report, Vandarkuzhali et al. [177] synthesized green fluorescent, 2.5 nm sized carbon quantum dots using a pseudo stem of the banana plant through the hydrothermal method. The study showcased that the quantum dot exhibits low cytotoxicity against HeLa cells with the cell viability of about 95% at 0.5 mg/ml of concentration. It is noteworthy that even at 1 mg/ml of concentration, the dots showed 85% of viability. These quantum dots were also used as a potential materials for bioimaging and for the detection of Fe^{3+} [177]. Likewise, Borovaya et al. [21] described a novel green fabrication technique for CdS quantum dot fabrication of size 5–7 nm using hairy *Linaria maroccana L.* root culture [21]. These studies show that the plant mediated synthesis methods for quantum dots are environment friendly with less toxicity towards healthy human cells, which have promising benefits in cell biology, bio imaging and drug delivery. However, the stability of synthesized quantum dots in culture medium and body fluids remains as a limitation, which has led to hybrid methods of incorporating microwave or ultrasound. In future studies, it will be possible to overcome the stability issue while fabricating quantum dots via plant extracts using these hybrid synthesis approaches.

2.3.2 Microbial Synthesis

Quantum dot synthesis via biological entities could be another promising method for producing clean, environment friendly and non-toxic zero-dimensional particles. This method has emerged as an alternative to plant mediated quantum dot fabrication approach for elevating their stability [158]. Prokaryotic microbes have advantages of growing at a faster rate, which has potential in yielding less or non-toxic quantum dots to overcome the issues of toxicity in conventional methods. In this regard, Bao et al. [12] recently developed biocompatible CdTe quantum dots through *Escherichia coli.* The quantum dots were also synthesized using conventional hydrothermal methods and tested against PC 12 cells to analyze their cytotoxicity. The results indicated that the hydrothermal synthesized quantum nanosized dots at >500 nM concentration has shown significant toxicity towards PC 12 cells, whereas, the *E. coli* synthesized quantum dots have upheld viability of cells to 92.9%, even at a concentration of 2 μM. This shows that the *E. coli* derived quantum dots possess better biocompatibility, compared to hydrothermally synthesized quantum dots [12]. Likewise, Bruna et al. [25] synthesized fluorescent CdS by *Halobacillus* sp. DS2 with high NaCl resistance. The synthesized quantum nanosized dots with a range of sizes (2–5 nm) and enhanced stability was tested against high NaCl concentration, and the findings illustrated that the *Halobacillus* synthesized quantum nanosized dots uphold their fluorescence after 4% NaCl exposure [25].

Other than bacteria, fungi, yeast and algae such as *Pleurotus ostreatus*, *Saccharomyces cerevisiae* and *Candida albicans* are also capable of producing quantum dots. It was reported that quantum CdS dots were fabricated by incubating the $CdSO_4$ with the fungus *Pleurotus ostreatus* at 26 °C for 10 days. The produced quantum CdS dots was in 4–5 nm size with wurtzite crystalline structure [19]. Likewise, 30–40 nm sized, ZnS quantum dots in sphalerite phase were synthesized by optimizing the concentration of $ZnSO_4$ and *Saccharomyces cerevisiae* yeast strain at different time periods. This study shows that the yeasts are also a potential source for the bio production of quantum dots [123]. Furthermore, microalgal biomass possess significant characteristics which comprises of efficient photo catalysts, high ability to capture carbon and for food production [57]. These microalgal extracts can serve as a source of carbon for quantum carbon dot fabrication, due to its worldwide availability. The 8 nm sized quantum carbon dots were produced via eutrophic algal blooms. The cytotoxicity screening was carried out against MCF-7 cells, in which 60% viability was observed, even at high dosage (1 mg/ml). In addition, the in vitro bio imaging ability of CDs were also investigated using confocal images by treating 0.2 mg/ml of CDs with breast cancer MCF-7 cell lines. A durable green photoluminescence on the membrane of cells was observed which indicates that the CDs possess cell penetration ability and exhibited fluorescent property enhancement capability in the cell environment [145]. The precursor concentration, extract volume, reaction time and temperature are the essential parameters, that must be optimized during biosynthesis process to yield size reduced quantum dots [24, 163]. In summary, the biosynthesized quantum dots are highly biocompatible, efficient in cellular uptake and have great potential in high contrast bio imaging applications. However, stability, long duration of microbial culture and the difficulty in the optimization of synthesis mechanisms are the drawbacks which hinder the biogenic approach for scaled-up quantum dots production.

3 Nanocomposites of Quantum Dots

Composites are a combination of two or more different materials that are blended to produce drastically enhanced electrical, optical and mechanical strength properties that are not possible to be attained via individual materials. The nanocomposites are novel composites, in which at least one of the component must be in the nano-size regime [75, 125]. Nanocomposites of quantum dots can be composited with polymers, metal/metal oxides, carbon materials and biomolecules to enhance their unique properties for desired applications such as non-linear optical devices, computing and electro-optical devices.

3.1 Polymer—Quantum Dot Composites

Quantum dots are found to possess numerous applicational significance in biosensing, bioanalysis, image probing, multicolor imaging, viral capsids, lasers, photovoltaic cells and diodes with light emitting capability [190]. Due to charge mobility, indissolubility and photoluminescence instability of quantum dots, the passivation process are needed to be modulated for increasing the luminescence efficiency of the device [146]. Such modifications can be achieved by encapsulating quantum dots in polymers with superior charge mobility and high stability to improve their photoluminescence. It is noteworthy that encapsulating QDs with certain polymers may result in either gain or loss in their photoluminescence emission [190]. Also, Bobrovsky et al. [17] demonstrated a novel nanocomposite of CdSe/ZnS quantum dot and liquid crystal with nanoporous polypropylene polymer at different spin ratios. The polypropylene served as a polymer matrix and quantum CdSe/ZnS dots as an inorganic fluorescent component. The results demonstrated that the filling of liquid crystals provided a low light scattering ability and high optical transparency of the composites. The spin draw ration of the polypropylene had strong optical effects towards composite films. Hence, introduction of quantum dots into polypropylene with a low spin draw ratio has been maintained by a noticeable dichroism decline of the resultant liquid crystal composite film [17]. Likewise, polymer-quantum dot nanocomposites of highly ultra-stable and luminescent quantum ZnS/CdSe@ZnS, ZnS/CdS and CdS/CdSe dots with polydimethylsiloxane (PDMS) was developed by Kong et al. [95] via in situ hydrosilation based strategy of manipulating the surface. Solvent dosage effect, time required for purifying QDs, curing temperature, concentration of quantum dots in PDMS and oxidation processes were analyzed for the optimization of properties. The quantum dots emitted blue, green and red fluorescence lights. The resistance of quantum dots on various solvents namely deionized water, acetone and ethanol, was examined by immersing the composites into it. After 2 h of treatment with ultrasonic waves, the intensities of blue, green and red fluorescence, were diminished by 0.12, 0.1 and 0.06, correspondingly, which indicates that the fluorescent material possess enhanced solvent resistance property. Finally, the quantum yield of 82.03% was obtained by exploring the solvent dosage, curing temperature, purification time and concentration which proved that the composite material has excellent solvent resistance with high stability [95].

3.2 Carbon Material—Quantum Dot Composites

Quantum carbon dots have fascinated greater consideration in recent years due to their size dependent photoluminescence, exceptional up conversion ability, inertness, stability towards photobleaching, aqueous solubility and low cytotoxicity [205]. These advantages have made them a widely used choice in bioimaging, optical sensing, photocatalysis and fast energy conversation [204]. Lim et al. [118] fabricated quantum carbon dots using citric acid and ethylenediamine (EDA), and investigated their optical, electrical and structural properties. The obtained CQDs exhibited cyan color under ultraviolet irradiation at 365 nm with the size range of 10 nm. Moreover, the electron extraction characteristics were investigated by fabricating CQDs as an interfacial layer to act as photovoltaic cells. It can be noted that the CQDs possess enhanced charge transport properties due to the addition of EDA and citric acid [118]. In recent years, quantum dots are combined with carbon materials have gained attention due to their remarkable optical, photovoltaic and enhanced photocatalytic properties. Wang et al. [188] synthesized a series of metal (zinc, cobalt, bismuth, cadmium, titanium) doped carbon quantum dots (CDs) via pyrolysis to evaluate their photocatalytic properties. The CDs with metal dopants were further joint with co-catalyst such as nanowires of CdS for photocatalyst mediated production of hydrogen. The composite of CdS/CDs with dopants namely titanium, cadmium and bismuth, showed improved production of hydrogen than composites of undoped CdS/CDs. The hydrogen evaluation of composites formed by CdS/CDs with bismuth as dopant is 4.2 times and slightly higher than individual CdS. This property of quantum dot can be attributed to the outstanding electron transfer property which was achieved due to the nanocomposite formation of two materials [188]. In another study, CdSe quantum dots are combined with nanotubes of carbon having multi-walls (MWCNT) and CdS with graphene by Chen et al. [28] via hydrothermal approach. The 'as prepared' composites were evaluated for its photocatalytic activity against methylene blue dye under irradiation of visible light. The graphene-CdS and MWCNT-CdSe composite showed excellent photocatalytic activity than pure CdSe [28]. Hence, the combination of carbon material with quantum dots will provide new insights in future for quantum carbon dot nanocomposite preparation with exclusive properties.

3.3 *Metal—Quantum Dot Composites*

The combination of metal/metal oxide nanoparticles with quantum dots is vital for fabricating solar cells for next generation, diodes light emitting capability and arrays of nanostructured electronics. The quantum nanosized dots with semiconductor properties including CdS, CdSe, InAs, PbS and organic-inorganic hybrid perovskite have been utilized in solar cells formed of dye-sensitization as a photosensitizer due to their adjustable band gap depending on the QDs size, stable chemistry, improved optical absorption, coefficient of extinction with ability to form numerous excitons [171]. Leschkies et al. [109] combined CdSe semiconductor quantum dots with ZnO nanowire to develop an innovative solar cell by sensitizing quantum nanosized dots. The quantum cadmium selenide dots were mercaptopropionic acid encapsulated and were fixed over nanowires of ZnO surface. The injections of electrons were carried out across quantum dot-ZnO nanowire nanocomposite interface via electronic state overlaps of ZnO conduction band and nanosized dots. These solar cells with dye-sensitized nanowire in quantum regime displayed 1–2 mA/Cm^2 of short circuit and 100mW/cm^2 of open circuit after illumination and simulated by AM 1.5 spectrums. Moreover, the fabricated composite has exhibited 50–60% higher quantum efficiencies at core [109]. Likewise, Kulakovich et al. [99] has developed quantum ZnS/CdSe dots on gold colloids for enhancing their luminescent property. Quantum ZnS/CdSe dots and gold colloids were deposited with well-defined spacer layers using polyelectrolyte deposition in layers to observe enrichment related to distance and quantum dot quenching of photoluminescence. The core-shell nanocomposites of ZnS/CdSe were fabricated via reactions at elevated temperatures in the mixture of TOPO/HAD. The obtained quantum dots were 8 nm in size and colloids of gold with a diameter of 12–15 nm was fabricated via $HAuCl_4$ reduced by citrate approach. The combination of gold colloid with quantum CdSe/ZnS dots resulted in further effectiveness of QDs excitation property, compared to the quantum dots without gold colloids. The efficient excitation of quantum dot-gold colloidal nanocomposites was proved by surface plasmon resonance observation in the quantum CdSe/ZnS dot spectrum of photoluminescence excitation [99]. Moreover, Galyomedinov et al. [51] also studied the quantum CdSe dot luminescence in the PMMA matrix that are deposited on the ion implanted silver nanoparticles containing sapphire substrate. The results showed that the photoluminescence enhancement in the quantum dots is due to their excitation in the spectral plasmon absorption band region of silver nanoparticles. Also, luminescence quenching may also occur in a number of quantum dots, when they are in contact with silver nanoparticles [51].

3.4 Novel Quantum Dot Nanocomposites

Quantum dots have distinct advantages in biological field as a biosensor, bio markers and image probing [31, 82]. Currently, heavy hybrid quantum dot composites are restricted in biological applications due their potential toxicity that are identified via in vivo studies. The limitations of toxicity can be avoided by bioconjugation of quantum nanosized dots using biorecognition elements such as aptamers and nucleic acids through covalent and non-covalent can be considered [126]. He et al. [64] functionalised $Zn_xHg_{1-x}Se$ quantum dots with proteins via one spot single step reaction at room temperature. Initially, the 'unfunctionalized' $Zn_xHg_{1-x}Se$ quantum dots were prepared using MPA as ligands. A maximum quantum yield of 25.6% was obtained at 10% molar ratio with 4 ± 0.6 nm of size with extreme emission spectra at 704 nm. The 'as prepared' quantum dots exhibited bright red lights near the infrared photoluminescence that are highly beneficial for bioimaging applications. The bioconjugated quantum dots were prepared by using various protein molecules such as BSA, Lysozyme, Trypsin, Haemoglobin and Transferrin. The bioimaging ability of the 'as prepared' protein functionalized $Zn_xHg_{1-x}Se$ quantum dots were compared with the traditional CdSe QDs. The results revealed that the CdSe QDs were photo bleached swiftly after 30 min of excitation, while the quantum $Zn_xHg_{1-x}Se$ dot photoluminescence was retained even after the excitation for 60 min. These results indicate that the bioconjugation of protein-quantum dot has led to high quantum yield, high colloidal photostability and high functionalization efficiency [64]. Likewise, Tikhomirov et al. [172] developed a novel, self-assembled quantum cadmium telluride dots with specific sequences of DNA along with mercaptoacetic acid (MPA). Guanine oligomers were used for phosphorothioate sequence anchoring to attach with the ligand of DNA. The results showed that the G-oligomers with short sequence produced quantum dots of smaller size with yield of improved emission, compared to longer G-oligomers. This indicates that the 5–20 range of nucleotides were suitable for quantum dot binding efficiencies. The absorption and emission spectra were observed on quantum dots growth in the presence of oligonucleotides. Moreover, X-ray photoelectron spectroscopy analysis revealed the binding at surface and quantum nanosized dot composition. These results established that MPA bonded to quantum dot via thiol moiety by replacing telluride, which is ascribed to the passivation of CdS-like shell upgradation. In addition, the photophysical properties were investigated using the red, orange and green quantum dots. The red dot with smaller band gap has symmetric dual to it. Furthermore, the orange colour emitting dots were altered to have irregular valency in the centre, so that two orange dots would bind dual valency to attach with a red and green dot site. As a result, the contribution of red dot to the total absorption spectra was identical to one-fifth and the green and orange was two-fifth. Hence, this work demonstrated that the tuneable quantum dots assemblies via bonding materials such as DNA can be used in nanoscale optoelectronic applications [172]. In general, all these nanocomposites are synthesized to overcome the limitations of the conventional quantum dots by blending two materials. Thus, nanocomposites are extensively used in recent times for improving the properties of quantum nanosized dots and will replace simple quantum dots for desired applications in future.

4 Physicochemical Properties of Quantum Dots

The physiochemical entities of the quantum dots are defined by their shape, structure, composition and size [58], which eventually modifies their crystallinity and electronic configurations as illustrated in Fig. 1. The wide application of quantum nanosized dots, including biomedicine, LEDs, solar cells, sensing applications and lasers are due to their exclusive properties [161]. Recent studies illustrated that each QDs contain exclusive optical, electrical, thermal and biological properties. In fact, the same quantum dot material with different sizes showed distinct variations in their properties [207].

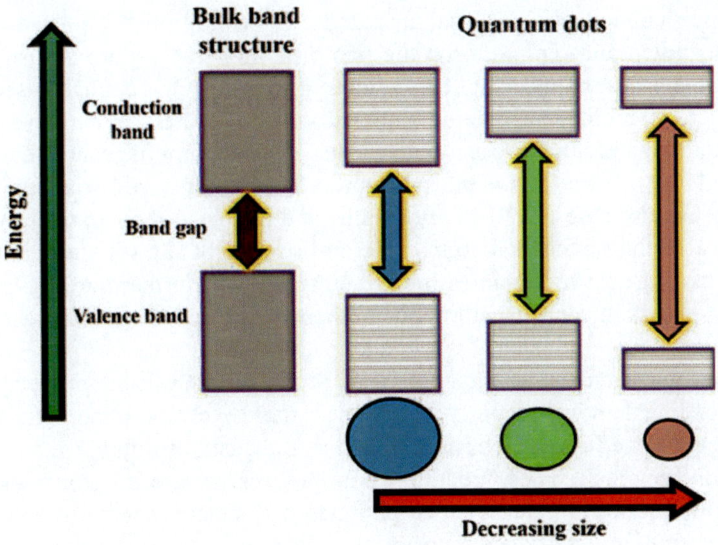

Fig. 1 Electronic configuration and band gaps in bulk materials and quantum dots

4.1 Optical Properties

The effects due to confinement of electrons in quantum regime determines the optical QDs properties and it makes them an ideal material for bioimaging, LEDs, and solar cell applications [153, 206]. Most of the QDs possess the ability of emitting light upon excited by energy sources such as light and electricity. Specifically, the color which is emitted by QDs is determined by several parameters such as their size, morphology, and structure [150]. Generally, when the electrons are excited from the electronic bands, formation of hole occurs and when the electron recombines with the valence band, it releases photons as energy and the phenomenon is known as fluorescence. It is possible to create a continuous valance band and conduction band via quantum dots, unlike bulk semiconductor materials [73, 164]. Additionally, 2–3 nm sized QDs emits shorter green, violet and blue wavelengths, and 5–6 nm sized QDs emits longer yellow, orange and red wavelength of colors. Hence, the fluorescence property of QDs is tunable by changing their size [220]. The ultimate brightness and stability of the quantum nanosized dots makes them an unavoidable entity in bioimaging and targeting applications. Li et al. [114] synthesized polyethylene glycol (PEG) coated quantum dots and used them to achieve targeted live animal images in mice. The visibility of QDs was detected in their bone marrow and lymph nodes, even after a few months due to the stability of the QDs [114, 128]. Recently, the phosphorescence property of carbon quantum dots was identified at room temperature by dispersing them into polyvinyl alcohol (PVA) matrix under UV condition [187]. Likewise, Gao et al. [53] studied and reported that the photostability of QDs are thousand times greater compared to organic dyes. This study suggests that QDs could be used in long term cellular and molecular imaging applications [53]. Moreover, carbon QDs were used in photodynamic therapy as photosensitizers to prevent the cancer cell growth (Du145 and PC3). Nitrogen and phosphorus co-doped QDs were also prepared using Adenosine triphosphate (ATP) biomolecules and analyzed their optical properties which revealed their enhanced ability to emit bright light [2]. Fluorescent QDs are often used as drug delivery vehicles due to their small size. Furthermore, QDs are used to achieve success in cancer therapy due to their nontoxic behavior and exceptional PL property.

Quantum confinement of quantum dots is also a significant property, which provides beneficial optical property alterations to these exclusive nanoparticles. Dong et al. [46] synthesized cesium lead halide nanocrystal quantum dots via a size reducing precipitation approach. These quantum dots possess precise control towards size with superior equilibrium and yielded particles with heterogenous free broadening to exhibit exciton luminescence. Further, these quantum dot in the size below 7 nm exhibited superior quantum confinement mediated luminescent optical property [46]. Likewise, Zhang et al. [210] fabricated lead sulfide-tungsten di-selenide as zero and two-dimensional hybrid quantum dot structure with tunable quantum confinement for ultrafast transfer of charges to be beneficial in optoelectronic devices. The study emphasized that the reduction in size of lead sulfide has supported the transition of weak to strong coupling hybrids, due to effect of quantum confinement [210].

Further, Liu et al. [119] reported that the coarse adjustment of quantum confinement in carbon quantum dots with fine surface trap states can improve their photoluminescence property [119]. All these recent studies showed that the reduction in size of quantum dots to provide quantum confinement will improve their optical property.

4.2 Thermal Properties

QDs are unique materials because their emission band is entirely dependent on temperature. The DC and AC conductivity of CdSe quantum dots are observed to be increasing with an increase in temperature. Due to the outstanding optical property of QDs, their contribution is essential in LED technologies (QLEDs). Recently, the thermal stability of QLEDs was studied at different temperatures and results showed that the QLEDs maintain about 97% of luminescence at 125 °C, while conventional LEDs maintain only 65% [103, 179]. In the electronic systems, QDs can control the transfer of heat and it is significant in fabricating sophisticated electronic devices. Indium arsenide (InAs) QDs was annealed at various temperatures and an elevation in their photoluminescence intensity was observed with a temperature elevation. Moreover, a decrement in the intensity of photoluminescence was observed after 850 °C, which indicates the quality degradation of QDs at high temperatures [8, 200]. Maestro et al. [121] also investigated the thermal sensitivity of CdSe and CdTe QDs for nanothermometer applications and the results suggested that CdTe QDs possess 0.2 °C of thermal sensitivity. This study further demonstrated that CdTe QDs are five times greater than CdSe QDs in exhibiting thermal sensitivity and are highly beneficial in thermal imaging purposes [121]. In biomedicine field, thermal imaging is necessary to understand the dynamics of tissues. For instance, accurate temperature of a single cell is required to recognize the complete mechanism and the cell metabolism. There is no technology available to monitor the temperature differences of a cell, during cell division and cell death. The identification of temperature differences in cell cycle will help in the early detection of diseases, including circulatory problems, cancer, inflammations. Thus, CdTe QDs possess high sensitive temperature sensing ability which is significant for developing early disease detection technologies in the future [59]. Further, Si/Ge quantum dots are also found to be suitable for thermoelectric applications due to their electron transmitting-phonon blocking behavior. Si/Ge superlattices were investigated for their thermal conductivity measurements using 3ω method and the observed average results were 10 W/mK and 3.5 W/mK at 300 and77 K respectively. The significant reduction in conductivity of temperature that was measured suggests that Si/Ge QDs are suitable for thermoelectric applications [1, 13]. In recent times, the thermal conductivity examination was conducted on graphene QDs which was synthesized in different conditions such as supercritical (400 °C), near critical (370 °C) and subcritical (250 °C) temperatures. For QDs synthesized using supercritical and subcritical temperatures, the thermal conductivity was significantly increased up to 60–65%, which is highly recommended for thermal therapies than graphene oxide sheets as they are highly safe to use without inhibiting cells due to high temperature [167]. Furthermore, nanoindentation is a valuable tool to understand the mechanical behavior of QDs. It was reported in a study that the incorporation of CdSe quantum dots has improved the elastic modules and hardness of the polymers [127]. However, there is no other literature to specifically mention that quantum dots can improve the mechanical property of materials as there is no proper characterization method to prove and nanoindentation may lead to errors in characterizing mechanical strength.

4.3 Electrical and Electronic Properties

The electrons and holes in a quantum dot are confined on all three directions, which makes it a unique electrical material. Their energy levels are discrete and comparable with Bhor's radius. The behavior of quantum dot electrons and holes with the surroundings are determined by their electrical property. Most of the studies showed that the electronic structure of QDs is based on their molecular orbital theory. In contrast, carbon quantum dots show a transition at $n \rightarrow \pi^*$ and $\pi \rightarrow \pi^*$, due to their enhanced energies at transition that are easily accessible. Recently, Imran et al. [71] used CdSe and CdTe QDs for energy harvesting applications and they were not able to completely understand their electrical properties. Even though, the experimental success rate is very limited in the energy harvesting field, their results suggested that an increment in atomic numbers may lead to an alteration in the property of the QDs [71]. The flow of electrons in graphene quantum dots is ten times greater than commercially available silicon materials and thus, GQDs are gaining popularity in fabricating next generation devices. Currently, flexible displays are receiving considerable interest among consumers due to their non-breakable property, lightweight, and easy wear-ability. In recent times, quantum dot based LEDs receive several commercial significance, due to their excellent color purity, elevated brightness, and low operating voltage [32]. Xie et al. [197]confirmed that addition of graphene QDs into SnO_2 significantly increases their electronic conductivity [197]. Moreover, QDs based memory devices are under extensive investigations which includes Si/Ge QDs. In quantum computation, QDs are considered as a potential material for the operation of spin qubits [39].

4.4 Other Properties

QD exhibited toxicity towards live cells in in vitro and in vivo conditions as certain QDs are prepared using ions of heavy metals such as cadmium, arsenic, lead and mercury [41]. To date, numerous experiments have been carried out to analyze the toxicity of QDs. Derfus et al. [41] examined the potential toxic effects of QDs and their finding suggested that the CdSe QDs are biocompatible for biological applications. However, they also revealed that the same QDs are highly toxic under UV conditions due to their photolysis ability. Likewise, CdSe QDs possess a tremendous advantage in wearable devices such as body sensors, smart phones, and LEDs. It is noteworthy that Cd^{2+} ions are highly hazardous to human and environment. Specifically, cadmium ions can cause cell lysis by binding with thiol group biomolecules in the mitochondria. A recent study demonstrated the toxic mechanism of Cd^{2+} ions using liver cell. It was reported that when UV radiation or air oxidation occurs in QDs, Cd^{2+} ions are released that acts as a carcinogen [174]. Several studies also suggested that continuous exposure of QDs may damage the lung tissues [193]. Hence, the European Union regulated the usage of cadmium QDs in commercial products. On the other hand, the release of Cd^{2+} ions into the biological medium can be controlled via surface modification such as coating and functionalizing them with biocompatible polymers. An interesting report reveals that approximately two billion QDs can be delivered into the single nucleus without toxicity [52]. In addition, a toxicity analysis study of carbon quantum dots (CDs) suggested that CDs can be used in bio imaging applications as PEGylated CDs and are less toxic to living cells even at higher concentrations up to 28 days. This study emphasize that CDs are safer than other semiconductor quantum dots. Hence, it is possible to decrease the toxicity of the QDs by fictionalization at surfaces using polymers and peptides [61].

Various semiconductor nanostructures were examined for water treatment applications, particularly TiO_2 nanoparticles, due to their low cost, chemical stability, bioavailability and enhanced optical absorption. However, the slow degradation kinetics of TiO_2 limits their application as photocatalyst [160]. It was recently demonstrated that the cadmium oxide QDs possess improved photocatalytic activity in the visible light region [149]. Likewise, organic functionalization is necessary for silicon QDs in photocatalytic applications as they are hydrophobic in nature. Kojima et al. [94] developed a novel silicon QDs without any functionalization process to avoid additional synthesis stage and toxicity [94]. Moreover, Kumar et al. [100, 101] studied the degradation duration of carbon QDs decorated TiO_2 nanoparticles and revealed a significant improvement in their photocatalytic degradation of organic pollutant due to the incorporation of CQDs [100]. All these unique and exclusive properties are the reason for using quantum dots in various commercial and engineering applications.

5 Quantum Dots in Engineering Applications

Quantum nanosized dots are extensively established to be beneficial in engineering applications that ranges from electronics to biomedical innovations [85, 126]. The most significant entity of quantum dots that grabbed the attention of several researchers is their electronic confinement due to their small size [60]. This property has been utilized in several aspects of storing drugs to confine electrons in electronics, which has led to a drastic change in the engineering of materials [124, 226]. In recent times, quantum dots are extensively employed to engineer materials in five major areas such as photocatalysts, solar cells, electronics, mechanical, bioimaging and pharmaceutical applications as mentioned in Fig. 2.

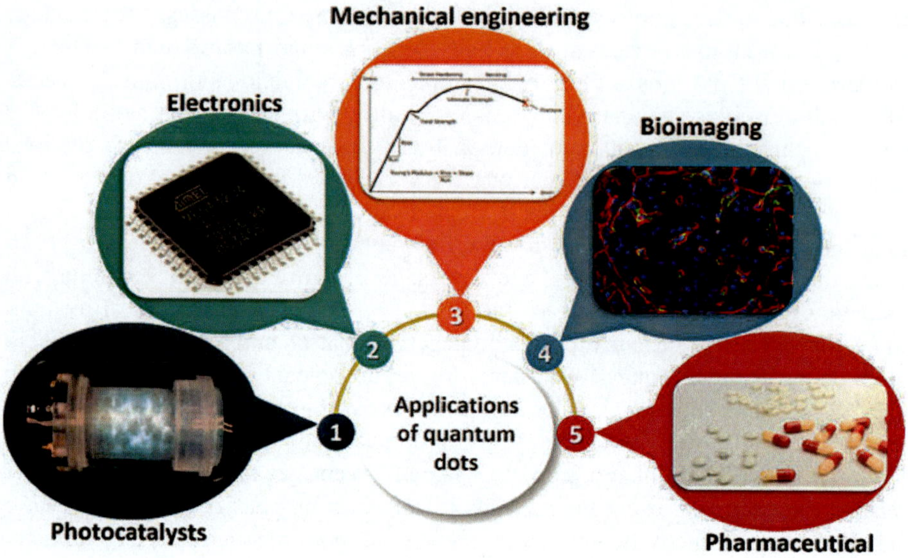

Fig. 2 Diverse applications of quantum dots

5.1 Photocatalysts and Solar Cells

Numerous quantum dots have been utilized as photocatalysts and to engineer normal nanomaterials to exhibit the enhanced photocatalytic ability. The size of nanomaterials usually contributes to the electron-hole pair formation that helps in their catalytic ability to combine with neighboring elements as shown in Fig. 3, which is essential for water splitting, wastewater treatment and applications in solar cells [11]. Li et al. [111] synthesized titanium dioxide nanoparticles and deposited carbon quantum dots over their surface using one-step alkali metal-assisted electrochemical method. This material is known to exhibit strong and stable photoluminescence with 12% of quantum yield. The results also revealed that these quantum dots elevate the optical and upconverted photoluminescence properties depending on their size. These quantum nanosized dot-metal oxide nanocomposites were proposed to be useful as a highly efficient photocatalyst and also a marker with fluorescent ability in fabricating materials for energy production [111]. Similarly, quantum cadmium selenide dots were embedded on the surface of graphene nanosheets that are bound together with poly (allylamine hydrochloride) via sequential self-assembly approach in layers. These quantum nanosized dots are proven to be beneficial in graphene nanosheets, which eventually elevates their photoelectrochemical properties and photocatalytic abilities under irradiation of visible light [196]. Moreover, quantum carbon nitride graphitic dots were fabricated from their bulk counterparts via thermal chemical etching process. This study demonstrated that these quantum dots exhibit blue emissions and upconversion ability that are significant as general component that utilizes visible light energy transfer in metal free photocatalytic systems [186]. In addition,

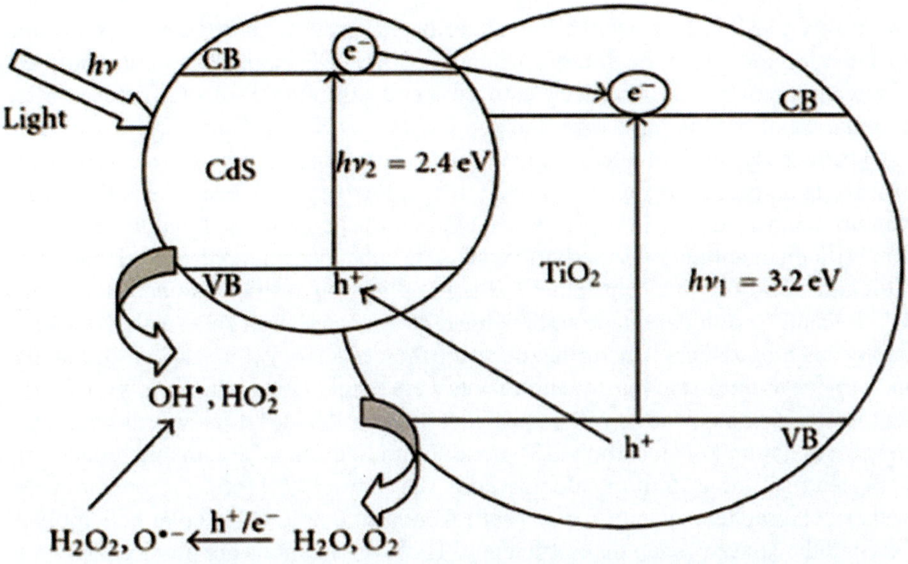

Fig. 3 Photocatalytic mechanism of a typical quantum dots. Reproduced with permission from Huang et al. [69] under the Creative Commons Attribution License, ©Hindawi publications

p- and n-type semi-conductor properties of quantum graphene oxide dots doped with nitrogen were reported to possess exclusive photocatalytic properties which are significant in catalyzing water splitting activity under visible light exposure [202]. Furthermore, zinc selenide quantum dots were doped with ions of transition metals such as manganese, cobalt and nickel to improve their photocatalytic efficiency. The synthesized quantum dots were 1–3 nm in size and exhibited exclusive abilities to remove methyl violet cationic dye using photocatalytic ability under visible light irradiation, which are highly useful in wastewater treatment [144]. In recent times, silver [44], carbon [213, 216], composites of graphite-like carbon nitride nanosheets and quantum carbon dots [113], carbon quantum dots with high reactive titanium oxides [185], graphene oxide composite and cesium-lead-bromide perovskite quantum dots [201] were fabricated to exhibit photocatalytic ability which is possible due to their capability of engineering already existing nanomaterials.

The formation mechanism of electron-hole pairs via external stimulation such as photons (light) in ultra-small sized quantum dots were also in engineering materials for solar cell applications. Nozik et al. [130] reported that quantum dots have exciton multiplication and relaxation potential dynamics which opens its door for the application of solar photon conversion with ultrahigh efficiency [130]. Further, Nozik et al. [131] reported in his review article that the quantum dots with semiconductor and array properties are highly beneficial in third-generation photovoltaic solar cells by forming multiple excitons [131]. Likewise, Chang and Lee [26] established that the cadmium selenide quantum dots can be deposited over mesoporous titanium dioxide films via chemical bath deposition process. These quantum dots possess a low surface tension, high wettability, superior penetration ability over metal oxide films and inhibit recombination of injected electrons. Thus, these quantum dots are significant as solar cells sensitized with dyes and exhibits 1.84% of efficiency after illumination of visible light from the sun (AM1.5, 100 mW/cm^2) [26]. Later, Lee et al. [106] used cadmium selenide quantum dot over titanium dioxide material as dye sensitized photoelectrode and are proved to be effective as solar cells sensitized with quantum nanosized dots [106]. Jiang and Green [81] emphasized that three dimensional silicon quantum dot superlattices possess enhanced energy bands, densities of states and mobilities that contributes to their application in silicon tandem solar cells [81]. Recently, colloidal quantum dots have been coated with solar cells via spray-coating and hierarchical structuring to elevate their efficiency [98, 104]. Additionally, zinc-copper-indium-selenium quantum dots were employed in solar cells with an efficient and certified power conversion of 11.6% [47]. Besides, titanium dioxide with amorphous nature [148], carbon [23], neodymium (sulfur, selenium and tellurium)$_3$ [115], lead sulfide-cadmium selenide core shell structure [166] and corn powder synthesized graphene quantum dots [168] were also employed in solar cell applications and they have proven their efficiency. These applications are possible only due to the ability of quantum dots in engineering nano, micro and bulk structures.

5.2 *Electronics*

Quantum dots are gaining much interest among researchers in electronic field due to their ability to confine electrons in their zero-dimensional structures and release them via quantum tunneling effect [170]. These semiconductor quantum dots are extensively under research to be used in integrated circuits [129]. Generally, integrated circuits consist of transistors that are made up of source, gate and drain [36]. The gate in the transistor is the key to reducing the travelling length of electrons to reach drain rapidly, which will increase the performance of transistors as well as reduce their size [231]. The utilization of quantum nanosized dots as a gate in transistors will eventually reduce the electron movement from source to drain, time taken, and space required for transistors in integrated circuit [35]. This will encourage several numbers of transistors to be placed in the given centimeter of integrated circuits and follow Moore's law [173]. In addition, the quantum tunneling effect of zero-dimensional quantum dots plays a crucial role in transferring electrons without losing energy and thus, they are proposed to be highly significant as the material that has potential for ballistic transport [70]. Ballistic transport is the transfer of electrons at the speed of light [107] and conventional materials loses energy during ballistic transport due to the movement of electrons in three-dimensions [50]. However, quantum dots are zero dimensions where electrons are confined [105], and the transfer of electron is possible only through the tunneling effect which makes ballistic transport possible in circuits [70]. Ravindran et al. [147] stated that the thiol-stabilized zinc sulfide capped cadmium selenide quantum dots coated over multiwalled carbon nanotubes via covalent coupling process possess enhanced electronic properties that helps them to be used in electronic device applications [147]. Moreover, Hayashi et al. [63] revealed that the double quantum dots are able to coherently manipulate electronic states of materials [63]. Korala et al. [97] also revealed that selenides zinc-cadmium core-shell quantum dots possess superior charge transfer, enhanced electrochemical photocurrent production under white light illumination, elevated photovoltage and photophysical ability to be useful in optoelectronics [97]. In recent times, core-shell [178], glowing graphene [221], black phosphorous [215], wearable diode arrays [33], blue emitting perovskite of bismuth halide [108] and pixelated multi-color light emitting diode fabricated using quantum dots [89] are gaining popularity in electronic applications. Even in biosensors, the quantum tunneling effect of quantum dots are been used recently which enhances their property [38, 87] and paves way for a new field called bioelectronics in future [165].

5.3 Mechanical Engineering Applications

The quantum state of the zero-dimensional quantum dots is highly significant in enhancing properties related to mechanics of devices. Childress et al. [30] described the coherent, quantum-mechanical coupling of quantum dots and separated them spatially in a microchip. This coupling will enhance the mechanical property of the microchip which will eventually help in the integration of quantum systems such as superconducting qubits [30]. Moreover, Kirschbaum et al. [92] established the fabrication of an integrated nanoelectromechanical circuit for ultrasensitive displacement detection. This circuit contains a two-dimensional electron system with suspended quantum dots of aluminum-gallium-arsenic/gallium-arsenic heterostructure and a mechanical resonator. The result revealed that the system improves the mechanical and transport properties of the resonator due to the presence of quantum dots along with its enhanced coulomb blockade [92]. These nanoelectromechanical systems show promise in enhancing nano-mechanics in the future. Likewise, Jeong et al. [78] reported that the crystalline indium-manganese-arsenic quantum dots are formed as symmetric single spin domain particles on the gallium-arsenic substrates via molecular beam epitaxy. The result revealed that the quantum size of these materials elevates their structural, mechanical and magnetic properties to support above room temperature ferromagnetic transition [78]. Similarly, carbon quantum dot hybrids were fabricated along with inorganic nanoparticles such as oxides of iron, zinc, silicon and titanium. These quantum dot nanohybrids demonstrated that they possess integrated fluorescent, magnetic, optical and mechanical properties. These nanohybrids of quantum dots with enhanced mechanical strength are significant as magneto-optical biolabeling agents and as photocatalysts [189]. Recently, tin dioxide quantum dots with graphene oxide were prepared to be beneficial in lithium-ion batteries as an anode. The study demonstrated that the addition of a quantum dot with graphene oxide exhibited quantum mechanical effects which increase the mechanical strength, life of batteries and their capacity retention up to 86% even after 2000 cycles [218]. Konwar et al. [96] fabricated a novel chitosan-carbon dot nanocomposite hydrogel film using tea as precursor for carbon dots. It was proved from this study that the mechanical property of a material can be engineered using quantum dots. The obtained hydrogel film was soft and tough, enhanced with mechanical, thermal, swelling, ultra-violet and visible light blocking properties as compared to chitosan hydrogel films [96]. Additionally, Qian et al. [140] reported that the titanium dioxide nanotube arrays were sensitized using graphene quantum dots. The sensitization of the nanotube with quantum dots supports the enhanced performance of photocathodic protection in 304 stainless steel and also increases their mechanical strength [140]. Similarly, Zhang et al. [221] fabricated quantum core-shell cadmium telluride-zinc selenide dots and are used to sensitize titanium dioxide nanotubes. This study also confirms that the quantum dot sensitization were used to improve the photocathodic protection and mechanical strength of stainless steel [211]. Contrarily, molybdenum disulfide-tungsten disulfide quantum dots were fabricated by combining sonication and solvothermal procedures. These quantum dots exhibited exclusive polyalkylene glycol (PAG) stability in base oil and are utilized as a high-performance lubricant additives to reduce friction and wear during steel-steel contact at high temperatures. The lubricant will enhance the process of coupling steels which eventually increases their mechanical strength [194].

Gobi et al. [54] reported that the graphene quantum dots can be infused with polymer to form stronger, brighter and tougher composites. In this study, the graphene quantum dots were infused with epoxy polymer matrix, which increased the toughness up to 2.6 fold, tensile strength up to 2.25 fold and Young's modulus up to 2.5 times of the polymer, improved their uniform loading capacity up to 10% and elevated their tensile strain up to 18% [54]. Further, Hawkins et al. [62] demonstrated the tensile properties of zinc oxide quantum dot and multiwalled carbon nanotube incorporated epoxy composite films. The study showed that the quantum dot decorated carbon nanotube in epoxy films possess 51% of Young's modulus and 20% of tensile strength with enhanced electrical conductivity [62]. Furthermore, Zhu et al. [225, 230] fabricated a hydrophobic electrospun carbon nanofiber fabric with high toughness and flexibility via cross-linking coal-based engineered graphene quantum dots. The results revealed that the Young's modulus of the nanofiber is increased more than 7 times due to addition of quantum dots, compared to pure polyacrylonitrile fabrics [225]. Thus, all these studies report the promising ability of quantum dots in engineering mechanical strength of a material which will help in several applications.

5.4 Bioimaging Applications

Generally, quantum dots are smaller in size, zero-dimensional and electron-hole pair formation ability and they exhibit photoluminescent, fluorescence or luminescent properties [112, 117] as shown in Fig. 4. Zhu et al. [228] prepared graphene quantum dots using one-step solvothermal method that exhibited strong green photoluminescent. The quantum photoluminescent yield of these graphene dots is 11.4%

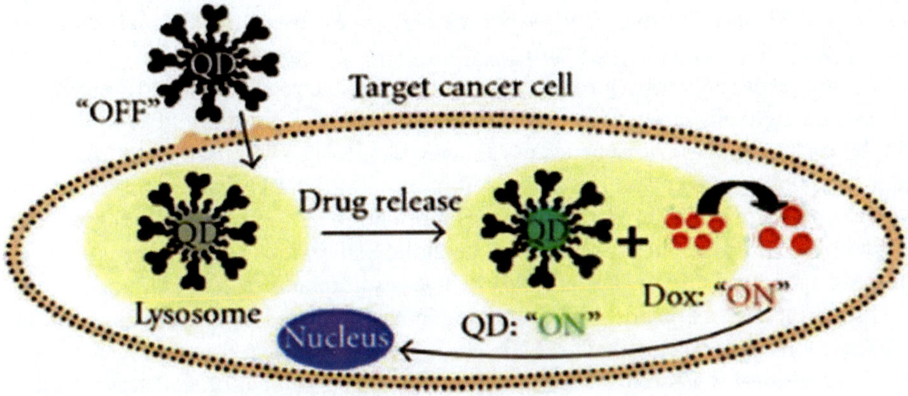

Fig. 4 Typical bioimaging property of a quantum dots. Reproduced with permission from Jin et al. [83] under the Creative Commons Attribution License, ©Hindawi publications

with high stability, ability to dissolve in most polar solvents, high biocompatibility and less toxicity. These qualities of quantum dots make them an unavoidable entity in bioimaging applications [228]. Likewise, Zhu et al. [229] proved that the surface chemistry of the quantum graphene dots can be modulated to exhibit enhanced photoluminescence and fluorescence mechanism. Such quantum dots were further reported to be highly beneficial in the bioimaging up-conversion purposes [229]. In addition, Kairdolf et al. [86] reported that semiconductor quantum dots are zero-dimensional, nano-sized, light-emitting particles that can be significant as labels of fluorescence for molecular imaging, chemical analysis and in biomedical diagnostic fields. The symmetric and narrow emission spectra, emission of light via tuning of size, broad absorption mediated multiple fluorescence excitation are the properties that gains much attention of researchers on these materials to be used in bioimaging applications [86]. Similarly, carbon dots with several surface passivation levels have been used as an emerging class of fluorescent quantum dots. The non-toxic nature, biocompatibility, structure and mechanical properties as well as ability to engineer conventional bioimaging materials to exhibit fluorescence are the advantageous factors for carbon dots to be beneficial to retrieve optical bio images with futuristic approach [120]. Tu et al. [175] stated that silicon quantum dots can be useful in obtaining images from Positron Emission Tomography (PET) with enhanced bio distribution in mice [175]. Silicon doped carbon [141], carbon dots synthesized from orange juice [151], paramagnetic cadmium sulfide: manganese-zinc sulfide [152], glutathione-capped cadmium telluride [223] and zinc oxide [195] are the other types of quantum dots that are used as dominant imaging probe classes and universal engineering platform of multifunctional nanodevices and bioimaging applications [232].

Recently, Zhu et al. [227] revealed the photoluminescent mechanism of three distinct quantum dots, such as graphene, carbon and polymer nanodots. The surface quantum confinement effect, the molecule state and the emission effect enhanced by crosslinking are the possible mechanism that are described to be accountable for the quantum dots' photoluminescent properties that are essential for bioimaging applications [227]. Du and Guo [48] further stated that heteroatom doping of fluorescent carbon dots, especially nitrogen will enhance their optical property which is beneficial in bio-imaging, sensing and photoelectronic devices [48]. Similarly, Qu et al. [142, 143] fabricated quantum graphene dots doped with nitrogen was used to improve their color emission property and are employed in bioimaging applications [143]. Moreover, Kumawat et al. [102] reported a novel fabrication method of quantum graphene dots with red luminescent properties via ethanolic leaf extract of *Mangifera indica* as reducing agent and microwave as catalyst. The result quantum dots exhibited exclusive properties that are useful for bioimaging in near-infrared region and intracellular nanothermometry [102]. Likewise, Xu et al. [199] synthesized monolayer molybdenum disulfide-tungsten disulfide quantum dots using sonication and solvothermal combination which forms exfoliated and structure with numerous active sites and defects. The resultant quantum nanosized dots are proven to have bioimaging probes ability and effectual electrocatalysts for hydrogen evolution reaction [199]. Other than these materials, molybdenum disulfide [45], sulfur and nitrogen co-doped graphene [142], paramagnetic graphene [67], lanthanide hybridized carbon [192] and glutathione capped copper-indium disulfide quantum dots [224] were also used as enhanced engineered materials with bioimaging property.

5.5 *Pharmaceutical Applications*

The zero-dimensional and quantum tunneling effects also helps quantum dots to exhibit exclusive drug delivery properties and other pharmaceutical abilities. Bagalkot et al. [10] synthesized an innovative quantum dot-aptamer-doxorubicin conjugate to exhibit self-quenching properties in fluorescence resonance energy transfer model. This system delivers doxorubicin to the targeted prostate cancer cells via RNA A10aptamer, since the delivery of doxorubicin by activating quantum dot fluorescence also assists in cancer cell imaging processes [10]. Probst et al. [137] expressed in their review that the quantum dots as a core-shell can exhibit intracellular and systemic distribution as nanocarrier, degrades, releases drugs and undergo clearance, while maintaining properties of the shell which is the nanoparticle drug delivery vehicle. Such quantum dot core-shell drug delivery systems are under extensive research to be used in photothermal therapy and magneto-transfection [137]. Zinc oxide [208], multifunctional graphene [187] and luminescent carbon dots are the other quantum dots that are useful in drug delivery applications. Klostranec et al. [93] stated that quantum nanosized dots can be utilized as convergent barcodes with microfluidics that can help in multiplexed signal processing for high-throughput diagnosis of infectious diseases [93]. Likewise, Derfus et al. [42] revealed that PEGylated quantum nanosized dots were highly significant as conjugates for siRNA delivery [42]. Moreover, zinc oxide quantum dots exhibited enhanced antimicrobial activity against *E. coli*, *Salmonella enteritidis* and *Listeria monocytogenes* [84]. In addition, quantum graphene dots proved as a common fluorophores in exposing the controlled trafficking of adipocyte insulin receptors [222]. In recent times, numerous quantum dots, such as super cationic carbon [80], nanoparticle conjugated graphene oxide [34], black phosphorus [203], peptide-based [27] and dopamine functionalized quantum dots [213, 216] are widely utilized in several disease treatments, delivery of drugs, antimicrobial agents and other pharmaceutical applications. The ability to engineer the quantum dot materials and using quantum dots to engineer the properties of other micro or bulk materials via unique material engineering methods has paved way for all these applications.

6 Conclusion, Limitation and Future Perspective

This chapter lists the evidences for the synthesis procedures and unique properties of quantum dots, which eventually helps in their material engineering and diversified applications. The exclusive properties of quantum dots are different from other nanomaterials due to their zero-dimensionality and small size related quantum properties including quantum tunneling effect. Even though, they are widely used to engineer bulk materials to exhibit quantum effects and to be utilized in solar cells, photocatalysts, electronics, mechanical engineering, bioimaging and pharmaceutical applications, there exist certain limitations. Individual quantum dots are stable and evaluating their mechanics and certain other properties are quite tedious. The stability of quantum dots is also questionable as they tend to agglomerate each other due to effects of quantum confinement. Once agglomerated, the quantum dot size may increase, which will reduce their exclusive properties and make them similar to nanosized particles. Further, there is no standard protocol to define quantum dots. Certain studies term 1–3 nm sized particles as quantum dots, others term 1–5 nm sized particles as quantum dots and few term particles with quantum effects with zero dimensionality as quantum dots. This has led to confusions in categorizing quantum dots under the class of nanoparticles or as a separate entity. Also, the quantum effects are still under research which is not understood properly. The ballistic transport and quantum confinement effects are theoretically proven and quantum dots have been proposed to possess those effects due to their size, however, there is no instrument to visualize and characterize those effects. Thus, the field of quantum dots, their synthesis approach, property enhancements and applications in engineering will develop in future, if a perfect instrument is invented to visualize and characterize quantum effects.

In the future, quantum dots will replace all the existing transistor models with its enhanced quantum tunneling effect to transfer electrons. This will enhance integrated circuits to transfer electrons without loss of energy and help electronic devices to be more advanced with more battery life, memory and performance. Further, it is possible to use quantum dots as self-powering devices which is the key to develop bioelectronics. This will eventually lead to the development of nanorobots which can detect disease or metabolic alterations in a patient's body, carry drugs to the target site and exhibit controlled release. These self-powered nanorobots made up of quantum dots will be highly useful in personalized patient treatment. Also, it is possible to design quantum dot based nanorobots to sense metabolic alterations in patients' body and signal them to smart phones which can indicate doctors. Even in photocatalyst applications, quantum dots synthesized using biogenic methods will be beneficial which will be less toxic to the environment and help in wastewater treatment, bioremediation, splitting of water into hydrogen to use as fuel. The photocatalytic quantum dots will also be beneficial in carbon capture and carbon sequestration application. Certain quantum dots can also bind with the proteins or activate enzymes in algae that can capture carbon and utilize for carbon sequestration application. It is noteworthy that only carbon dots and graphene quantum dots are under extensive research in several applications. In the future, research should be extended and synthesized in using several quantum dots in various applications. The introduction of novel quantum dots in the future will further widen their efficiencies as well as their properties and create options among researchers to widen their attention towards quantum dots for various applications.

References

1. Alvarez-Quintana, J., et al.: Cross-plane thermal conductivity reduction of vertically uncorrelated Ge/Si quantum dot superlattices. Appl. Phys. Lett. **93**, 013112 (2008)
2. Ananthanarayanan, A., et al.: Nitrogen and phosphorus co-doped graphene quantum dots: synthesis from adenosine triphosphate, optical properties, and cellular imaging. Nanoscale **7**, 8159–8165 (2015)
3. Andra, S., Balu, S.K., Jeevanandham, J., Muthalagu, M., Vidyavathy, M., San Chan, Y., Danquah, M.K.: Phytosynthesized metal oxide nanoparticles for pharmaceutical applications. Naunyn-Schmiedeberg's archives of pharmacology, pp. 1–17 (2019)
4. Anikin, K.V., Melnik, N.N., Simakin, A.V., Shafeev, G.A., Voronov, V.V., Vitukhnovsky, A.G.: Formation of ZnSe and CdS quantum dots via laser ablation in liquids. Chem. Phys. Lett. **366**, 357–360 (2002)
5. Antolini, F., Orazi, L.: Quantum dots synthesis through direct laser patterning: a review. Front. Chem. **7**, 252 (2019)
6. Arul, V., Sethuraman, M.G.: Facile green synthesis of fluorescent N-doped carbon dots from Actinidia deliciosa and their catalytic activity and cytotoxicity applications. Opt. Mater. **78**, 181–190 (2018)
7. Ayyub, P., Chandra, R., Taneja, P., Sharma, A., Pinto, R.: Synthesis of nanocrystalline material by sputtering and laser ablation at low temperatures. Appl. Phys. A **73**, 67–73 (2001)
8. Babiński, A., Jasiński, J., Bożek, R., Szepielow, A., Baranowski, J.: Rapid thermal annealing of InAs/GaAs quantum dots under a GaAs proximity cap. Appl. Phys. Lett. **79**, 2576–2578 (2001)
9. Bae, W.K., Char, K., Hur, H., Lee, S.: Single-step synthesis of quantum dots with chemical composition gradients. Chem. Mater. **20**, 531–539 (2008)
10. Bagalkot, V., Zhang, L., Levy-Nissenbaum, E., Jon, S., Kantoff, P.W., Langer, R., Farokhzad, O.C.: Quantum dot—aptamer conjugates for synchronous cancer imaging, therapy, and sensing of drug delivery based on bi-fluorescence resonance energy transfer. Nano Lett. **7**, 3065–3070 (2007). https://doi.org/10.1021/nl071546n
11. Banerjee, S., Pillai, S.C., Falaras, P., O'shea, K.E., Byrne, J.A., Dionysiou, D.D.: New insights into the mechanism of visible light photocatalysis. J. Phys. Chem. Lett. 5, 2543–2554 (2014)

12. Bao, H., Lu, Z., Cui, X., Qiao, Y., Guo, J., Anderson, J.M., Li, C.M.: Extracellular microbial synthesis of biocompatible CdTe quantum dots. Acta Biomater. **6**, 3534–3541 (2010)
13. Bao, Y., Liu, W., Shamsa, M., Alim, K., Balandin, A., Liu, J.: Electrical and thermal conductivity of Ge/ Si quantum dot superlattices. J. Electrochem. Soc. **152**, G432–G435 (2005)
14. Baptista, A., Silva, F., Porteiro, J., Míguez, J., Pinto, G.: Sputtering Physical Vapour Deposition (PVD) coatings: a critical review on process improvement and market trend demands. Coatings **8**, 402 (2018)
15. Bera, D., Kuiry, S.C., Seal, S.: Synthesis of nanostructured materials using template-assisted electrodeposition. JOM **56**, 49–53 (2004)
16. Bera, D., Qian, L., Tseng, T.-K., Holloway, P.H.: Quantum dots and their multimodal applications: a review. Materials **3**, 2260–2345 (2010)
17. Bobrovsky, A., Shibaev, V., Abramchuk, S., Elyashevitch, G., Samokhvalov, P., Oleinikov, V., Mochalov, K.: Quantum dot–polymer composites based on nanoporous polypropylene films with different draw ratios. Eur. Polymer J. **82**, 93–101 (2016)
18. Bonilla, C.A.M., Kouznetsov, V.V.: "Green" quantum dots: basics, green synthesis, and nanotechnological applications. In: Green Nanotechnology-Overview and Further Prospects. IntechOpen (2016)
19. Borovaya, M., Burlaka, O., Yemets, A., Blume, Y.B.: Biosynthesis of quantum dots and their potential applications in biology and biomedicine. In: Nanoplasmonics, Nano-Optics, Nanocomposites, and Surface Studies, pp 339–362. Springer (2015a)
20. Borovaya, M., Pirko, Y., Krupodorova, T., Naumenko, A., Blume, Y., Yemets, A.: Biosynthesis of cadmium sulphide quantum dots by using Pleurotus ostreatus (Jacq.) P. Kumm. Biotechnology & Biotechnological Equipment **29**, 1156–1163 (2015)
21. Borovaya, M.N., Naumenko, A.P., Matvieieva, N.A., Blume, Y.B., Yemets, A.I.: Biosynthesis of luminescent CdS quantum dots using plant hairy root culture. Nanoscale Res. Lett. **9**, 686 (2014)
22. Brichkin, S.B., Razumov, V.F.: Colloidal quantum dots: synthesis, properties and applications. Russ. Chem. Rev. **85**, 1297 (2016)
23. Briscoe, J., Marinovic, A., Sevilla, M., Dunn, S., Titirici, M.: Biomass-derived carbon quantum dot sensitizers for solid-state nanostructured solar cells. Angew. Chem. Int. Ed. **54**, 4463–4468 (2015)
24. Brooks, J., Lefebvre, D.D.: Optimization of conditions for cadmium selenide quantum dot biosynthesis in Saccharomyces cerevisiae. Appl. Microbiol. Biotechnol. **101**, 2735–2745 (2017)
25. Bruna, N., et al.: Synthesis of salt-stable fluorescent nanoparticles (quantum dots) by polyextremophile halophilic bacteria. Scientific reports **9**, 1953 (2019)
26. Chang, C.-H., Lee, Y.-L.: Chemical bath deposition of CdS quantum dots onto mesoscopic TiO2 films for application in quantum-dot-sensitized solar cells. Appl. Phys. Lett. **91**, 053503 (2007). https://doi.org/10.1063/1.2768311
27. Chen, H., Zhang, M., Li, B., Chen, D., Dong, X., Wang, Y., Gu, Y.: Versatile antimicrobial peptide-based ZnO quantum dots for in vivo bacteria diagnosis and treatment with high specificity. Biomaterials **53**, 532–544 (2015). https://doi.org/10.1016/j.biomaterials.2015.02.105
28. Chen, M.-L., et al.: Synthesis of carbon nanomaterials-CdSe composites and their photocatalytic activity for degradation of methylene blue. J. Nanomater. **2012**, 21 (2012)
29. Chen, X., Peng, L., Huang, K., Shi, Z., Xie, R., Yang, W.: Non-injection gram-scale synthesis of cesium lead halide perovskite quantum dots with controllable size and composition. Nano Res. **9**, 1994–2006 (2016)
30. Childress, L., Sørensen, A.S., Lukin, M.D.: Mesoscopic cavity quantum electrodynamics with quantum dots. Phys. Rev. A **69**, 042302 (2004). https://doi.org/10.1103/PhysRevA.69.042302
31. Chinnathambi, S., Shirahata, N.: Recent advances on fluorescent biomarkers of near-infrared quantum dots for in vitro and in vivo imaging. Sci. Technol. Adv. Mater. **20**, 337–355 (2019)

32. Choi, M.K., Yang, J., Hyeon, T., Kim, D.-H.: Flexible quantum dot light-emitting diodes for next-generation displays. NPJ Flex. Electron. **2**, 10 (2018)
33. Choi, M.K., et al.: Wearable red–green–blue quantum dot light-emitting diode array using high-resolution intaglio transfer printing. Nat. Commun. **6**, 7149 (2015)
34. Choi, S.Y., Baek, S.H., Chang, S.-J., Song, Y., Rafique, R., Lee, K.T., Park, T.J.: Synthesis of upconversion nanoparticles conjugated with graphene oxide quantum dots and their use against cancer cell imaging and photodynamic therapy. Biosens. Bioelectron. **93**, 267–273 (2017). https://doi.org/10.1016/j.bios.2016.08.094
35. Crippa, A., et al.: Level spectrum and charge relaxation in a silicon double quantum dot probed by dual-gate reflectometry. Nano Lett. **17**, 1001–1006 (2017)
36. Crone, B., et al.: Large-scale complementary integrated circuits based on organic transistors. Nature **403**, 521 (2000)
37. Dahi, A., Colson, P., Jamin, C., Cloots, R., Lismont, M., Dreesen, L.: Radio-frequency magnetron sputtering: a versatile tool for CdSe quantum dots depositions with controlled properties. J. Mater. Environ. Sci. **7**, 2277–2287 (2016)
38. Dai, W.-X., Zhang, L., Zhao, W.-W., Yu, X.-D., Xu, J.-J., Chen, H.-Y.: Hybrid PbS quantum dot/nanoporous NiO film nanostructure: preparation, characterization, and application for a self-powered cathodic photoelectrochemical biosensor. Anal. Chem. **89**, 8070–8078 (2017)
39. Darma, Y., Rusydi, A.: Quantum dot based memory devices: Current status and future prospect by simulation perspective. In: AIP Conference Proceedings, vol 1, pp. 20–23. AIP (2014)
40. de Mello, Donegá C., Liljeroth, P., Vanmaekelbergh, D.: Physicochemical evaluation of the hot-injection method, a synthesis route for monodisperse nanocrystals. Small **1**, 1152–1162 (2005)
41. Derfus, A.M., Chan, W.C., Bhatia, S.N.: Probing the cytotoxicity of semiconductor quantum dots. Nano Lett. **4**, 11–18 (2004)
42. Derfus, A.M., Chen, A.A., Min, D.-H., Ruoslahti, E., Bhatia, S.N.: Targeted Quantum Dot Conjugates for siRNA Delivery. Bioconjug. Chem. **18**, 1391–1396 (2007). https://doi.org/10.1021/bc060367e
43. Dhawan, S., Dhawan, T., Vedeshwar, A.G.: Growth of Nb2O5 quantum dots by physical vapor deposition. Mater. Lett. **126**, 32–35 (2014)
44. Di, J., et al.: New insight of Ag quantum dots with the improved molecular oxygen activation ability for photocatalytic applications. Appl. Catal. B **188**, 376–387 (2016)
45. Dong, H., et al.: Fluorescent MoS2 quantum dots: ultrasonic preparation, up-conversion and down-conversion bioimaging, and photodynamic therapy. ACS Appl. Mater. Interfaces **8**, 3107–3114 (2016). https://doi.org/10.1021/acsami.5b10459
46. Dong, Y., Qiao, T., Kim, D., Parobek, D., Rossi, D., Son, D.H.: Precise control of quantum confinement in cesium lead halide perovskite quantum dots via thermodynamic equilibrium. Nano Lett. **18**, 3716–3722 (2018). https://doi.org/10.1021/acs.nanolett.8b00861
47. Du, J., et al.: Zn–Cu–In–Se quantum dot solar cells with a certified power conversion efficiency of 11.6%. J. Am. Chem. Soc. **138**, 4201–4209 (2016)
48. Du, Y., Guo, S.: Chemically doped fluorescent carbon and graphene quantum dots for bioimaging, sensor, catalytic and photoelectronic applications. Nanoscale **8**, 2532–2543 (2016). https://doi.org/10.1039/C5NR07579C
49. Fan, L., et al.: Direct synthesis of graphene quantum dots by chemical vapor deposition. Part. Part. Syst. Charact. **30**, 764–769 (2013)
50. Ferry D (2016) Semiconductor Transport. CRC Press
51. Galyametdinov, Y.G., Shamilov, R., Nuzhdin, V., Valeev, V., Stepanov, A.: Luminescence of CdSe quantum dots near a layer of silver nanoparticles ion-synthesized in sapphire. Tech. Phys. Lett. **42**, 1067–1070 (2016)
52. Gao, X., Cui, Y., Levenson, R.M., Chung, L.W.K., Nie, S.: In vivo cancer targeting and imaging with semiconductor quantum dots. Nat. Biotechnol. **22**, 969 (2004)
53. Gao, X., Yang, L., Petros, J.A., Marshall, F.F., Simons, J.W., Nie, S.: In vivo molecular and cellular imaging with quantum dots. Curr. Opin. Biotechnol. **16**, 63–72 (2005)

54. Gobi, N., Vijayakumar, D., Keles, O., Erogbogbo, F.: Infusion of graphene quantum dots to create stronger, tougher, and brighter polymer composites. ACS Omega **2**, 4356–4362 (2017). https://doi.org/10.1021/acsomega.6b00517
55. Groiss, H., et al.: Size control and midinfrared emission of epitaxial Pb Te/ Cd Te quantum dot precipitates grown by molecular beam epitaxy. Appl. Phys. Lett. **91**, 222106 (2007)
56. Guo, J., et al.: Nuclear quantum effects of hydrogen bonds probed by tip-enhanced inelastic electron tunneling. Science **352**, 321–325 (2016)
57. Guo, L.-P., Zhang, Y., Li, W.-C.: Sustainable microalgae for the simultaneous synthesis of carbon quantum dots for cellular imaging and porous carbon for CO2 capture. J. Colloid Interface Sci. **493**, 257–264 (2017)
58. Hardman, R.: A toxicologic review of quantum dots: toxicity depends on physicochemical and environmental factors. Environ. Health Perspect. **114**, 165–172 (2005)
59. Haro-González, P., Martínez-Maestro, L., Martín, I., García-Solé, J., Jaque, D.: High-sensitivity fluorescence lifetime thermal sensing based on CdTe quantum dots. Small **8**, 2652–2658 (2012)
60. Harris, R.D., et al.: Electronic processes within quantum dot-molecule complexes. Chem. Rev. **116**, 12865–12919 (2016)
61. Havrdova, M., et al.: Toxicity of carbon dots–effect of surface functionalization on the cell viability, reactive oxygen species generation and cell cycle. Carbon **99**, 238–248 (2016)
62. Hawkins, S.A., Yao, H., Wang, H., Sue, H.-J.: Tensile properties and electrical conductivity of epoxy composite thin films containing zinc oxide quantum dots and multi-walled carbon nanotubes. Carbon **115**, 18–27 (2017). https://doi.org/10.1016/j.carbon.2016.12.058
63. Hayashi, T., Fujisawa, T., Cheong, H.-D., Jeong, Y.H., Hirayama, Y.: Coherent manipulation of electronic states in a double quantum dot. Phys. Rev. Lett. **91**, 226804 (2003)
64. He, X., Gao, L., Ma, N.: One-step instant synthesis of protein-conjugated quantum dots at room temperature. Sci. Rep. **3**, 2825 (2013)
65. Horoz, S., et al.: CdSe quantum dots synthesized by laser ablation in water and their photovoltaic applications. Appl. Phys. Lett. **101**, 223902 (2012)
66. Hu, M.Z., Zhu, T.: Semiconductor nanocrystal quantum dot synthesis approaches towards large-scale industrial production for energy applications. Nanoscale Res. Lett. **10**, 469 (2015)
67. Huang, C.-L., et al.: Application of paramagnetic graphene quantum dots as a platform for simultaneous dual-modality bioimaging and tumor-targeted drug delivery. J. Mater. Chem. B **3**, 651–664 (2015). https://doi.org/10.1039/C4TB01650E
68. Huang, C., Dong, H., Su, Y., Wu, Y., Narron, R., Yong, Q.: Synthesis of carbon quantum dot nanoparticles derived from byproducts in bio-refinery process for cell imaging and in vivo bioimaging. Nanomaterials **9**, 387 (2019)
69. Huang, K., Chen, L., Deng, J., Xiong, J.: Enhanced visible-light photocatalytic performance of nanosized anatase TiO_2 doped with cds quantum dots for cancer-cell treatment. J. Nanomater. **2012**, 720491 (2012). https://doi.org/10.1155/2012/720491
70. Iftikhar, Z., et al.: Tunable quantum criticality and super-ballistic transport in a "charge" Kondo circuit. Science **360**, 1315–1320 (2018)
71. Imran, M., Saif, M.J., Kuznetsov, A.E., Idrees, N., Iqbal, J., Tahir, A.A.: Computational investigations into the structural and electronic properties of Cd n Te n (n = 1–17) quantum dots. RSC Advances **9**, 5091–5099 (2019)
72. Irshad, M., Ahmad, F., Mohamed, N., Abdullah, M.: Preparation and structural characterization of template assisted electrodeposited copper nanowires. Int. J. Electrochem. Sci. **9**, 2548–2555 (2014)
73. Jasieniak, J., Califano, M., Watkins, S.E.: Size-dependent valence and conduction band-edge energies of semiconductor nanocrystals. ACS Nano **5**, 5888–5902 (2011)
74. Jeevanandam, J., Barhoum, A., Chan, Y.S., Dufresne, A., Danquah, M.K.: Review on nanoparticles and nanostructured materials: history, sources, toxicity and regulations. Beilstein J. Nanotechnol. **9**, 1050–1074 (2018)
75. Jeevanandam, J., Chan, Y.S., Pan, S., Danquah, M.K.: Metal oxide nanocomposites: Cytotoxicity and targeted drug delivery applications. In: Hybrid Nanocomposites: Fundamentals, Synthesis and Applications, pp 111–147. Pan Stanford Publishing (2019)

76. Jeevanandam, J., San Chan, Y., Danquah, M.K.: Nano-formulations of drugs: recent developments, impact and challenges. Biochimie **128**, 99–112 (2016)
77. Jeevanandam, J., Sundaramurthy, A., Sharma, V., Murugan, C., Pal, K., Kodous, M.H.A., Danquah, M.K.: Sustainability of one-dimensional nanostructures: fabrication and industrial applications. In: Sustainable Nanoscale Engineering, pp 83–113. Elsevier (2020)
78. Jeon, H.C., et al.: (In1–xMnx)As diluted magnetic semiconductor quantum dots with above room temperature ferromagnetic transition. Adv. Mater. **14**, 1725–1728 (2002). https://doi.org/10.1002/1521-4095(20021203)14:23%3c1725:AID-ADMA1725%3e3.0.CO;2-Q
79. Jha, S., Mathur, P., Ramteke, S., Jain, N.K.: Pharmaceutical potential of quantum dots. Artif. Cells Nanomed. Biotechnol. **46**, 57–65 (2018)
80. Jian, H.-J., et al.: Super-cationic carbon quantum dots synthesized from spermidine as an eye drop formulation for topical treatment of bacterial keratitis. ACS Nano **11**, 6703–6716 (2017). https://doi.org/10.1021/acsnano.7b01023
81. Jiang, C.-W., Green, M.A.: Silicon quantum dot superlattices: modeling of energy bands, densities of states, and mobilities for silicon tandem solar cell applications. J. Appl. Phys. **99**, 114902 (2006). https://doi.org/10.1063/1.2203394
82. Jin, S., Hu, Y., Gu, Z., Liu, L., Wu, H.-C.: Application of quantum dots in biological imaging. J. Nanomater. **2011**, 13 (2011)
83. Jin, S., Hu, Y., Gu, Z., Liu, L., Wu, H.-C.: Application of quantum dots in biological imaging. J. Nanomater. **2011**, 834139 (2011). https://doi.org/10.1155/2011/834139
84. Jin, T., Sun, D., Su, J.Y., Zhang, H., Sue, H.J.: Antimicrobial efficacy of zinc oxide quantum dots against listeria monocytogenes, salmonella enteritidis, and escherichia coli O157:H7. J. Food Sci. **74**, M46–M52 (2009). https://doi.org/10.1111/j.1750-3841.2008.01013.x
85. Kagan, C.R., Lifshitz, E., Sargent, E.H., Talapin, D.V.: Building devices from colloidal quantum dots. Science **353**, aac5523 (2016)
86. Kairdolf, B.A., Smith, A.M., Stokes, T.H., Wang, M.D., Young, A.N., Nie, S.: Semiconductor quantum dots for bioimaging and biodiagnostic applications. Ann. Rev. Anal. Chem. **6**, 143–162 (2013)
87. Kang, T., Um, K., Park, J., Chang, H., Lee, D.C., Kim, C.-K., Lee, K.: Minimizing the fluorescence quenching caused by uncontrolled aggregation of CdSe/CdS core/shell quantum dots for biosensor applications. Sens. Actuators B Chem. **222**, 871–878 (2016)
88. Karmakar, R.: Quantum dots and it method of preparations-revisited. Prajnan O Sadhona A Sci. Ann. **116** (2015)
89. Kim, B.H., et al.: Multilayer transfer printing for pixelated, multicolor quantum dot light-emitting diodes. ACS Nano **10**, 4920–4925 (2016). https://doi.org/10.1021/acsnano.5b06387
90. Kim, M., Osone, S., Kim, T., Higashi, H., Seto, T.: Synthesis of nanoparticles by laser ablation: a review. KONA Powder Part. J. 2017009 (2017)
91. Kim, Y.T., Han, J.H., Hong, B.H., Kwon, Y.U.: Electrochemical synthesis of CdSe quantum-dot arrays on a graphene basal plane using mesoporous silica thin-film templates. Adv. Mater. **22**, 515–518 (2010)
92. Kirschbaum, J., Höhberger, E.M., Blick, R.H., Wegscheider, W., Bichler, M.: Integrating suspended quantum dot circuits for applications in nanomechanics. Appl. Phys. Lett. **81**, 280–282 (2002). https://doi.org/10.1063/1.1492302
93. Klostranec, J.M., et al.: Convergence of quantum dot barcodes with microfluidics and signal processing for multiplexed high-throughput infectious disease diagnostics. Nano Lett. **7**, 2812–2818 (2007). https://doi.org/10.1021/nl071415m
94. Kojima, T., Sugimoto, H., Fujii, M.: Size-dependent photocatalytic activity of colloidal silicon quantum dot. J. Phys. Chem. C **122**, 1874–1880 (2018)
95. Kong, L., Zhang, L., Meng, Z., Xu, C., Lin, N., Liu, X.-Y.: Ultrastable, highly luminescent quantum dot composites based on advanced surface manipulation strategy for flexible lighting-emitting. Nanotechnology **29**, 315203 (2018)
96. Konwar, A., Gogoi, N., Majumdar, G., Chowdhury, D.: Green chitosan–carbon dots nanocomposite hydrogel film with superior properties. Carbohyd. Polym. **115**, 238–245 (2015)

97. Korala, L., Wang, Z., Liu, Y., Maldonado, S., Brock, S.L.: Uniform thin films of CdSe and CdSe(ZnS) Core(Shell) quantum dots by sol-gel assembly: enabling photoelectrochemical characterization and electronic applications. ACS Nano **7**, 1215–1223 (2013). https://doi.org/10.1021/nn304563j
98. Kramer, I.J., et al.: Efficient spray-coated colloidal quantum dot solar cells. Adv. Mater. **27**, 116–121 (2015)
99. Kulakovich, O., et al.: Enhanced luminescence of CdSe quantum dots on gold colloids. Nano Lett. **2**, 1449–1452 (2002)
100. Kumar, M.S., Yasoda, K.Y., Kumaresan, D., Kothurkar, N.K., Batabyal, S.K.: TiO2-carbon quantum dots (CQD) nanohybrid: enhanced photocatalytic activity. Mater. Res. Express **5**, 075502 (2018)
101. Kumar, S., Aziz, S.T., Girshevitz, O., Nessim, G.D.: One-step synthesis of N-doped graphene quantum dots from chitosan as a sole precursor using chemical vapor deposition. J. Phys. Chem. C **122**, 2343–2349 (2018)
102. Kumawat, M.K., Thakur, M., Gurung, R.B., Srivastava, R.: Graphene quantum dots from mangifera indica: application in near-infrared bioimaging and intracellular nanothermometry. ACS Sustain. Chem. Eng. **5**, 1382–1391 (2017). https://doi.org/10.1021/acssuschemeng.6b01893
103. Kurtin, J.: Quantum dot LED phosphors: performance and reliability improvements. In: Light-Emitting Diodes: Materials, Devices, and Applications for Solid State Lighting XVII, p. 86411D. International Society for Optics and Photonics (2013)
104. Labelle, A.J., et al.: Colloidal quantum dot solar cells exploiting hierarchical structuring. Nano Lett. **15**, 1101–1108 (2015). https://doi.org/10.1021/nl504086v
105. Lee, J., et al.: Imaging electrostatically confined Dirac fermions in graphene quantum dots. Nat. Phys. **12**, 1032 (2016)
106. Lee, Y.-L., Huang, B.-M., Chien, H.-T.: Highly efficient CdSe-sensitized TiO2 photoelectrode for quantum-dot-sensitized solar cell applications. Chem. Mater. **20**, 6903–6905 (2008)
107. Lee, Y.K., Lee, H., Lee, C., Hwang, E., Park, J.Y.: Hot-electron-based solar energy conversion with metal–semiconductor nanodiodes. J. Phys. Condens. Matter **28**, 254006 (2016)
108. Leng, M., et al.: Lead-free, blue emitting bismuth halide perovskite quantum dots. Angew. Chem. Int. Ed. **55**, 15012–15016 (2016)
109. Leschkies, K.S., et al.: Photosensitization of ZnO nanowires with CdSe quantum dots for photovoltaic devices. Nano Lett. **7**, 1793–1798 (2007)
110. Li, G., et al.: Solvent-polarity-engineered controllable synthesis of highly fluorescent cesium lead halide perovskite quantum dots and their use in white light-emitting diodes. Adv. Func. Mater. **26**, 8478–8486 (2016)
111. Li, H., et al.: Water-soluble fluorescent carbon quantum dots and photocatalyst design. Angew. Chem. Int. Ed. **49**, 4430–4434 (2010). https://doi.org/10.1002/anie.200906154
112. Li, H., He, X., Liu, Y., Huang, H., Lian, S., Lee, S.-T., Kang, Z.: One-step ultrasonic synthesis of water-soluble carbon nanoparticles with excellent photoluminescent properties. Carbon **49**, 605–609 (2011)
113. Li, K., Su, F.-Y., Zhang, W.-D.: Modification of g-C3N4 nanosheets by carbon quantum dots for highly efficient photocatalytic generation of hydrogen. Appl. Surf. Sci. **375**, 110–117 (2016). https://doi.org/10.1016/j.apsusc.2016.03.025
114. Li, L., Daou, T.J., Texier, I., Kim Chi, T.T., Liem, N.Q., Reiss, P.: Highly luminescent CuInS2/ZnS core/shell nanocrystals: cadmium-free quantum dots for in vivo imaging. Chem. Mater. **21**, 2422–2429 (2009)
115. Li, Q., et al.: Nd2 (S, Se, Te) 3 colloidal quantum dots: synthesis, energy level alignment, charge transfer dynamics, and their applications to solar cells. Adv. Func. Mater. **26**, 254–266 (2016)
116. Li, Y., et al.: A template/electrochemical deposition method for fabricating silver nanorod arrays based on porous anodic alumina. Nanomater. Nanotechnol. **7**, 1847980417717543 (2017)

117. Liang, Q., Ma, W., Shi, Y., Li, Z., Yang, X.: Easy synthesis of highly fluorescent carbon quantum dots from gelatin and their luminescent properties and applications. Carbon **60**, 421–428 (2013)
118. Lim, H., Liu, Y., Kim, H.Y., Son, D.I.: Facile synthesis and characterization of carbon quantum dots and photovoltaic applications. Thin Solid Films **660**, 672–677 (2018)
119. Liu, Z., et al.: Photoluminescence of carbon quantum dots: coarsely adjusted by quantum confinement effects and finely by surface trap states. Sci. China Chem. **61**, 490–496 (2018). https://doi.org/10.1007/s11426-017-9172-0
120. Luo, P.G., et al.: Carbon "quantum" dots for optical bioimaging. J. Mater. Chem. B **1**, 2116–2127 (2013). https://doi.org/10.1039/C3TB00018D
121. Maestro, L.M., Jacinto, C., Silva, U.R., Vetrone, F., Capobianco, J.A., Jaque, D., Solé, J.G.: CdTe quantum dots as nanothermometers: towards highly sensitive thermal imaging. Small **7**, 1774–1778 (2011)
122. Mahajan, S., Rani, M., Dubey, R., Mahajan, J.: Synthesis of CdSe crystal using hot injection method. Int. J. Latest Res. Sci. Technol. **2**, 518–521 (2013)
123. Mala, J.G.S., Rose, C.: Facile production of ZnS quantum dot nanoparticles by Saccharomyces cerevisiae MTCC 2918. J. Biotechnol. **170**, 73–78 (2014)
124. Malgras, V., Tominaka, S., Ryan, J.W., Henzie, J., Takei, T., Ohara, K., Yamauchi, Y.: Observation of quantum confinement in monodisperse methylammonium lead halide perovskite nanocrystals embedded in mesoporous silica. J. Am. Chem. Soc. **138**, 13874–13881 (2016)
125. Mansur, H.S.: Quantum dots and nanocomposites. Wiley Interdiscip. Rev. Nanomed. Nanobiotechnol. **2**, 113–129 (2010)
126. Martynenko, I., Litvin, A., Purcell-Milton, F., Baranov, A., Fedorov, A., Gun'ko, Y.: Application of semiconductor quantum dots in bioimaging and biosensing. J. Mater. Chem. B **5**, 6701–6727 (2017)
127. McCumiskey, E.J., Chandrasekhar, N., Taylor, C.R.: Nanomechanics of CdSe quantum dot–polymer nanocomposite films. Nanotechnology **21**, 225703 (2010)
128. Michalet, X., et al.: Quantum dots for live cells, in vivo imaging, and diagnostics. Science **307**, 538–544 (2005)
129. Midolo, L., et al.: Electro-optic routing of photons from single quantum dots in photonic integrated circuits (2017). arXiv preprint arXiv:170706522
130. Nozik, A.J.: Exciton multiplication and relaxation dynamics in quantum dots: applications to ultrahigh-efficiency solar photon conversion. Inorg. Chem. **44**, 6893–6899 (2005)
131. Nozik, A.J., Beard, M.C., Luther, J.M., Law, M., Ellingson, R.J., Johnson, J.C.: Semiconductor quantum dots and quantum dot arrays and applications of multiple exciton generation to third-generation photovoltaic solar cells. Chem. Rev. **110**, 6873–6890 (2010)
132. Oh, E., Liu, R., Nel, A., Gemill, K.B., Bilal, M., Cohen, Y., Medintz, I.L.: Meta-analysis of cellular toxicity for cadmium-containing quantum dots. Nat. Nanotechnol. **11**, 479 (2016)
133. Osman, H., et al.: One-step hot injection synthesis of gradient alloy CdxZn1-xSySe1-y quantum dots with large-span self-regulating ability. J. Lumin. **206**, 565–570 (2019)
134. Owen, J., Brus, L.: Chemical synthesis and luminescence applications of colloidal semiconductor quantum dots. J. Am. Chem. Soc. **139**, 10939–10943 (2017)
135. Panagiotopoulou, M., et al.: Molecularly imprinted polymer coated quantum dots for multiplexed cell targeting and imaging. Angew. Chem. Int. Ed. **55**, 8244–8248 (2016)
136. Peyvast, N., Zhou, K., Hogg, R.A., Childs, D.T.D.: Dominant role of many-body effects on the carrier distribution function of quantum dot lasers. Appl. Phys. Express **9**, 032705 (2016)
137. Probst, C.E., Zrazhevskiy, P., Bagalkot, V., Gao, X.: Quantum dots as a platform for nanoparticle drug delivery vehicle design. Adv. Drug Deliv. Rev. **65**, 703–718 (2013). https://doi.org/10.1016/j.addr.2012.09.036
138. Pu, Y., Cai, F., Wang, D., Wang, J.-X., Chen, J.-F.: Colloidal synthesis of semiconductor quantum dots toward large-scale production: a review. Ind. Eng. Chem. Res. **57**, 1790–1802 (2018)
139. Qi, B.-P., Zhang, X., Shang, B.-B., Xiang, D., Zhang, S.: Solvothermal tuning of photoluminescent graphene quantum dots: from preparation to photoluminescence mechanism. J. Nanopart. Res. **20**, 20 (2018)

140. Qian, B., Dai, H., Tang, S., Song, Z.: Enhanced photocathodic protection performance of graphene quantum dots sensitized TiO2 nanotube arrays for 304 stainless steel. Optik **178**, 128–134 (2019). https://doi.org/10.1016/j.ijleo.2018.10.027
141. Qian, Z., Shan, X., Chai, L., Ma, J., Chen, J., Feng, H.: Si-doped carbon quantum dots: a facile and general preparation strategy, bioimaging application, and multifunctional sensor. ACS Appl. Mater. Interfaces. **6**, 6797–6805 (2014). https://doi.org/10.1021/am500403n
142. Qu, D., et al.: Three colors emission from S, N Co-doped graphene quantum dots for visible light H2 production and bioimaging. Adv. Opt. Mater. **3**, 360–367 (2015). https://doi.org/10.1002/adom.201400549
143. Qu, D., Zheng, M., Li, J., Xie, Z., Sun, Z.: Tailoring color emissions from N-doped graphene quantum dots for bioimaging applications. Light Sci. Appl. **4**, e364 (2015b)
144. Rajabi, H.R., Farsi, M.: Effect of transition metal ion doping on the photocatalytic activity of ZnS quantum dots: synthesis, characterization, and application for dye decolorization. J. Mol. Catal. A: Chem. **399**, 53–61 (2015). https://doi.org/10.1016/j.molcata.2015.01.029
145. Ramanan, V., Thiyagarajan, S.K., Raji, K., Suresh, R., Sekar, R., Ramamurthy, P.: Outright green synthesis of fluorescent carbon dots from eutrophic algal blooms for in vitro imaging. ACS Sustain. Chem. Eng. **4**, 4724–4731 (2016)
146. Ramkumar, V., Ju, S.: Quantum-dot and polychalcone mixed nanocomposites for polymer light-emitting diodes. J. Nanomater. 2017 (2017)
147. Ravindran, S., Chaudhary, S., Colburn, B., Ozkan, M., Ozkan, C.S.: Covalent coupling of quantum dots to multiwalled carbon nanotubes for electronic device applications. Nano Lett. **3**, 447–453 (2003). https://doi.org/10.1021/nl0259683
148. Ren, Z., et al.: Amorphous TiO2 buffer layer boosts efficiency of quantum dot sensitized solar cells to over 9%. Chem. Mater. **27**, 8398–8405 (2015)
149. Reshak, A.H.: Quantum dots in photocatalytic applications: efficiently enhancing visible light photocatalytic activity by integrating CdO quantum dots as sensitizers. Phys. Chem. Chem. Phys. **19**, 24915–24927 (2017)
150. Rossi, D., Wang, H., Dong, Y., Qiao, T., Qian, X., Son, D.H.: Light-induced activation of forbidden exciton transition in strongly confined perovskite quantum dots. ACS Nano **12**, 12436–12443 (2018)
151. Sahu, S., Behera, B., Maiti, T.K., Mohapatra, S.: Simple one-step synthesis of highly luminescent carbon dots from orange juice: application as excellent bio-imaging agents. Chem. Commun. **48**, 8835–8837 (2012). https://doi.org/10.1039/C2CC33796G
152. Santra, S., Yang, H., Holloway, P.H., Stanley, J.T., Mericle, R.A.: Synthesis of water-dispersible fluorescent, radio-opaque, and paramagnetic CdS: Mn/ZnS quantum dots: a multifunctional probe for bioimaging. J. Am. Chem. Soc. **127**, 1656–1657 (2005)
153. Schiffman, J.D., Balakrishna, R.G.: Quantum dots as fluorescent probes: synthesis, surface chemistry, energy transfer mechanisms, and applications. Sens. Actuators B Chem. **258**, 1191–1214 (2018)
154. Schneider, R., Balan, L., Aldeek, F.: Synthesis, characterization and biological applications of water-soluble ZnO quantum dots. In: Nanomaterials. IntechOpen (2011)
155. Sekhar, M.C., Reddy, B.P., Mallikarjuna, K., Krishna, G.G., Park, S.-H.: Biogenic Fabrication of Au/Pd bimetallic quantum dots from mushroom extract and their application to organic dye pollutant reduction. Curr. Nanosci. **14**, 313–318 (2018)
156. Şenel, B., Demir, N., Büyükköroğlu, G., Yıldız, M.: Graphene quantum dots: synthesis, characterization, cell viability, genotoxicity for biomedical applications. Saudi Pharm. J. (2019)
157. Shah, K.A., Bhat, B.M.U.D.: Effect of magnetic field on quantum state energies of an electron confined in the core of a double walled carbon nanotube. Physica B **498**, 55–58 (2016)
158. Shah, M., Fawcett, D., Sharma, S., Tripathy, S., Poinern, G.: Green synthesis of metallic nanoparticles via biological entities. Materials **8**, 7278–7308 (2015)
159. Shamsi, J., et al.: Colloidal synthesis of quantum confined single crystal CsPbBr 3 nanosheets with lateral size control up to the micrometer range. J. Am. Chem. Soc. **138**, 7240–7243 (2016)

160. Shamsipur, M., Rajabi, H.R.: Study of photocatalytic activity of ZnS quantum dots as efficient nanoparticles for removal of methyl violet: effect of ferric ion doping. Spectrochim. Acta Part A Mol. Biomol. Spectrosc. **122**, 260–267 (2014)
161. Shen, J., Zhu, Y., Yang, X., Li, C.: Graphene quantum dots: emergent nanolights for bioimaging, sensors, catalysis and photovoltaic devices. Chem. Commun. **48**, 3686–3699 (2012)
162. Silveiro, I., Ortega, J.M.P., De Abajo, F.J.G.: Quantum nonlocal effects in individual and interacting graphene nanoribbons. Light Sci. Appl. **4**, e241 (2015)
163. Singh, A.K., Pal, P., Gupta, V., Yadav, T.P., Gupta, V., Singh, S.P.: Green synthesis, characterization and antimicrobial activity of zinc oxide quantum dots using Eclipta alba. Mater. Chem. Phys. **203**, 40–48 (2018)
164. Smith, A.M., Nie, S.: Semiconductor nanocrystals: structure, properties, and band gap engineering. Acc. Chem. Res. **43**, 190–200 (2009)
165. Sobolev, A.M., Byzova, N.A., Goryacheva, I.Y., Zherdev, A.V.: Silanized quantum dots as labels in lateral flow test strips for C-reactive protein. Anal. Lett. **52**, 1874–1887 (2019)
166. Spangler, L.C., Lu, L., Kiely, C.J., Berger, B.W., McIntosh, S.: Biomineralization of PbS and PbS–CdS core–shell nanocrystals and their application in quantum dot sensitized solar cells. J. Mater. Chem. A **4**, 6107–6115 (2016)
167. Tayyebi, A., Akhavan, O., Lee, B.-K., Outokesh, M.: Supercritical water in top-down formation of tunable-sized graphene quantum dots applicable in effective photothermal treatments of tissues. Carbon **130**, 267–272 (2018)
168. Teymourinia, H., Salavati-Niasari, M., Amiri, O., Farangi, M.: Facile synthesis of graphene quantum dots from corn powder and their application as down conversion effect in quantum dot-dye-sensitized solar cell. J. Mol. Liq. **251**, 267–272 (2018). https://doi.org/10.1016/j.molliq.2017.12.059
169. Thambiraj, S., Shankaran, R.: Green synthesis of highly fluorescent carbon quantum dots from sugarcane bagasse pulp. Appl. Surf. Sci. **390**, 435–443 (2016)
170. Thierschmann, H., et al.: Three-terminal energy harvester with coupled quantum dots. Nat. Nanotechnol. **10**, 854 (2015)
171. Tian, J., Cao, G.: Control of nanostructures and interfaces of metal oxide semiconductors for quantum-dots-sensitized solar cells. J. Phys. Chem. Lett. **6**, 1859–1869 (2015)
172. Tikhomirov, G., Hoogland, S., Lee, P., Fischer, A., Sargent, E.H., Kelley, S.O.: DNA-based programming of quantum dot valency, self-assembly and luminescence. Nat. Nanotechnol. **6**, 485 (2011)
173. Toumey, C.: Less is moore. Nat. Nanotechnol. **11**, 2 (2016)
174. Tsoi, K.M., Dai, Q., Alman, B.A., Chan, W.C.: Are quantum dots toxic? Exploring the discrepancy between cell culture and animal studies. Acc. Chem. Res. **46**, 662–671 (2012)
175. Tu, C., Ma, X., House, A., Kauzlarich, S.M., Louie, A.Y.: PET imaging and biodistribution of silicon quantum dots in mice. ACS Med. Chem. Lett. **2**, 285–288 (2011). https://doi.org/10.1021/ml1002844
176. Valizadeh, A., Mikaeili, H., Samiei, M., Farkhani, S.M., Zarghami, N., Akbarzadeh, A., Davaran, S.: Quantum dots: synthesis, bioapplications, and toxicity. Nanoscale Res. Lett. **7**, 480 (2012)
177. Vandarkuzhali, S.A.A., Jeyalakshmi, V., Sivaraman, G., Singaravadivel, S., Krishnamurthy, K.R., Viswanathan, B.: Highly fluorescent carbon dots from pseudo-stem of banana plant: applications as nanosensor and bio-imaging agents. Sens. Actuators B Chem. **252**, 894–900 (2017)
178. Vasudevan, D., Gaddam, R.R., Trinchi, A., Cole, I.: Core–shell quantum dots: properties and applications. J. Alloy. Compd. **636**, 395–404 (2015)
179. Vu, H.-T., Chiang, R.-K., Huang, C.-Y., Chen, C.-J., Yu, H.-C., Lien, J.-Y., Su, Y.-K.: Enhanced thermal stability of green-emission quantum-dot light-emitting diodes via composition-gradient thick-shell quantum dots. Appl. Phys. Express **9**, 082101 (2016)
180. Wagner, A.M., Knipe, J.M., Orive, G., Peppas, N.A.: Quantum Dots in Biomedical Applications (2019). Available at SSRN 3321961

181. Wang, C., Jiang, Z., Wei, L., Chen, Y., Jiao, J., Eastman, M., Liu, H.: Photosensitization of TiO2 nanorods with CdS quantum dots for photovoltaic applications: a wet-chemical approach. Nano Energy **1**, 440–447 (2012)
182. Wang, R., Lu, K.-Q., Tang, Z.-R., Xu, Y.-J.: Recent progress in carbon quantum dots: synthesis, properties and applications in photocatalysis. J. Mater. Chem. A **5**, 3717–3734 (2017)
183. Wang, S., Louie, A.Y., Kauzlarich, S.M.: Synthesis and properties of water-soluble CdSe/Zn 1-x Mn x S semiconductor quantum dots using an amphiphilic polymer. In: Colloidal Quantum Dots for Biomedical Applications II, p. 64480A. International Society for Optics and Photonics (2007)
184. Wang, T., Hsieh, H., Hsieh, Y., Chiang, C., Sun, Y., Wang, C.: The in vivo biodistribution and fate of CdSe quantum dots in the murine model: a laser ablation inductively coupled plasma mass spectrometry study. Anal. Bioanal. Chem. **404**, 3025–3036 (2012)
185. Wang, W., Ni, Y., Xu, Z.: One-step uniformly hybrid carbon quantum dots with high-reactive TiO2 for photocatalytic application. J. Alloy. Compd. **622**, 303–308 (2015). https://doi.org/10.1016/j.jallcom.2014.10.076
186. Wang, W., Yu, J.C., Shen, Z., Chan, D.K.L., Gu, T.: g-C3N4 quantum dots: direct synthesis, upconversion properties and photocatalytic application. Chem. Commun. **50**, 10148–10150 (2014). https://doi.org/10.1039/C4CC02543A
187. Wang, X., et al.: Multifunctional graphene quantum dots for simultaneous targeted cellular imaging and drug delivery. Colloids Surf. B **122**, 638–644 (2014). https://doi.org/10.1016/j.colsurfb.2014.07.043
188. Wang, Y., et al.: Novel metal doped carbon quantum dots/CdS composites for efficient photocatalytic hydrogen evolution. Nanoscale **11**, 1618–1625 (2019)
189. Wang, Y., Hu, A.: Carbon quantum dots: synthesis, properties and applications. J. Mater. Chem. C **2**, 6921–6939 (2014). https://doi.org/10.1039/C4TC00988F
190. Weaver, J., Zakeri, R., Aouadi, S., Kohli, P.: Synthesis and characterization of quantum dot–polymer composites. J. Mater. Chem. **19**, 3198–3206 (2009)
191. Werner, M., Gresty, N., Pickett, N., Chalker, P., Harris, J., Naasani, I.: Multi-layer-coated quantum dot beads. Google Patents (2016)
192. Wu, F., Su, H., Zhu, X., Wang, K., Zhang, Z., Wong, W.-K.: Near-infrared emissive lanthanide hybridized carbon quantum dots for bioimaging applications. J. Mater. Chem. B **4**, 6366–6372 (2016). https://doi.org/10.1039/C6TB01646D
193. Wu, T., Tang, M.: Toxicity of quantum dots on respiratory system. Inhal. Toxicol. **26**, 128–139 (2014)
194. Wu, X., Gong, K., Zhao, G., Lou, W., Wang, X., Liu, W.: MoS2/WS2 quantum dots as high-performance lubricant additive in polyalkylene glycol for steel/steel contact at elevated temperature. Adv. Mater. Interfaces **5**, 1700859 (2018). https://doi.org/10.1002/admi.201700859
195. Wu, Y.L., Lim, C.S., Fu, S., Tok, A.I.Y., Lau, H.M., Boey, F.Y.C., Zeng, X.T.: Surface modifications of ZnO quantum dots for bio-imaging. Nanotechnology **18**, 215604 (2007)
196. Xiao, F.-X., Miao, J., Liu, B.: Layer-by-layer self-assembly of CdS quantum dots/graphene nanosheets hybrid films for photoelectrochemical and photocatalytic applications. J. Am. Chem. Soc. **136**, 1559–1569 (2014). https://doi.org/10.1021/ja411651e
197. Xie, J., et al.: Enhanced electronic properties of SnO2 via electron transfer from graphene quantum dots for efficient perovskite solar cells. ACS Nano **11**, 9176–9182 (2017)
198. Xu, M., Li, Z., Zhu, X., Hu, N., Wei, H., Yang, Z., Zhang, Y.: Hydrothermal/solvothermal synthesis of graphene quantum dots and their biological applications. Nano Biomed. Eng. **5** (2013)
199. Xu, S., Li, D., Wu, P.: One-pot, facile, and versatile synthesis of monolayer MoS2/WS2 quantum dots as bioimaging probes and efficient electrocatalysts for hydrogen evolution reaction. Adv. Func. Mater. **25**, 1127–1136 (2015). https://doi.org/10.1002/adfm.201403863
200. Xu, S., Wang, X., Chua, S., Wang, C., Fan, W., Jiang, J., Xie, X.: Effects of rapid thermal annealing on structure and luminescence of self-assembled InAs/GaAs quantum dots. Appl. Phys. Lett. **72**, 3335–3337 (1998)

201. Xu, Y.-F., Yang, M.-Z., Chen, B.-X., Wang, X.-D., Chen, H.-Y., Kuang, D.-B., Su, C.-Y.: A CsPbBr 3 perovskite quantum dot/graphene oxide composite for photocatalytic CO2 reduction. J. Am. Chem. Soc. **139**, 5660–5663 (2017). https://doi.org/10.1021/jacs.7b00489
202. Yeh, T.-F., Teng, C.-Y., Chen, S.-J., Teng, H.: Nitrogen-doped graphene oxide quantum dots as photocatalysts for overall water-splitting under visible light illumination. Adv. Mater. **26**, 3297–3303 (2014). https://doi.org/10.1002/adma.201305299
203. Yin, F., et al.: Black phosphorus quantum dot based novel siRNA delivery systems in human pluripotent teratoma PA-1 cells. J. Mater. Chem. B **5**, 5433–5440 (2017). https://doi.org/10.1039/C7TB01068K
204. Yu, H., Shi, R., Zhao, Y., Waterhouse, G.I., Wu, L.Z., Tung, C.H., Zhang, T.: Smart utilization of carbon dots in semiconductor photocatalysis. Adv. Mater. **28**, 9454–9477 (2016)
205. Yu, H., et al.: Carbon quantum dots/TiO 2 composites for efficient photocatalytic hydrogen evolution. J. Mater. Chem. A **2**, 3344–3351 (2014)
206. Yuan, F., et al.: Engineering triangular carbon quantum dots with unprecedented narrow bandwidth emission for multicolored LEDs. Nat. Commun. **9**, 2249 (2018)
207. Yuan, G., Gómez, D., Kirkwood, N., Mulvaney, P.: Tuning single quantum dot emission with a micromirror. Nano Lett. **18**, 1010–1017 (2018)
208. Yuan, Q., Hein, S., Misra, R.D.K.: New generation of chitosan-encapsulated ZnO quantum dots loaded with drug: synthesis, characterization and in vitro drug delivery response. Acta Biomater. **6**, 2732–2739 (2010). https://doi.org/10.1016/j.actbio.2010.01.025
209. Zeng, Z., et al.: Unraveling the cooperative synergy of zero-dimensional graphene quantum dots and metal nanocrystals enabled by layer-by-layer assembly. J. Mater. Chem. A **6**, 1700–1713 (2018)
210. Zhang, C., Lian, L., Yang, Z., Zhang, J., Zhu, H.: Quantum confinement-tunable ultrafast charge transfer in a PbS quantum dots/WSe2 0D–2D hybrid structure: transition from the weak to strong coupling regime. J. Phys. Chem. Lett. **10**, 7665–7671 (2019). https://doi.org/10.1021/acs.jpclett.9b03293
211. Zhang, J., Hu, J., Zhu, Y.-F., Liu, Q., Zhang, H., Du, R.-G., Lin, C.-J.: Fabrication of CdTe/ZnS core/shell quantum dots sensitized TiO2 nanotube films for photocathodic protection of stainless steel. Corros. Sci. **99**, 118–124 (2015)
212. Zhang, P.: Scaling for quantum tunneling current in nano-and subnano-scale plasmonic junctions. Sci. Rep. **5**, 9826 (2015)
213. Zhang, W.-H., Ma, W., Long, Y.-T.: Redox-mediated indirect fluorescence immunoassay for the detection of disease biomarkers using dopamine-functionalized quantum dots. Anal. Chem. **88**, 5131–5136 (2016). https://doi.org/10.1021/acs.analchem.6b00048
214. Zhang, W., Zhang, H., Feng, Y., Zhong, X.: Scalable single-step noninjection synthesis of high-quality core/shell quantum dots with emission tunable from violet to near infrared. ACS Nano **6**, 11066–11073 (2012)
215. Zhang, X., et al.: Black phosphorus quantum dots. Angew. Chem. Int. Ed. **54**, 3653–3657 (2015)
216. Zhang, Z., Zheng, T., Li, X., Xu, J., Zeng, H.: Progress of carbon quantum dots in photocatalysis applications. Part. Part. Syst. Charact. **33**, 457–472 (2016)
217. Zhao, C., Bai, Z., Liu, X., Zhang, Y., Zou, B., Zhong, H.: Small GSH-capped CuInS2 quantum dots: MPA-assisted aqueous phase transfer and bioimaging applications. ACS Appl. Mater. Interfaces **7**, 17623–17629 (2015). https://doi.org/10.1021/acsami.5b05503
218. Zhao, K., et al.: SnO2 quantum dots@graphene oxide as a high-rate and long-life anode material for lithium-ion batteries. Small **12**, 588–594 (2016). https://doi.org/10.1002/smll.201502183
219. Zhao, P., Li, X., Baryshnikov, G., Wu, B., Ågren, H., Zhang, J., Zhu, L.: One-step solvothermal synthesis of high-emissive amphiphilic carbon dots via rigidity derivation. Chem. Sci. **9**, 1323–1329 (2018)
220. Zheng, J., Zhang, C., Dickson, R.M.: Highly fluorescent, water-soluble, size-tunable gold quantum dots. Phys. Rev. Lett. **93**, 077402 (2004)

221. Zheng, X.T., Ananthanarayanan, A., Luo, K.Q., Chen, P.: Glowing graphene quantum dots and carbon dots: properties, syntheses, and biological applications. Small **11**, 1620–1636 (2015)
222. Zheng, X.T., Than, A., Ananthanaraya, A., Kim, D.-H., Chen, P.: Graphene quantum dots as universal fluorophores and their use in revealing regulated trafficking of insulin receptors in adipocytes. ACS Nano **7**, 6278–6286 (2013). https://doi.org/10.1021/nn4023137
223. Zheng, Y., Gao, S., Ying, J.Y.: Synthesis and cell-imaging applications of glutathione-capped CdTe quantum dots. Adv. Mater. **19**, 376–380 (2007)
224. Zhou, J., Yang, Y., Zhang, C-y: Toward biocompatible semiconductor quantum dots: from biosynthesis and bioconjugation to biomedical application. Chem. Rev. **115**, 11669–11717 (2015)
225. Zhu, J., et al.: Engineering cross-linking by coal-based graphene quantum dots toward tough, flexible, and hydrophobic electrospun carbon nanofiber fabrics. Carbon **129**, 54–62 (2018). https://doi.org/10.1016/j.carbon.2017.11.071
226. Zhu, S., Song, Y., Wang, J., Wan, H., Zhang, Y., Ning, Y., Yang, B.: Photoluminescence mechanism in graphene quantum dots: Quantum confinement effect and surface/edge state. Nano Today **13**, 10–14 (2017)
227. Zhu, S., Song, Y., Zhao, X., Shao, J., Zhang, J., Yang, B.: The photoluminescence mechanism in carbon dots (graphene quantum dots, carbon nanodots, and polymer dots): current state and future perspective. Nano Research **8**, 355–381 (2015). https://doi.org/10.1007/s12274-014-0644-3
228. Zhu, S., et al.: Strongly green-photoluminescent graphene quantum dots for bioimaging applications. Chem. Commun. **47**, 6858–6860 (2011). https://doi.org/10.1039/C1CC11122A
229. Zhu, S., et al.: Surface chemistry routes to modulate the photoluminescence of graphene quantum dots: from fluorescence mechanism to up-conversion bioimaging applications. Adv. Func. Mater. **22**, 4732–4740 (2012)
230. Zhu, T., Wang, A., Kirsten, K., Cleaver, G., Sheng, Q.: Primordial non-Gaussianity and power asymmetry with quantum gravitational effects in loop quantum cosmology. Phys. Rev. D **97**, 043501 (2018)
231. Zhuang, L., Guo, L., Chou, S.Y.: Silicon single-electron quantum-dot transistor switch operating at room temperature. Appl. Phys. Lett. **72**, 1205–1207 (1998)
232. Zrazhevskiy, P., Sena, M., Gao, X.: Designing multifunctional quantum dots for bioimaging, detection, and drug delivery. Chem. Soc. Rev. **39**, 4326–4354 (2010). https://doi.org/10.1039/B915139G

Synthesis of Thin Film and Its Application

Sohail Ahmed and Shahzad Abu Bakar

Abstract Thin film technology is a major area of scientific research in the modern world because of its fascinating surface properties and wide range of applications from microelectronics to optics, space science to aircraft, and superconductivity to photovoltaic and solar cells. The performance of the thin films depends on the atomic structure, composition, microstructure, defects and interfaces that are controlled by thermodynamics and kinetics of the synthesis. The major advantage of thin film fabrication is its lower production temperature, facil fabrication process and offer to produce flexible and transparent films. High quality, compact and multi shades crystalline thin films can be fabricated using low cost, robust and highly efficient deposition techniques. However, there are lot of challenges to fabricate desired thins films for the latest technological and industrial applications. This chapter will cover different thin film technologies, challenges and future prospects.

Keywords Techniques · Nano structured · Thin films · Fabrication · AACVD

1 Introduction

A thin film is a layer of material on a surface ranging from fractions of nanometer to several micrometers in thickness. Thin film is a two dimensional form of deposited solid material, whose one dimension, called the thickness, is much smaller than the other two dimensions. The thin film is formed by atom to atom or molecule to molecule deposition through various condensation process. Historically, in 1852 Bunsen and Grove produced the first metallic thin film in a vacuum system. In the literature, research work has been focused on the development of thin films of various materials with different deposition techniques. The advantages in the development of

S. Ahmed (✉)
Faculty of Sciences, The University of Haripur, KPK 22620, Haripur, Pakistan
e-mail: Sohailamalik@hotmail.com

S. A. Bakar
National Center of Physics, Quaid-I-Azam University, Islamabad, Pakistan

N. M. Mubarak et al. (eds.), *Contemporary Nanomaterials in Material Engineering Applications*, Engineering Materials,
https://doi.org/10.1007/978-3-030-62761-4_10

thin film are its superior shifting of materials properties to the surface and enable the basic characteristic of materials properties. The enhancement of different physical and chemical properties are linked with the formation of a system comprising a substrate material and thin film. The development of fabrication instrument is the key to success in the technology development to control the three fundamental properties of the thin films; (a) thickness, (b) nanometer range of crystalline structure, chemical composition and (c) surface geometry [1].

In past decades, the research on thin films, particularly for their remarkably different properties at nanoscale from bulk material, was believed out of scientific curiosity. Subsequently, the desired capability of controlling properties of thin films helped electronics, optoelectronics and other technological devices to boost their performance immensely. Since the thin film technology contributes toward the development of microelectronics, the electronics industry start depending on thin film technology for minimising the size of semiconductor devices to two dimensions. In thin film, effective characteristics of the materials are classified as; resistance to wear, anti-scratch stiffness, strain-to-failure and abrasion for mechanical properties [2–4]. Similarly, optical properties are related to transparency, optical entrapment, fluorescence, light valves, opaque, waveguides, and antireflection [1].

Similarly, chemical properties are anti-fog, water repellence, chemical barriers, moisturising and oxygen shields on polymers, antimicrobial being chemically neutral, and surface compaction[1, 5–7]. Moreover, the binding energy, insulation, potential energy, and conduction are related to electronic properties [5, 8]. Magnetic and thermal properties are related to either saving data and/or an obstacle in multiple layers thin films for atomic vibration control expansion to produce thermal transmission and thermoelectric tool usage due to the flow of electrons, respectively [1].

Scientists have performed tremendous work in the development of thin film materials for different applications. These fields include the development of thin films for the electronic industry to use either as a light barrier with low dielectric constant of metallic dielectrics or semiconductor wafers production in integrated circuits. Further work has been focused on the construction of organic light emitting or anticorrosive coating indicators. The advantage of organic thin film fabrication as compared to mineral or ceramic materials is its lower production temperature and more facile fabrication. Following are some of the more common methods adopted for the fabrication of thin films;

- Physical vapour deposition (PVD)
- Chemical vapour deposition (CVD)
- Electronic precipitation/electronic coating
- Sol–gel technique
- Rotating coating
- Spray coating
- Electro-deposition
- Self assembling.

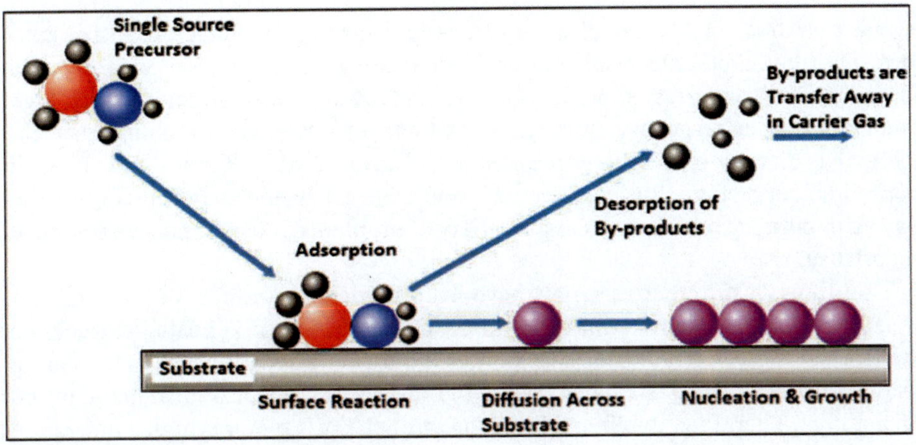

Fig. 1 Illustration of thin films composition

Some of the fabrication techniques are described in the preparation section. These thin film fabrication techniques are different from each other with respect to the mode of operation and experimental setup. Typically, these techniques performed the fabrication of the thin film according to the nature of precursors and substrate. Sol–gel method enabled the synthesis and fabrication of powder and thin films at atmospheric pressure. Nowadays, high quality, compact and multi shades crystalline thin films are fabricated using modern chemical vapour deposition (CVD) and physical vapour deposition (PVD) techniques operating at optimum conditions. A general CVD process is illustrated in Fig. 1.

Another important aspect of thin film fabrication is the preparation of the substrate. This includes the use of extensive methods such as surface cleaning and chemical modification to prepare thin films with desired stiffness. The preparation of substrate is a crucial parameter for thin film growth and cannot be ignored. Some of the commonly used substrate materials are conductive glass coating with fluorinated tin oxide, indium oxide and aluminium oxide, etc.

The wide range of thin film applications extends from micrometer dots to several square meters. Metals and metallic oxides polycrystalline films were the first films that found industrial applications, mainly in the field of optical devices and electronics. At present, thin film technology is a major area of scientific research because of its wide range of applications, from microelectronics to optics, space science to aircraft, and superconductivity to photovoltaics. One of the most important applications of thin films is in photovoltaic devices.

Thin film in photovoltaic devices reduces material cost and also the fabrication of large area devices at a comparatively low cost. Thin films have several applications in various fields, such as anit-reflecting (AR) coating, interference filters, polarisers, narrowband filters, solar cells, photoconductors, IR detectors, waveguide coatings, temperature control of satellites, photo thermal solar coatings such as black chrome, nickel, cobalt etc., magnetic films, superconducting films, anti-corrosive films, microelectronics devices, diamond films, reduction of fabrication through coating or surface modification, i.e., epitaxy and heterostructure films, high temperature wear resistance films, hard coatings, etc. Besides, the thin film technology has directly or indirectly resulted in the advancement of many new areas of research in solid state physics, chemistry and material sciences. It will continue to play an important role in addressing a variety of problems of basic and technological importance.

Thin films of different materials have been fabricated through various conventional techniques as shown in Fig. 2. In CVD, the precursor is heated to generate vapours and to convert it into the gaseous phase ultrasonically and carry to the heating chamber using carrier gas to deposited at the surface of the substrate at pre adjusted temperature. Therefore, thin film fabrication through CVD requires high temperature in a vacuum chamber. CVD fabrication is regarded as an expensive technique for the industrial production of large surface thin films.

Similarly, PVD technique requires heating of precursor material through electron beam or resistive heating to convert it into the gas phase, which is deposited on the substrate forming thin film. Generally, these methods are classified in different categories, such as pulsed laser deposition, magnetron sputtering and thermal evaporation. These methods have demerits to control the size of the fabricated grain at the surface of the thin films and need improvement in the operating procedures. Other requirements need the use of an expensive single crystal substrate to help the growth of grain along specific orientation and structure at the surface of thin films. Keeping in mind the high cost of single crystal substrate, it is costly to grow such kind of thin film on a large surface area such as wafers and required expensive equipment

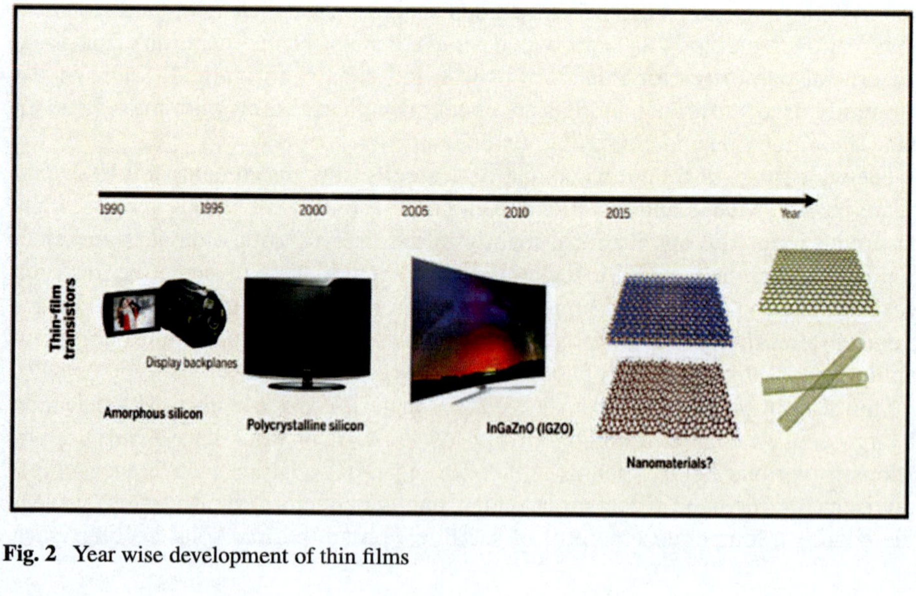

Fig. 2 Year wise development of thin films

for semiconducting industry processes. Such type of deposition is performed in a vacuum chamber and hinders the surface coating [4, 5].

The development of new technology for the growth of thin films with nanostructure features is gaining interest among the scientific community for the quality production of industrial products. This provides an excellent opportunity for the industry to develop products with enhanced efficiency and quality at a low price. The development of new techniques with better control over the quality of products having higher efficiency is the key moto for the growth of any R&D organisation. Therefore, companies and industries that are investing more in R&D will have a better future. The development of new technology has been attracting industries of advanced countries to invest in the development of new procedures for the deposition of materials on large surface areas.

2 Nanostructured Thin Films

Nanostructured thin films of good quality, compact and uniform thin layer is required in the industry for preparing and packaging of the products for all market supply. The basic purpose of thin film fabrication is to provide resistance against corrosion from environmental agents, enhance surface compatibility, hardness and more importantly, to improve physical properties like electrical and magnetic [6]. The quality of most of the industrial products degrades from 4 to 6% due to corrosion. The significance of thin film fabrication is itself important to protect the surface from corrosion. The enhanced characteristics of nanostructured thin films as compared to conventional coatings are: increase wear strength, hardness, abrasion, environmental pollutions, and decadence, etc. This section will cover the attempts which have been made to fabricated nanostructured thin films in the modern industry. We have discussed the importance of nanostructured thin films surface engineering and competence. A detailed summary is provided to compare the advantages and applications of nanostructured thin films with traditional coatings [9].

The applications of TiO_2 nanostructured thin films have many advantages of low fabrication price and potential use for mass use for indoor applications in electrical and electronic appliances. Similarly, the growth of economical nanostructured carbon nanotubes have potential use in solar cells, transparent layer conductive of magnetic and polarisers. Therefore, the focus is on nanostructured thin films for industries. Currently, the modern industries which are harnessing the benefits of nanostructured thin films are namely; glasswork industry, automobile industry, packaging industry, military toiletries, and transportation industry. Widely grown nanostructured thin films for industrial productions are listed as; ZnO_2, TiO_2, CuO_2, and Al_2O_3 [7, 10, 11].

3 Importance of Nanostructured Thin Films

The objective of nanostructured thin film fabrication is to promote the quality of thin films and having long lifetime. Another aspect is related to low cost of fabrication for industrial production, maintenance and repair, easy to adapt to the fabrication environment. The development of nanotechnology has made the fabrication of nanostructured thin film as one of the leading fields for industries. The nanostructured thin film fabrication has advanced the production of materials with better control of surface quality and efficiency as compared to the previously manufactured with micrometric layers. The development of nanostructured thin films encouraged the surface engineering of the product to compete with the market requirement and produce a healthy environment of competition among the industries. The development of thin film technology for optoelectronic applications is summarised in Fig. 2.

Moreover, the surface engineering of nanostructured thin films is considered an economical approach for key industries and explains the mechanism of different processes during the growth of the product layer. More research work is underway to tune the future of surface engineering for future aspects. The market for surface engineering has a fair share in modern countries. In the UK, a recent survey estimated an approximate share of above 50 billion pounds of surface engineering and more than 50% of this amount had been utilised towards enhancing and improving the surface resistance and properties against surface corrosion of products. The production of nanostructured thin films is focused on meeting the requirement originating from agriculture, electronic consumers, aerospace, automobile, and electronic sectors [12, 13]. Therefore, surface engineering of nanostructured thin films has been regarded as a pivotal tool to design the surface of substrate with enhanced performance and economical aspects. The objective of surface engineering is directly related to the surface features for obtaining a particular application. Hence, surface engineering is acted as a bridge between the developments of technology for final consumers.

4 Techniques in Thin Film Fabrication

Thin films are prepared onto the substrate's surface to study the properties of nanomaterials which are unachievable in bulk materials. Multiple-layer deposition of different materials is often used to attain functionality in thin films. Optical interference filters, for instance, consist of tens or even hundreds of layers alternating between high and low indexes of refraction. When nanometre thicknesses of semiconducting materials such as GaAs and AlGaAs, are used to form alternate layers. This results in a superlattice that has electrical properties controlled by the constructed periodicity rather than by the atomic periodicity.

Multilayer thin films can behave as new materials in the bulk form. The combination of multilayer with lithographic patterning in designing thin films results in the construction of microstructures of endless variety. The combination of multi-layer with lithographic patterning comprises the technology of the integrated-circuit industry, the optical waveguide industry and micromechanical fabrication. There exists a wide variety of thin film deposition methods and techniques. Each technique has its own merits and demerits.

4.1 Physical Vapour Deposition

Currently, most of the metallisation for microelectronics is carried out by physical vapour deposition (PVD), which consists of evaporation and sputtering processes. The formation of a layer on a substrate using either of the two methods involves three steps: (1) converting a condensed phase material (generally a solid) into the gaseous or vapour phase, (2) transporting the gaseous phase from the source to the substrate surface, followed by (3) nucleation and growth of a new layer.

4.2 (a) Physical Vapour Deposition (PVD) by Evaporation

In PVD, the material to be deposited can be heated directly by an electron beam (or laser beam) by directing a stream of high-energy electrons at the target material to create a molten region at the surface and vaporise the material. The deposition of the evaporated material occurs on a substrate which is maintained at a lower temperature than that of the vapour. In a conventional evaporation process, the material to be deposited is heated under high vacuum conditions (10^{-5}–10^{-8} torr) [14]. Thin film deposition of alloy that consists of two or more components can be complicated due to differences in vapour pressures, which cause difficulty in maintaining stoichiometry of both the target and deposited films since the target becomes richer in the less volatile species. The evaporation process has some advantages like simplicity and reliability, along with high deposition rates and high film purity.

4.3 (b) Physical Vapour Deposition (PVD) by Sputtering

PVD by Sputtering is a process to deposit a film on the substrate surface via a physical mechanism. In this process, the atoms from the surface of the target material are ejected by striking with energetic ions which are accelerated in a blow discharge or plasma (10^{-2} torr) towards the target having a negative potential. Subsequently, the surface atoms of the target material are vaporised towards the substrate surface where they condense to form a film, in contrast to the evaporation process, were target material is evaporated by heating under a high vacuum (10^{-7}–10^{-8} torr) [15, 16].

Heating can be achieved by resistive, inductive, electron beam and arc techniques. Then, the vaporised species condense on the substrate surface.

4.4 Chemical Vapour Deposition (CVD)

CVD has been developed into a highly advanced and efficient thin film growth technology over the last twenty years [17, 18]. Many synthetic techniques have been tried in CVD due to which they have different types. Metal–organic chemical vapour deposition (MOCVD) is the most successful and widely used technique to produce metal oxide thin films of high and reproducible quality.

MOCVD is utilised for the preparation of solid films on a substrate by the gas-phase and surface reactions. This technique depends upon the volatility of the precursor material which is sublimed from solid phase into the vapour phase (without chemically decomposing), followed by reaction over a heated substrate, finally decomposing to deposit a solid species. The process proceeds under non-equilibrium conditions by generating chemically active species, which induce the growth of the solid network. At the middle stage of this deposition process, the parent or carrier gas contains the element(s) required as the film constituent(s).

$$[\mathrm{Cu(OAc)(dmae)(H_2O_2)}]_2 + \mathrm{He(carrier\ gas)} \rightarrow \mathrm{Cu_xO_y} + \mathrm{by-products}$$

The target of modern MOCVD is to provide films with good thickness and uniformity, controlled composition and stoichiometry, high purity, good adhesion onto the substrate and good mechanical, optical and electrical properties. In conventional MOCVD (atmospheric pressure), a carrier gas is bubbled through the liquid and/or solid precursors, when the compound is reaching to its equilibrium vapour pressure, then precursors are carried into the reactor.

4.5 Conventional CVD Process

The volatile precursor is a primary requirement of a conventional CVD process. The second important requirement is a clean thermal decomposition pathway, which results in the deposition of fine and good quality thin films. In a conventional CVD, a volatile precursor is deposited on the substrate surface in the reactor by a carrier gas (1). The precursor sorbs at the substrate surface, (2) and reacts (3) to liberate the by-products that subsequently desorbs (4) and are transported out of the reactor (5). The target atoms then diffuse (6) to form nuclei of the materials (e.g., CuO), where subsequent growth occurs (7) [14].

4.6 CVD Reactor Design

A conventional CVD reactor system consists of; (i) A reagent handling arrangement for the delivery of source compounds, (ii) A reactor unit (or reaction chamber), (iii) An exhaust system. The design of the reagent handling system depends on the source compounds. The source compounds, which form either liquid or solid, are mostly placed in a bubbler where other parameters such as source temperature, the flow rate of the carrier gas. The total pressure determines the amount of reagent transported from the bubbler over the source. A typical CVD reactor configuration has been demonstrated in Fig. 3, which can be applied to accommodate many CVD applications.

The horizontal and vertical reactors are employed for atmospheric and reduced pressure growth. Horizontal reactor systems can be divided into two categories, hot- and cold-wall reactors. Hot wall reactors are mostly used for academic research based on small scale, whereas in the industry, most of the CVD work has been carried out in single-wafer based on cold-wall reactors. The hot-wall reactor has

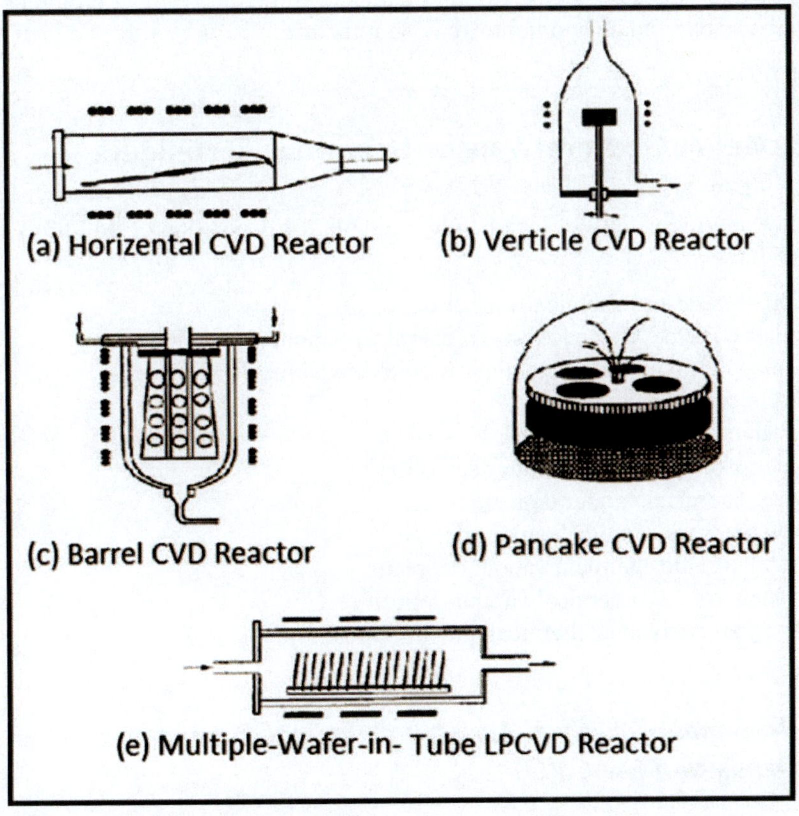

Fig. 3 **a** Horizontal CVD reactor; **b** Vertical CVD reactor; **c** Barrel CVD reactor **d** Pancake CVD reactor **e** Multiple-wafer-in-tube reactor [19]

many advantages, including the growth temperature, which can be kept constant as compared to cold-wall reactor. They are also simple to operate under various ranges of pressures and growth temperatures. There are some demerits of hot-wall reactor as well, critical one of which is that deposition can occur not only on the substrate but also on the reactor walls during the CVD process. Afterward, these deposits on the walls can fall off and contaminate the substrate surface. Cold-wall reactors have been used in the laboratory as well as industrial scale. Although cold-wall reactor only accommodates a few wafers (usually a single wafer) yet no chances of deposition on the reactor wall zone and growth temperature can be maintained in a certain area of the reactor zone, which results in less homogeneous reactions during the CVD process due to partial heating area in the reactor chamber. Higher deposition rates are possible with hot-wall reactor systems.

The barrel reactor is being extensively used for silicon epitaxy. The pancake reactor is also widely used for silicon technology but can be used with multi-wafers. The horizontal multiple-wafer-in-tube low pressure CVD reactor has been used for polycrystalline silicon, dielectric and passivation films. Hot-wall reactor should necessarily be used for the preparation of uniform films on substrate because it maintains growth temperature in the reactor chamber. Still, some critical problems may arise, such as particulate and impurity contamination.

5 Variant of Chemical Vapour Deposition Techniques

Nowadays a wide variety of techniques available for deposition of thin films using CVD such as,

1. Aerosol assisted chemical vapour deposition
2. Electric Field Directed Assisted chemical vapour deposition
3. Atmospheric pressure chemical vapour deposition
4. Chemical beam epitaxy
5. Chemical vapour infiltration
6. Electrostatic assisted vapour deposition
7. Laser chemical vapour deposition
8. Low pressure chemical vapour deposition
9. Metal organic chemical vapour deposition
10. Plasma assisted chemical vapour deposition
11. Thermally activated chemical vapour deposition.

5.1 Principal of Aerosol Assisted Chemical Vapour Deposition (AACVD)

This technique is versatile, straightforward to operate and handle [20]. The schematic AACVD is shown in Fig. 4. In AACVD process, single source molecular precursor

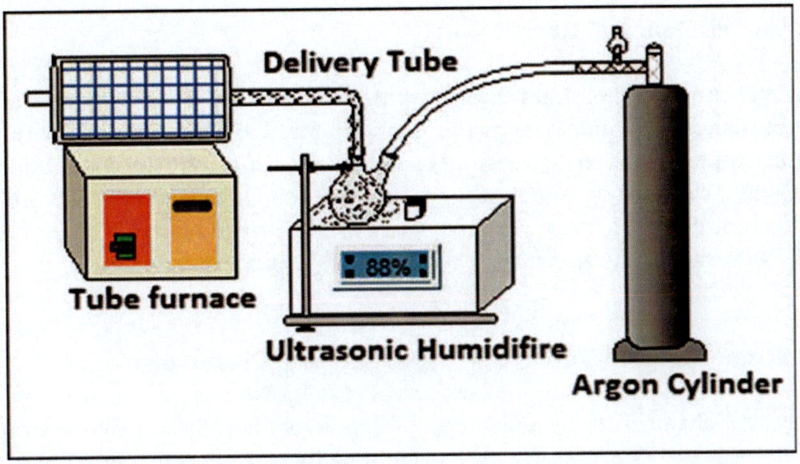

Fig. 4 Diagram of Advanced AACVD apparatus

is dissolved in an organic solvent and atomisation of the precursor solution take place at ambient temperature to produce sub-micro meter sized aerosol droplets that are distributed through a carrier gas which carries the generated aerosols into the thermal chamber, where the solvent undergoes rapid evaporation and the final product is deposited on the surface of substrate [14, 21].

5.2 *Technical Aspects of AACVD Process*

Following are the important technical aspects of AACVD process for the deposition of thin films.

5.2.1 Operational Pressure

The process of AACVD can be performed under atmospheric or low pressure conditions, while the deposition of oxides or sulphides thin films is carried out in air without reaction chamber.

5.2.2 Precursor

Metal alkyls and main group hydrides are mostly used as precursors since metal alkyls are often pyrophoric and the hydrides are highly toxic. Therefore, there is a need for special equipment to guard against safety and environmental hazards for such precursors.

5.2.3 Atomisation and Carrier Gas

In AACVD process, the atomisation of precursor solution can be carried out by using ultrasonic humidifier. The carrier gas facilitates the generation of aerosol and its subsequent transport to the reacting chamber. For deposition of oxide thin films, compressed air is used as career gas. At the same time, nitrogen or argon are used as inert carrier gases. However, reactive gases like hydrogen may also be introduced together with carrier gas as required by the AACVD process.

5.2.4 Evaporation of Solvent and Vaporisation of Precursor

In the heated chamber, the vaporisation of precursors is followed by the evaporation of the solvent. The aerosol droplets are directly converted to vapor phase. The precursor's vaporisation is the key feature of the AACVD process. Therefore, the selection of molecular precursor and the control over processing parameters are important to ensure the correct CVD process. Before complete evaporation of the solvent and vaporisation of precursor, if aerosol droplets reach the heated substrate, a spray pyrolysis process will take place instead of a true AACVD process.

5.2.5 Heterogeneous Reaction

In the gas phase, the partial decomposition of vaporised precursor may occur. The vaporised precursors and their gaseous intermediate undergoes substantial decomposition and chemical reaction when absorbed into the surface of substrate thus giving desired materials. For the heterogeneous chemical vapour deposition, reactive gases are introduced into the reaction chamber to fabricate high quality thin films with good adhesive properties.

5.2.6 Homogeneous Reaction

At high temperature, most of the decomposition reactions may occur much earlier in the vapour phase leading to nucleation homogeneously, thus forming fine metal particles of uniform composition. It is assumed that the temperature is uniform, and the velocity as well as the concentration of the reactive gaseous species tends to zero [22]. Reactive gases can also be introduced in homogeneous reaction externally. If these fine particles are absorbed on to the substrate surface, the heterogeneous reaction will lead to the formation of porous film. Alternatively, the powder in size ranges from nanometre to micrometer can be collected from the gas phase depending upon the reaction parameters.

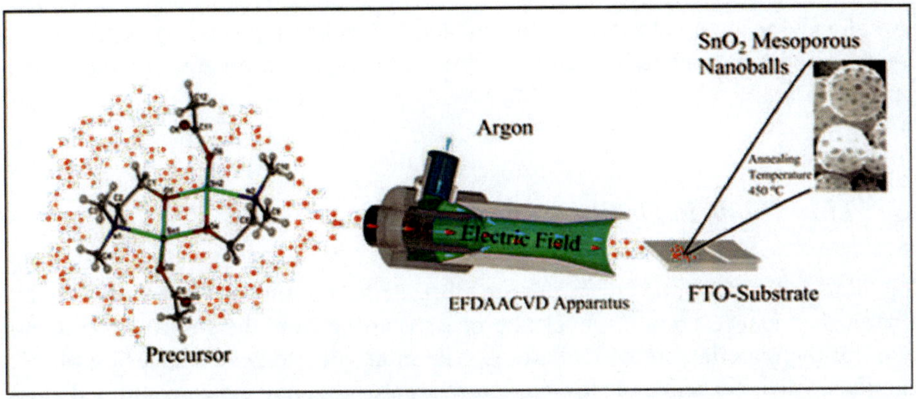

Fig. 5 The flow diagram of Electric Field Directed AACVD

6 Principal of Electric Field Directed Aerosol Assisted Chemical Vapour Deposition (EFDAACVD)

The aerosol under the influence of an electric field is sprayed on the surface of plain glass or fluorine doped tin oxide glass substrate at the desired temperature. A complete set up of an in-house built Electric Field Directed aerosol assisted chemical vapour deposition (EFDAACVD) apparatus is shown in Fig. 5 [23]. The synthetic strategy is mainly concerned with the synthesis of tin (IV) oxide porous nano-balls with diameter of 195–632 nm range sizes and with well-defined reproducible structure that has nano-sized pores in the range of 33 to 35 nm. The films deposited at 350 °C temperature and were annealed at 400 °C exhibited excellent adhesion properties to the substrate and were stable toward air and moisture contents.

7 Applications of Thin Films

The efficiency of the particular material for specific applications depends on the composition, atomic structure, microstructure, interfaces and defects, which are monitored due to the processing and synthesis by kinetics and thermodynamics of the production methods. The physical dimension of the material, i.e., either grain size or particle size less than 100 nm; plays important role for specific applications ranging from engineering and energy fields due to its unique characteristics as compared to conventional materials. Intensive study has been performed for almost a century in the development of a precise method for the fabrication of materials thin films with controlled characteristics as aforementioned. In this regard, many techniques have been improved and developed widely for industrial production of thin films [24, 25]. These have been providing excellent opportunities for further improvement and development of growth techniques. The well-known deposition techniques have already been summarised in the chapter. The following sections will emphasise the important applications of thin films in energy, coating and optoelectronic applications.

7.1 Thin Films in Optics

Thin films of the developed materials are fabricated at the surface of the substrate by either a chemical reaction or physical process, starting from the controlled condensation of the materials individual atoms. The more precise benefit of using plastic materials with an excellent coating of materials are replacing the conventional glass optics due to their unique characteristics such as lightweight, low breakage and low cost. More importantly, polymer based plastic optics can be easily moulded due to their flexibility according to the shape and dimension requirements. The coating of plastic optics is a very critical phenomenon and needs early attention during the development process. Antireflection coating is considered a critical property of the optical devices and has been tested on the eyeglasses in the mid nineteenth century. With the development of optical devices, the high precision antireflection coating has been in high demand. The more commonly used polymer for better antireflection coating on the display covers and camera lenses are either polycarbonate (PC) or acrylic [poly(methyl methacrylate) (PMMA)] of thermoplastic polymers. More efforts are required to meet the emerging demands of antireflection coating on optical devices with prolonged environmental stability and high adhesion. These are still challenges in dealing with the problems associated with antireflection coating on plastic optics due to complex chemical reactions during the production process. Therefore, efforts are on the way to overcome the challenges and to produce optical devices with enhanced antireflection coating on the plastic displays and lenses.

Generally, the commonly available plastic optics in the market are in the different shapes ranging from aspherical and spherical lenses, diffractive, free form, mirror surfaces and prisms. Injection moulding is commonly employed to produce polymer based plastic optics. Other than this, a single-point prototyping stage manufacturing process is used to fabricated plastic optics with extended heat resistance, lower birefringence and improved stability. Optical devices that relay on plastic are extensively used in medical devices, eyeglasses, cameras, projectors, fiber optics, displays, and automotive devices. Eyeglass coatings are very important because of the use of thermosetting polymer used in the manufacturing process to meet the requirement of high index. The materials are expected to be flexible shaped, biodegradable and biobased.

For optics, new thermoplastic materials, e.g., Iupizeta EP PC resins having high refractive indices low birefringence and high-temperature stability, have been introduced in the market in the former few years. Commercially antireflective applications of glass coatings have been introduced in the market since the 1960s. The optical glass coatings are performed by wet-chemical method involve the heating up to 400 °C of metal alkoxides or inorganic salts that forms inorganic oxide network at the surface of substrate due to polycondensation and hydrolysis. Organically modified silanes are produced using an alternative way through organic–inorganic composite materials. These monomers are polymerised either by thermally/photochemically or a combination of both. Pencil hardness and scratch-resistant coatings were increased by the use of grade 2B to 5H having colloidal silica cross-linking agents and by adjusting the total layer composition thickness [23].

7.2 Thin-Film Photovoltaic Cells

The exponential consumption of fossil fuels dramatically increases environmental pollution and global warming. In the past few decades significant amount of research is dedicated in the development of alternative renewable energy resources. Among renewable energy resources, the sun is the biggest source of energy through which a sufficient amount of energy can be harvested by using photovoltaic devices. The solar cells used for the conversion of solar energy are silicon based. The solar cell modules are designed on bulk silicon wafers. In 2016, up-to 94% the Si-wafer-based technology was responsible for the overall production. But, now from the total production 70% of the part is handled by the multi-crystalline technology. In 2017, 300 GWp were characterised by all the installed photovoltaic power, which will extend up to 1 TWp in 2021 [24].

For the production of low-cost and sustainable sources, thin-films photovoltaic modules are getting popular.

The major advantages of thin-films solar cells compare to crystalline and polycrystalline solar cells are;

(1) In the production processes, fewer amounts of energy and materials are used.
(2) Due to the relevancy to the large area, it maintains low-cost production.

There are also some failures faced by solar-cell thin film technologies such as:

- Minor conversion efficiency of the module
- The degradation of the material in some cases by the induced-light. Concerning the price and the state of their presentation, the inorganic thin films solar cells are competitive to crystalline silicon solar cells [25].

At the industrial stage, two of the main classes have been made for thin film solar cells:

The first one is based on different phases of silicon like amorphous, nano-crystalline and polycrystalline, while the other is made on polycrystalline chalcogenide semiconductor compounds (CdTe, CI(G)S). Advancement in the chemical synthesis for the production of organic materials is the result of hard work done by different companies like Sumitomo Chemical, Heliatek, Mitsubishi, Konarka, Solarmer, and Philips 66; which is appreciable [15]. Finally, the third classification is made based on organic materials.

8 Conclusion

In summary, this chapter presented a comprehensive overview of nanostructured thin film fabrication techniques from the surface engineering and chemistry point of view to enlighten their potential applications. An inclusive discussion has been sorted to highlight the surface chemistry of thin films for controlled and efficient growth using different protocols. Special emphasis has been given to different nanostructured thin film techniques and the effect of growth parameters on the surface of the substrate. The role of chemical composition and mechanical properties of thin film techniques are linked with the optimising processing cost. Modification of the chemical surface of thin films for advanced application and surface interactions is discussed for industrial processing. Indeed, various new areas for potential usage of nanostructured thin films are presented with recent developments and uses in possible applications.

References

1. Lee, T.D., Ebong, A.U.: A review of thin film solar cell technologies and challenges. Renew. Sustain. Energy Rev. **70**, 1286–1297 (2017)
2. Bakar, S.A., Tajammul Hussain, S., Mazhar, M.: CdTiO3 thin films from an octa-nuclear bimetallic single source precursor by aerosol assisted chemical vapor deposition (AACVD). New J. Chem. **36**(9), 1844 (2012)
3. Bakar, S.A., Ribeiro, C.: Nitrogen-doped Titanium Dioxide: an overview of material design and dimensionality effect over modern applications. J. Photochem. Photobiol. C Photochem. Rev. (2016)
4. Rühle, S., Shalom, M., Zaban, A.: Quantum-dot-sensitised solar cells. ChemPhysChem **11**(11), 2290–2304 (2010)
5. Campbell, D.S.: Preparation methods for thin films. In: Physics of Nonmetallic Thin Films, pp. 9–48. Springer, Boston, MA, USA (1976)
6. Smith, H.M., Turner, A.F.: Vacuum deposited thin films using a ruby laser. Appl. Opt. **4**(1), 147 (1965)
7. Hussain, S.T., Bakar, S.A., Saima, B., Muhammad, B.: Low temperature deposition of silver sulfide thin films by AACVD for gas sensor application. Appl. Surf. Sci. **258**(24), 9610–9616 (2012)
8. Bakar, S.A., Hussain, S.T., Trimizi, S.A., Tahir, M.N.: Crack free and spherical shape copper–zinc composite oxide thin films fabrications by AACVD. Polyhedron **49**(1), 138–144 (2013)
9. Chang, J.A., et al.: High-performance nanostructured inorganic-organic heterojunction solar cells. Nano Lett. **10**(7), 2609–2612 (2010)
10. Ahmed, S. et al.: Semiconducting composite oxide Y<inf>2</inf>CuO<inf>4</inf>-5CuO thin films for investigation of photoelectrochemical properties. Dalt. Trans. **43**(22) (2014)
11. Sunada, K., Watanabe, T., Hashimoto, K.: Studies on photokilling of bacteria on TiO2 thin film. J. Photochem. Photobiol. A Chem. **156**(1–3), 227–233 (2003)
12. Szeifert, J.M., et al.: Ultrasmall titania nanocrystals and their direct assembly into mesoporous structures showing fast lithium insertion. J. Am. Chem. Soc. **132**(36), 12605–12611 (2010)
13. Ravindrakumar, J., Bavane, G., SOPS, NMU.: Chapter 3 Synthesis of Polyaniline (PANI). Synth. Charact. Thin Film. Conduct. Polym. Gas Sens. Appl. **2014**, 1–22 (2014)
14. Butler, S.Z., et al.: Progress, challenges, and opportunities in two-dimensional materials beyond graphene. ACS Nano **7**(4), 2898–2926 (2013)

15. Gizopoulos, D.: Advances in Electronic Testing : Challenges and Methodologies. Springer, Berlin (2006)
16. Deb, S.K.: A Novel Electrophotographic System. Appl. Opt. **8**(S1), 192 (1969)
17. Lin, Z., Orlov, A., Lambert, R.M., Payne, M.C.: New insights into the origin of visible light photocatalytic activity of nitrogen-doped and oxygen-deficient anatase TiO2. J. Phys. Chem. B **109**(44), 20948–20952 (2005)
18. Zhai, T., et al.: A comprehensive review of one-dimensional metal-oxide nanostructure photodetectors. Sensors (Basel) **9**(8), 6504–6529 (2009)
19. Avendaño, E., Berggren, L., Niklasson, G.A., Granqvist, C.G., Azens, A.: Electrochromic materials and devices: Brief survey and new data on optical absorption in tungsten oxide and nickel oxide films. Thin Solid Films **496**(1), 30–36 (2006)
20. Van Benschoten, J.E., Reed, B.E., Matsumoto, M.R., McGarvey, P.J.: Metal removal by soil washing for an iron oxide coated sandy soil. Water Environ. Res. **66**(2), 168–174 (1994)
21. Sutar, D.S., Padma, N., Aswal, D.K., Deshpande, S.K., Gupta, S.K., Yakhmi, J.V.: Preparation of nanofibrous polyaniline films and their application as ammonia gas sensor. Sensors Actuators B Chem. **128**(1), 286–292 (2007)
22. Tang, J., Wen, X., Liu, Z., Wang, J., Zhang, P.: Synthesis and electrorheological performances of 2D PANI/TiO2nanosheets. Colloids Surfaces A Physicochem. Eng. Asp. **552**(April), 24–31 (2018)
23. Huang, X., Dai, B., Ren, Y., Xu, J., Zhu, P.: Preparation and study of electromagnetic interference shielding materials comprised of Ni-Co coated on web-like biocarbon nanofibers via electroless deposition. J. Nanomater. **2015**, 1–7 (2015)
24. Kondo, Y., et al.: Preparation, photocatalytic activities, and dye-sensitised solar-cell performance of submicron-scale TiO2 hollow spheres. Langmuir **24**(2), 547–550 (2008)
25. Ayyan, J.P.R., Karmakar, S.: A Critical Review on Dye Sensitized Solar Cells. 8–10 (2011)

Synthesis, Spectroscopic Characterization and Applications of Tin Dioxide

Hawazin Alghamdi, Benjamin Concepcion, Shankar Baliga, and Prabhakar Misra

Abstract Metal oxides are useful for the detection and sensing of combustible and toxic gases, and for use in lithium batteries and solar cells. The present study focuses on the spectroscopic investigation of commercial and in-house laboratory synthesized tetragonal tin dioxide (SnO_2), aimed at studying its physical and chemical properties at nanoscale levels and in bulk. We have investigated the pure powder form and thin films prepared on two different types of substrate, silicon and UV-Quartz, each with five different thicknesses (i.e. 41, 78, 96.5, 373, and 908 nm). Raman spectroscopy with two different laser excitation wavelengths, namely 780 and 532 nm, has been used to investigate the various SnO_2 vibrational modes. Thermal effects on the primary vibrational features in the Raman spectra have been studied in the range 30–170 °C. X-ray diffraction (XRD) spectra have been recorded to confirm the rutile structure of tin dioxide and to obtain information on the spherical grain particle size of SnO_2 with EDS analysis for the thin film samples. Scanning Electron Microscope (SEM) images have been recorded in order to understand the morphology of the particles of SnO_2 at the nanoscale level. In addition, FT-IR spectra have been obtained to study the IR-active vibrational modes for the bulk and thin film samples on the two substrates. Moreover, UV-VIS spectra have been employed to determine the energy band gap for the SnO_2 film samples by an efficient process facilitated by a Tauc plot technique utilizing an in-house developed python script.

Keywords Tin dioxide · Raman scattering · FT-IR spectroscopy · X-ray diffraction · Scanning Electron Microscopy (SEM) · UV-VIS spectroscopy · Gas sensor · Solar cells · Li-ion batteries · Molecular Dynamics (MD) simulation · Tauc plot

H. Alghamdi · P. Misra (✉)
Laser Spectroscopy Laboratory, Department of Physics & Astronomy, Howard University, Washington, DC 20059, USA
e-mail: pmisra@howard.edu

B. Concepcion
Department of Physics, University of California, Berkeley, CA 94720, USA

S. Baliga
General Monitors, Lake Forest, CA 92630, USA

N. M. Mubarak et al. (eds.), *Contemporary Nanomaterials in Material Engineering Applications*, Engineering Materials,
https://doi.org/10.1007/978-3-030-62761-4_11

1 Introduction

1.1 Metal Oxides

Metal oxides play a crucial role in applications in a variety of diverse fields, such as material science, inorganic chemistry, condensed matter physics, geology, electrical engineering and mechanical engineering [19]. They also have exceptional potential as base materials in developing leading-edge technologies. In recent years, extensive research effort has been directed towards metal oxides covering a range of applications in frontier areas in optoelectronics and device engineering. Depending on the nature of the surface, size and shape of the active metal oxide material(s), the response characteristics may differ. For example, the optical and electronic properties of such materials rely on the structure and spatial dimensions involved, whereas the physical and chemical properties of low-dimensional oxide materials can be remarkably altered by the Coulombic, Van der Waals, and interatomic coupling, due to the large number of atoms present on the surface in comparison to its bulk form. Therefore, significant interest exists in gaining a better understanding of a variety of low-dimensional materials in different forms, such as nanoparticles, nanowires, nanotubes, nanospheres, nanorods, nanoribbons, nanodisks and nanosheets. Enhanced insight into the thermal, electrical, and optical properties of such nanomaterials significantly contributes to improved fabrication of these materials with enhanced magnetic, electronic, and optoelectronic properties for a wide range of device applications [12]. Metal oxides are considered flexible and have been found useful for a host of applications relating to medical technology, water treatment, air pollutant reduction, purification of industrial and military polluted soils, and in personal care products [9].

1.2 Tin Dioxide

Mineral tin dioxide, often called stannic oxide or cassiterite (SnO_2) has the tetragonal crystalline structure, as do many other metal oxides. The rutile structure of SnO_2 has a tetragonal unit cell with $P4_2/mnm$ space-group symmetry and lattice constants $a = b = 4.7374$ Å and $c = 3.1864$ Å [13]. This structure is the essential form among all other phases that can [14] be produced under high mechanical pressure from the commonly stable and available rutile phase. SnO_2 can have different phases of polymorphs, such as $CaCl_2$-type (Pnnm), α-PbO_2-type (Pbcn), ZrO_2-type orthorhombic phase I (Pbca), fluorite-type ($Fm\bar{3}m$), pyrite-type ($Pa\bar{3}$), and a nine-fold coordination cotunnite-type orthorhombic phase II (Pnam) [12]. The unit cell of rutile-type SnO_2 contains six atoms; two of which are six-fold coordinated Sn and four are three-fold coordinated O ions [13]. Tin (IV) oxide is an n-type semiconductor with the chemical structure

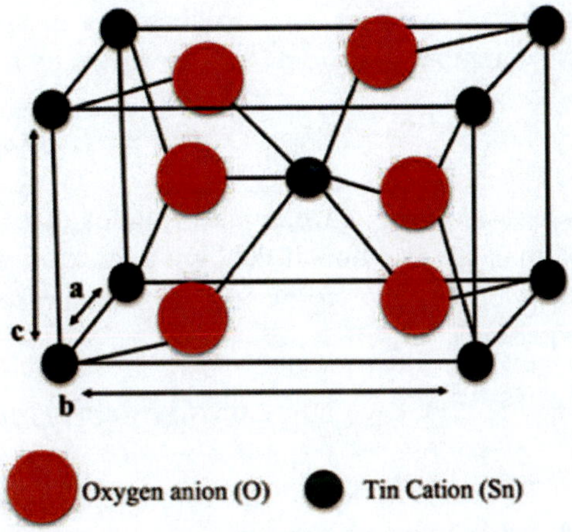

Fig. 1 Unit cell rutile crystalline structure of SnO_2 [4]

O=Sn=O. Figure 1 shows the rutile tetragonal structure of tin dioxide, with six-anion oxygen surrounding a tin cation in the middle to form an imperfect octahedron of rutile structured tin atoms. The ionic radius of O^{2-} is 1.40 Å, while that for Sn^{4+} is 0.71 Å [14]. Each of the Sn atoms is located at the lattice vertices corresponding to the neighboring six O's of the distorted octahedral. The Sn atoms are located at (0,0,0) and (½ a,½ a,½ c), while O atoms are located at ± (ua,-ua,0) and (½ a,½ a,½ c) ± (ua,ua,0), where u = 0.307 [3, 39]. SnO_2 is known for its unique properties, such as chemical and electrochemical stability, high optical transparency, nontoxicity, and efficient electron communication when doped [1]. It is a semiconductor with a wide direct band gap between the conduction and valence bands of approximately 3.6 eV [4]. It can be used effectively as a transparent conducting oxide gas sensor, a transparent electrode for lithium ion batteries, in solar cells, and finds potential applications in electrochromic devices [11, 58].

1.3 Methodology (Raman, FT-IR, XRD, SEM)

Raman Spectroscopy

The principle of Raman spectroscopy is based upon inelastic scattering of light incident on a sample. When a photon of frequency ν_0 and energy $h\nu_0$ impinges on molecules of a sample, the molecules are transported to a *virtual energy state* for a short time interval, followed by the emission of a photon of frequency ν_1 of energy $h\nu_1$. The electron cloud of the sample interacts with the electric field of the monochromatic light and as a result the molecules get excited to a new rovibronic state and undergo a variation in their electric dipole-electric dipole *polarizability* that results in the Raman scattering effect. The intensities of the Raman lines are proportional to the change in the polarizability that the sample molecules experience, whereby the Raman spectrum (a plot of the scattering intensity versus frequency shift) is a function of the rovibronic states of the molecules. The Raman shift $\Delta\nu$ in wavenumber units (cm^{-1}) is governed by the following equation:

$$\Delta\nu = [(1/\lambda_0) - (1/\lambda_1)] \quad (1)$$

where λ_0 is the wavelength of the excitation laser and λ_1 is the wavelength of the Raman spectral line. If the laser excitation wavelength (λ_0) and the Raman scattered wavelength (λ_1) are in nm, the Raman frequency shift (in cm^{-1}) is given by the equation:

$$\Delta\nu = [\{1/\lambda_0(\text{nm})\} - \{1/\lambda_1(\text{nm})\}] \times \left(10^7\,\text{nm}\right)/\,(\text{cm}) \quad (2)$$

The magnitudes of the Raman shifts are dictated by the polarizability of the electrons in the molecules.

Most of the excited molecules in the sample emit scattered radiation with a frequency that is equal to the incident radiation frequency and experience Rayleigh scattering (i.e. elastic scattering). However, a small portion of the scattered light also has a frequency different from the incident frequency and undergoes Raman scattering (i.e. inelastic scattering). If the final state of the molecule is higher than the initial energy state, the scattered photon will be shifted to a lower energy (lower wavenumber); this *downshift* defines the Stokes-shift. Conversely, when the final molecular state is lower in energy as compared to the initial state, the Raman scattered photon is shifted to a higher energy (higher wavenumber); the corresponding *upshift* is referred to as the anti-Stokes shift. The Raman spectra are therefore composed of both Stokes lines and anti-Stokes lines [7, 13, 38].

To characterize our tin dioxide samples, we used a DXR™ SmartRaman spectrometer with an excitation wavelength of 780 nm and $M^2 \leq 1.5$ beam quality. We used the 780 nm excitation to measure the Stokes Raman spectra of the SnO_2 in bulk form at different temperatures with a high brightness single-mode diode laser. Figure 2 shows the instrument's internal optical layout with 780 nm excitation, where the laser is focused on the sample by an objective lens resulting in scattered light that is filtered (to remove the Rayleigh component) and subsequently imaged on to the CCD detector, yielding the Raman spectrum of our sample [8].

A heated cell is accommodated within the Raman SmartDXR™ spectrometer sample compartment, in order to study the thermal effects on the recorded Raman spectra of the samples. We also used a Renishaw inVia Raman microscope with a laser wavelength of 532 nm to record the spectra of the tin dioxide samples. The powdered sample of SnO_2 (bulk) was compressed and transferred to the Ventacon™ model H4-200 cell (shown in Fig. 3) that can be heated up to 200 °C. We utilized the heated cell to control the environment, while recording the Raman spectra of the samples simultaneously. The H4 cell system comprises of an oven fitted with a window at the front and a sealing system at the back. The sample is gently pressed to the rear face of the window with a stainless-steel rod.

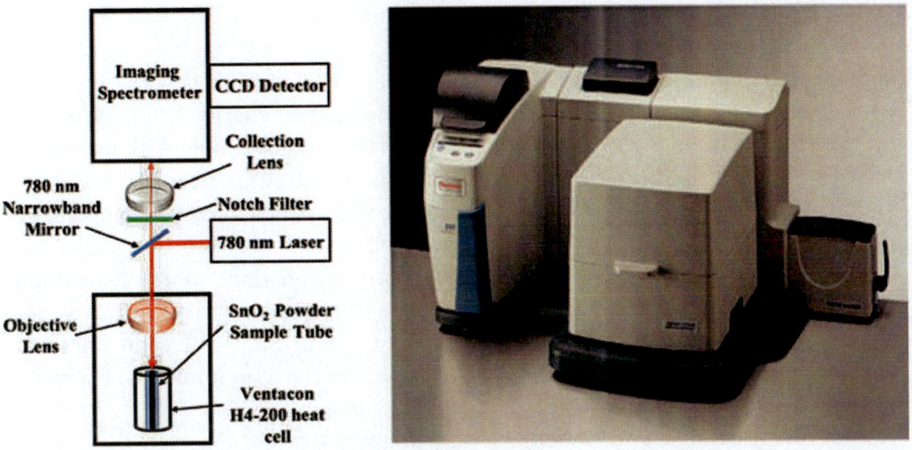

Fig. 2 Schematic of the Raman SmartDXR™ Spectrometer (left) and the Raman SmartDXR™ instrument (right). (Adapted from Casimir et al. [8])

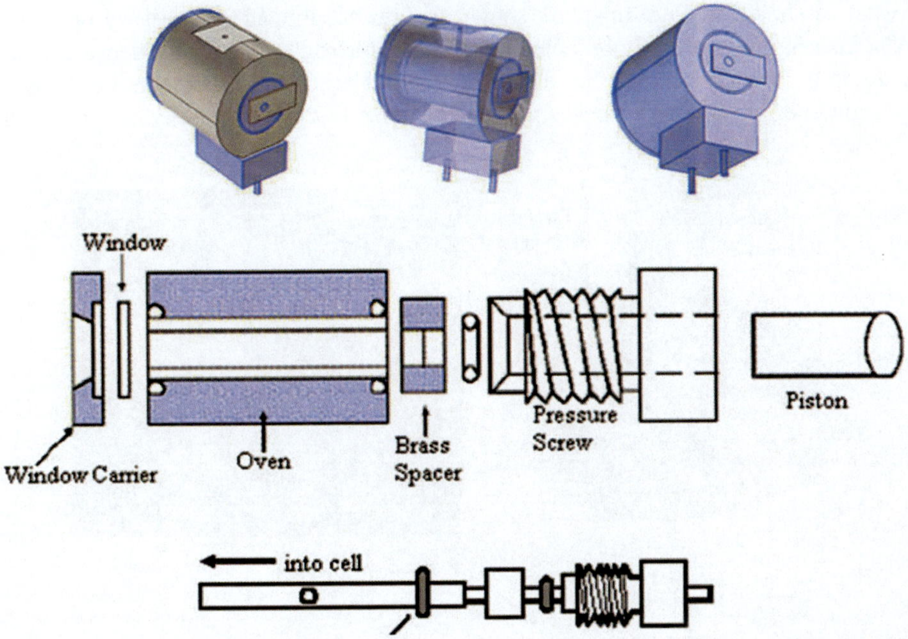

Fig. 3 (Top) An overview of the Ventacon™ H4-200 heated cell; (middle) schematic cross-section of the heating cell; and (bottom) schematic cross-section of the cylindrical container of powdered samples. (Adapted from Casimir et al. [8])

Fourier Transform-Infrared Spectroscopy

When infrared radiation interacts with a molecule, some of the light gets absorbed. The molecule then vibrates and gets excited to a higher energy level. When the photon energy between the two permitted energy levels is coincident with the energy difference between the pair of levels, an absorption spectrum is recorded [38]. As a result, the resultant modes of vibration are infrared vibrations due to a change in the dipole moment [5].

In the Fourier Transform (IR) technique, a broadband light source shines into a Michelson interferometer. Different wavelengths are modulated at different rates and the beam exiting the spectrometer is a snapshot in time called an interferogram. Fourier transformation is used to convert the interferogram into an absorption spectrum of the signal at a series of discrete wavelengths.

We have collected FT-IR spectra of the samples using a Thermo Scientific™ Nicolet™ iS50 FT-IR Spectrometer. The iS50 FT-IR has a tungsten-halogen white light source with a Polaris™ long-life IR source. The Attenuated Total Reflection (ATR) technique, which requires minimal sample preparation is used with a diamond crystal as shown in Fig. 4a–c. As shown in Fig. 4c, the ATR accessory operates by collecting the internally reflected infrared beam from an optically dense crystal (diamond). When the crystal comes into contact with the sample, the sample absorbs the light energy, and the evanescent wave is attenuated. Subsequently, the reflected

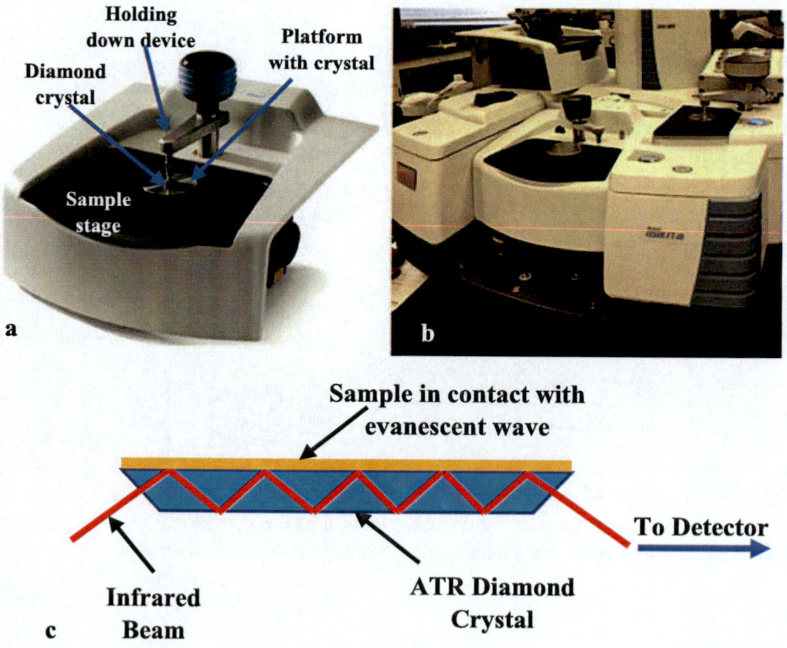

Fig. 4 **a** ATR sample stage; **b** Scientific™ Nicolet™ iS50 FT-IR Spectrometer; **c** Schematic of ATR top plate. Thermo Fisher Scientific™

energy goes to the IR spectrometer detector, which records the attenuated IR beam and generates the FT-IR spectra.

X-ray Diffraction

X-rays incident on a sample interact with the electrons of the array of atoms making up the material and experience *elastic scattering*. The resulting spherical waves undergo constructive interference to produce a diffraction pattern governed by Bragg's law:

$$2d \sin\theta = n\lambda \tag{3}$$

where d is the interplanar spacing in the nanomaterial sample, θ is the incident angle (as shown in Fig. 5) and λ is the wavelength of the incident X-ray radiation.

The X-Ray diffraction spectra of the SnO_2 samples were collected using a Thermo Scientific™ ARL™ EQUINOX 100 X-Ray diffractometer (XRD) (shown in Fig. 5). The size of the beam is approximately 5 mm × 300 μm, and a spinning stage for powder samples is used as illustrated in Fig. 5 (top). The EQUINOX curved detector can measure all diffraction peaks simultaneously across a wide angular range, which means it measures the whole 2θ range simultaneously and in real-time; no scanning is required. A regular X-ray diffraction instrument includes the X-ray source or tube,

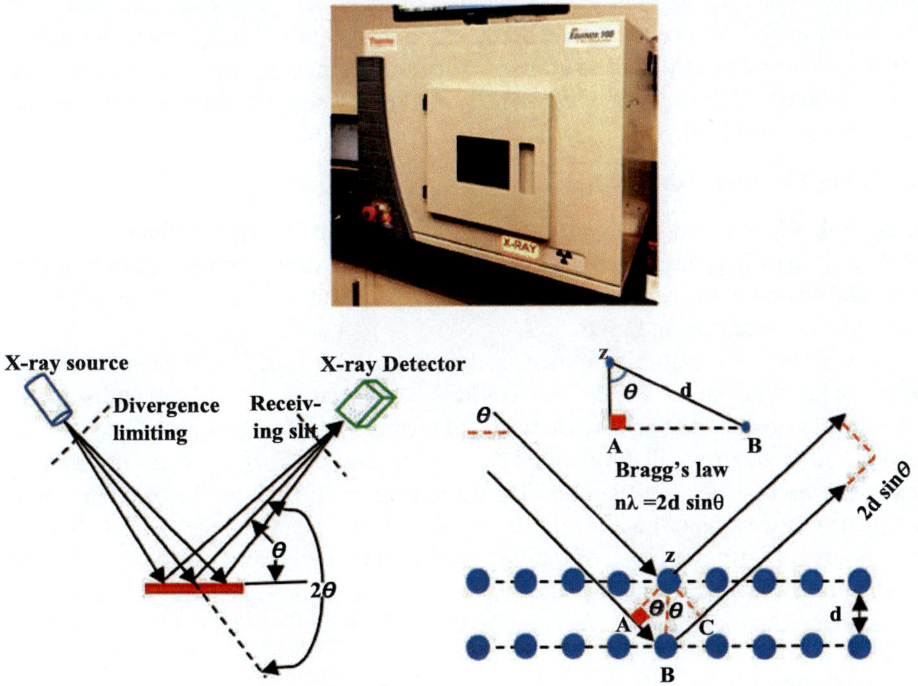

Fig. 5 (Top) Thermo Fisher Scientific™ ARL™ EQUINOX 100 X-Ray diffractometer; (bottom) Schematic of X-ray diffraction

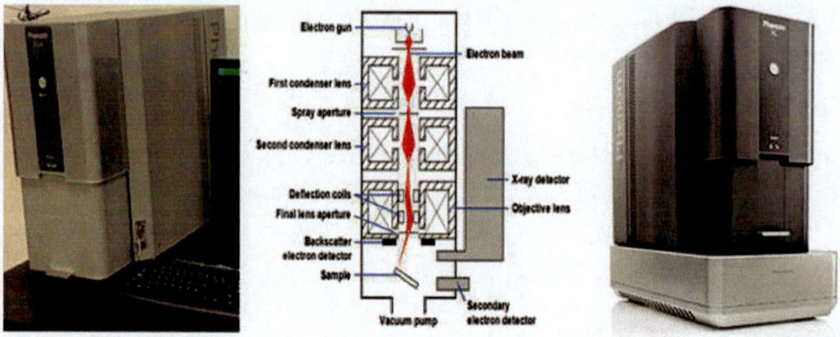

Fig. 6 (Left) Phenom Pure Scanning Electron Microscope (left); (Middle) Schematic of SEM adapted from Nanoscience Instruments, Inc.^{TM}; (Right) Phenom XL Desktop Scanning Electron Microscope, Thermo Fisher Scientific^{TM}

a sample stage, an optics receiver, and an X-ray detector. The ARL^{TM} EQUINOX has the X-ray tube, and the X-ray detector is stationary. The only moving part in the XRD is the sample stage. The foundational principle behind XRD analysis is Bragg's law, which relates X-ray diffraction to the crystalline structure, allowing it to be determined experimentally. X-ray diffraction is a powerful technique used to study solid matter materials. Solid materials can either be crystallized, with the atoms organized in a regular or ordered manner, or amorphous, where the atoms are organized randomly. As shown in the schematic diagram, the angle between the X-ray source and the sample is θ, and the angle between the X-ray projection and the detector is 2θ [49].

Scanning Electron Microscope

A Scanning Electron Microscope (SEM) was utilized in order to investigate the surface morphology and topography of the bulk and film samples. Both Phenom Pure and Phenom XL desktop instruments have been utilized to study the samples. Both SEMs are shown in Fig. 6.

An electron gun emits electrons from a cathode, with variation in electron energy between hundreds of eV and few tens of keV. The electrons are subsequently accelerated by the gradient of an electric field and the beam passes through two condenser lenses and through an electromagnetic scanning coil to be focused on the sample that is placed in a vacuum chamber. The scanning coil directs the electron beam into the plane of the sample (i.e., x and y directions). The accelerated electrons with high energy create different types of emitted and scattered electrons from inelastic and elastic collisions occurring near the surface of the sample. The inelastic scattering of incident electrons produces secondary electrons that are detected by an electron detector [36].

2 Synthesis Methods and Characterization of SnO_2 Films

2.1 SnO_2 Powder and Thin Films on Si and UV-Quartz

Commercially purchased SnO_2 powder from SIGMA-ALDRICH with 325 mesh and 99.9% purity was used as a standard and for spectroscopic characterization. The powder sample was annealed at 1000 °C for 1 h. Thin films of SnO_2 on UV quartz and silicon wafers were created by radio frequency (RF) sputter deposition in an oxygen/argon mixed atmosphere. The sputtering target was a 6-inch diameter, ¼ inch thick, ceramic target made from high purity tin dioxide. The deposited tin dioxide film thickness was monitored by sputter deposition onto a polished silicon substrate. A Filmetrics F20 instrument was used for curve fitting to calculate film thickness. Film thickness was varied by changing the sputter deposition time. UV-Quartz substrates were placed in the sputtering machine alongside the polished silicon thickness monitor substrate. UV-Quartz was chosen because it permitted transmission measurements in the UV at wavelengths as low as 190 nm with a UV spectrophotometer. The measurement of the SnO_2 band gap via UV transmission is described in Sect. 3.5.

3 Properties of SnO_2 via Spectroscopic Characterization

3.1 Temperature-Dependent Raman Spectroscopy of SnO_2

Raman spectroscopy is considered a versatile and sensitive technique to examine modifications in vibrational phonon modes crystallinity, surface and bulk atoms, disorder, and size effects in nanometric crystallites. The present research focuses on the vibrational modes of the rutile-structure of SnO_2, where the normal modes with space-group symmetry of $P4_2/mnm$ (point group symmetry D_{4h}^{14}) at the Brillouin zone are given by Eq. (4) [35]:

$$\begin{aligned}\Gamma =& 1\mathrm{A}_{1g} + 1\mathrm{A}_{2g} + 1B_{1g} + 1B_{2g} + 1E_g + 1\mathrm{A}_{2u} \\ &+ 2B_{1u} + 3E_u\end{aligned} \tag{4}$$

where A_{1g}, B_{1g}, B_{2g}, and doubly degenerate E_g vibrational modes are Raman active, with the feature that the tin atoms remain at rest. A_{2u} and the triply degenerate E_u modes are active in IR (i.e. InfraRed light). The A_{2g} and the B_{1u} modes are considered silent (i.e., inactive optically) [13]. The vibrational modes are illustrated in Fig. 7 [4].

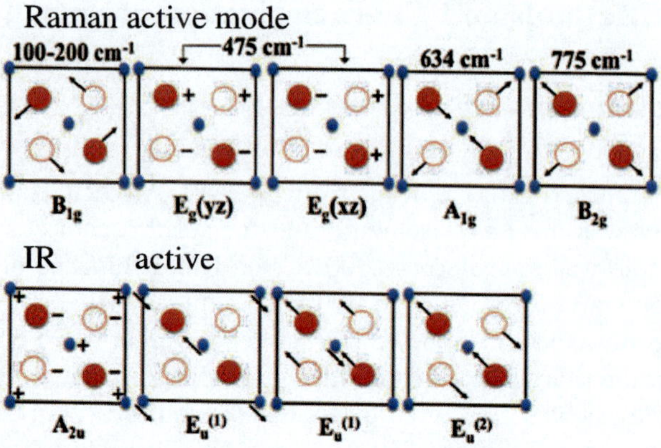

Fig. 7 Top view along the z-axis of rutile-structured SnO_2 optically active vibrational modes. The Sn atoms are indicated by (blue) dots, and O atoms are indicated by (red) dots: Raman active modes are shown with approximate wavenumber positions (top) and IR active modes (bottom), where the arrows represent the direction of movement in plane and (+ or −) represents the movement along the z-axis [13]

The B_{1g} Raman active mode, around 123 cm^{-1}, is not visible in many Raman spectrometers because its intensity is about three orders of magnitude less than that of the A_{2g} peak [31, 30]. The Raman spectra were collected by choosing a laser power of 10 mW and 25 μm aperture slit at different increasing temperatures from 30 to 170 °C, in steps of 20 °C (as illustrated in Fig. 8). Temperature variations were accomplished by use of the Ventacon H4-200 heated cell (shown in Fig. 3).

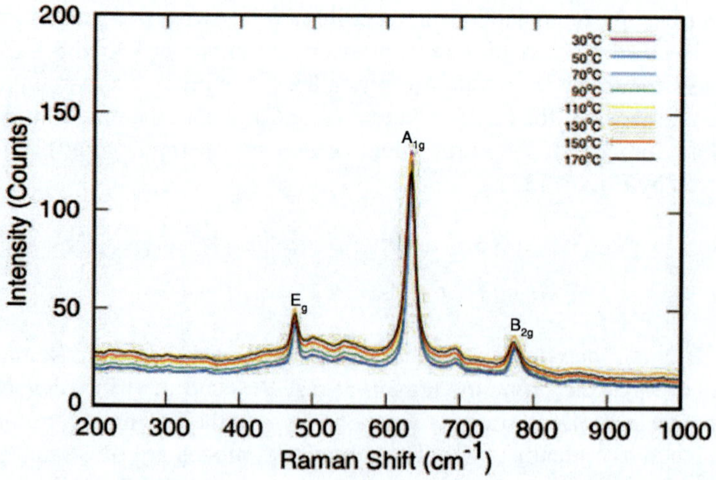

Fig. 8 Raman spectra of SnO_2 powder with 780 nm wavelength excitation as temperature varied between 30 and 170 °C

We have observed three of the four Raman active modes (E_g, A_{1g} and B_{2g}) in the SnO_2 powder sample, where $A_{1g} = 634\ cm^{-1}$ is the most intense peak and represents the symmetric O–Sn–O stretching, $B_{2g} = 775\ cm^{-1}$ represents the antisymmetric O–Sn–O stretching, and $E_g = 475\ cm^{-1}$, which is doubly degenerate, represents the translational mode of SnO_2 [18, 47]. The B_{1g} mode cannot be observed with the rest of the Raman active phonons. In contrast to the other Raman active modes, the B_{1g} band dramatically increases in wavenumber with increasing temperature [45]. Other weak abnormal Raman modes are noticed at approximately 501.7, 544.0, and 695.3 cm^{-1}, which are similar to the reported bands of SnO_2 nanoribbons [53].

A comparison is done below for the two Raman active bands A_{1g} and B_{2g} with temperature variation in the range 30–170 °C, as shown in Fig. 9. The A_{1g} and B_{2g} bands decrease in wavenumber with increasing temperature, while the E_g mode appears to have almost no change as the temperature increases from 30 to 170 °C, as demonstrated in Fig. 10. Raman spectra of SnO_2 are modified based on the size, shape, and temperature of the crystal.

Increasing crystal size leads to a dramatically decreased wave number for the A_{1g} and B_{2g} bands, while the E_g band slightly decreases [31, 30]. At smaller crystal sizes, three bands are noticeable in the spectrum of SnO_2 powder, denoted as S_1, S_2, and S_3 as shown in Fig. 11. S_1 and S_2 slightly decrease in peak wave number with increasing crystal size, while S_3 has no reported shift. The amplitudes of these bands decrease as particle size increases. Some values for these peaks, when examined using a 488 nm Raman spectrometer, are shown in Table 1. Average crystal sizes are denoted by L and were determined through TEM measurements [14].

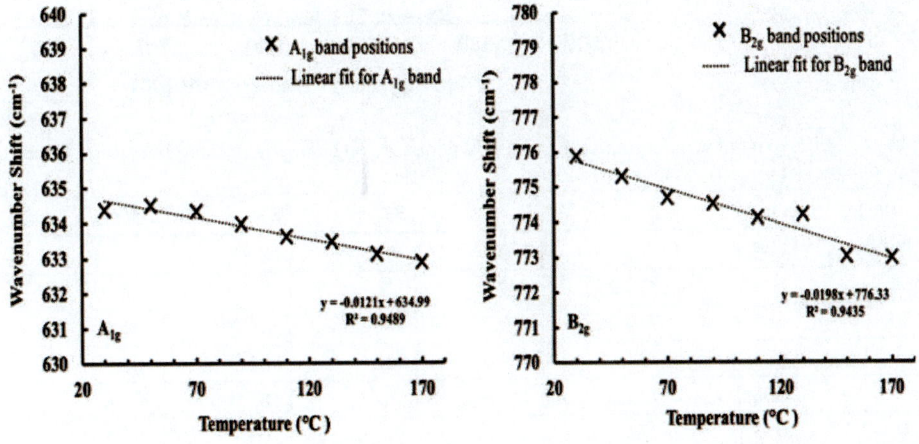

Fig. 9 Wavenumber shifts for the A_{1g} and B_{2g} vibration bands versus temperature in the range 30–170 °C

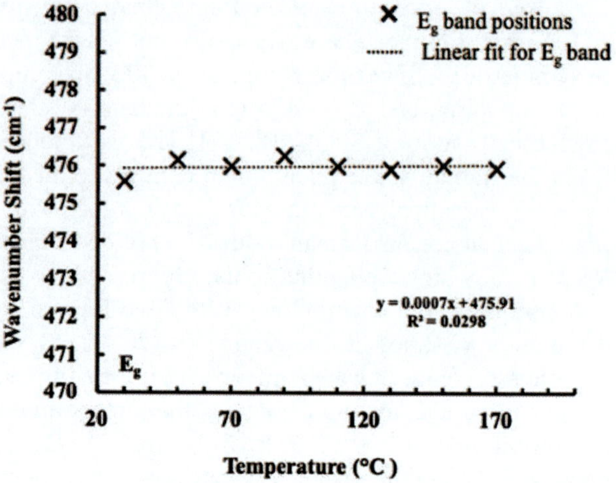

Fig. 10 Wavenumber variation for E_g vibration band versus temperature in the interval 30–170 °C

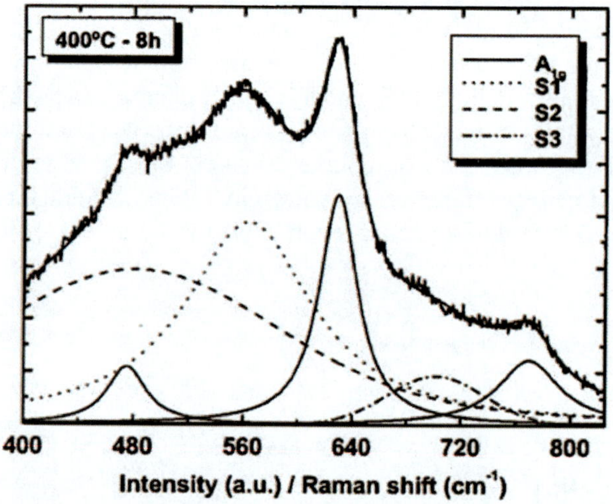

Fig. 11 Raman spectra of SnO_2 powder comprised of A_{1g}, B_{2g}, and E_g bands, along with S_1, S_2, and S_3 bands. Reproduced from Dieguez et al. [14], with permission of AIP Publishing

Table 1 Shifts in band wavenumber for SnO_2 powder based on changing particle size. Reproduced from Dieguez et al. [14], with permission of AIP Publishing

L (nm)	A_{1g}	B_{2g}	E_g	S1	S2	S3(A_{2u})
109.8	637.9	779.2	479.2	….	….	….
56.6	637.9	778.9	479.5	….	….	….
15.8	635.2	774.2	479.7	568.9	493.2	696.3
9.6	634.3	774.1	479.6	568.0	501.2	706.9
6.7	633.9	776.2	479.6	568.1	485.9	705.8
5.0	633.1	772.9	….	572.0	503.7	691.7
4.5	631.6	767.7	….	573.3	518.0	….
3.5	631.5	….	….	575.8	541.7	….

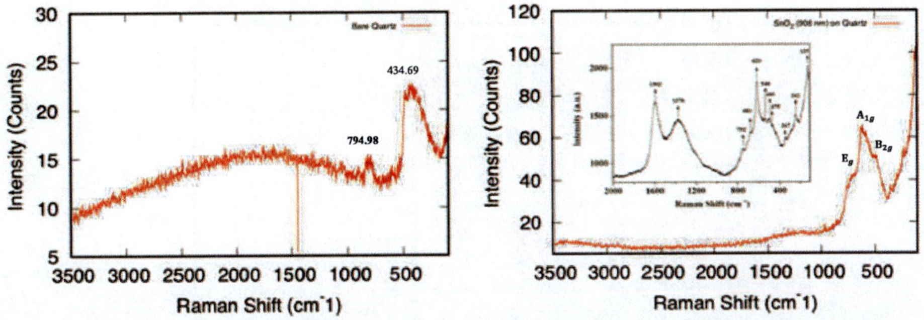

Fig. 12 DXR Raman spectra with 532 nm wavelength: Raman data of bare Quartz (left). Raman data of SnO_2 (908 nm) film thickness on Quartz compared with SnO_2/SiO_2 (right). The compared figure is retrieved from Ferreira et al. [18] with open access provided by SciELO Publishing

Thin films of SnO_2 will necessarily pick up spectroscopic characteristics from the media they are deposited on. Examples of such media are silicon and UV-Quartz; their pristine Raman spectra, along with their spectra with a 908 nm thick SnO_2 film, are shown in Fig. 12. A different wavelength of 532 nm was used because the 780 nm and the 514 nm wavelengths did not show any signs of the tin dioxide Raman bands.

The spectrum of the bare UV-Quartz has been collected with two pronounced features of quartz at (434.69 cm^{-1}) and in the interval (794.98–823.15 cm^{-1}), which approximately match the Raman peaks of borosilicate glass around (450–480 cm^{-1}) and (800 cm^{-1}), respectively [28]. The SnO_2/Quartz (908 nm) film exhibited evidence of the Raman active modes around (627.07, 490.19, and 745.42 cm^{-1}), which resemble A_{1g}, E_g, and B_{2g} features, but are at somewhat different wavenenumbers, possibly due to the presence of the quartz substrate, a similar effect to what has been observed previously [18].

3.2 *XRD Spectroscopy of* SnO_2

X-ray diffraction (XRD) spectra were measured by utilizing the Thermo Fisher Scientific ARLTM EQUINOX 100 X-Ray diffractometer with a Cu–K α monochromatic radiation source with wavelength $\lambda = 1.5406$ Å. The size of the beam is roughly 5 mm × 300 μm, with a spinning stage for powder sample loading. The spectra were collected in the range of $2\theta = 20$–$80°$. The XRD patterns for tin dioxide in powder form with 325 mesh, 99.9% trace metals, were recorded with 2 peak values and the associated numbers of Miller Indices (*hkl*) as 26.78° (110), 34.04° (101), 38.25° (200), 51.98° (211), 55.01° (220), 62.13° (310), 64.97° (112), and 66.15° (301), respectively, as shown in Fig. 13, which agree with Patil et al. [43] and Ferreira et al. [18].

The XRD spectra confirm the rutile tetragonal structure of SnO_2. XRD patterns for SnO_2 thin films were collected with thickness values of 41, 78, 96.5, 373, and

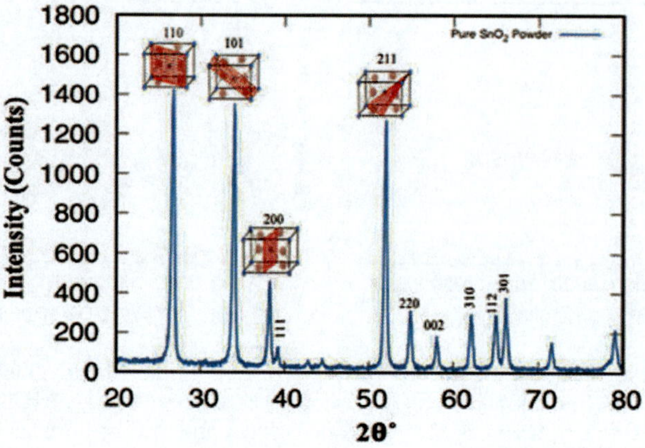

Fig. 13 XRD pattern of SnO_2 powder

908 nm, along with a bare double polished silicon wafer to act as a control. Figure 14 shows the XRD spectra of tin dioxide thin films over the range of $2\theta = 0$–$80°$. For the bare silicon wafer substrate, we can see the broad peak in the range $2\theta = 5$–$20°$ approximately. The peak looks shifted in comparison to Su et al. [51] where they have the range to be $2\theta = 10$–$20°$.

For the lowest ratio thickness of the film (i.e., 41 nm), we can observe properties similar to the silicon substrate, except for a small hump around 26.2° that matches

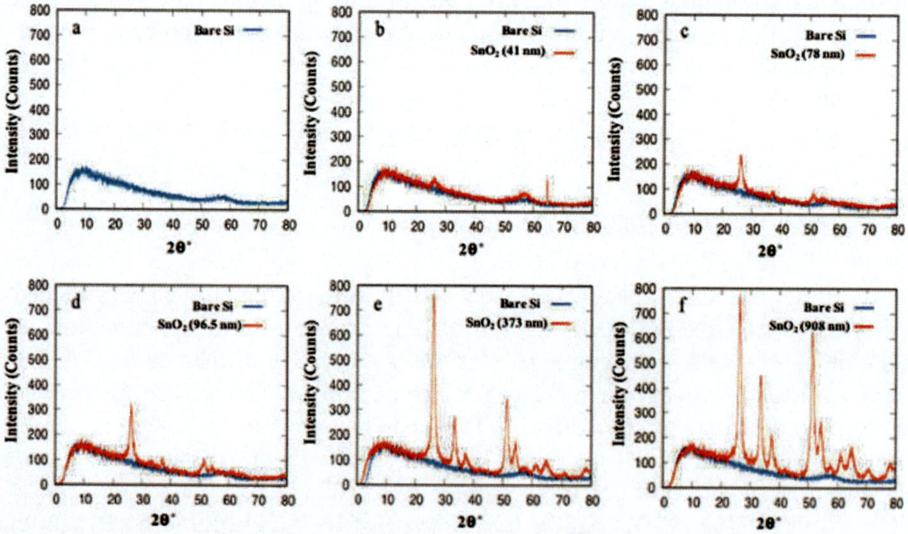

Fig. 14 XRD spectra of SnO_2 thin films on Si substrate: **a** bare double polished silicon wafer substrate, **b** comparison of silicon substrate with film of 41 nm thickness of SnO_2, **c** silicon substrate with 78 nm SnO_2 film thickness, **d** silicon substrate with 96.5 nm SnO_2 thickness, **e** silicon substrate with 373 nm SnO_2 film thickness, **f** silicon substrate with 908 nm SnO_2 film thickness

the (110) orientation. As the thickness increases to 78 nm, another feature appears around 37.1°, which could be associated with the (200) index, along with the peak 51.36°, which corresponds to the (211) index. When the thickness of the tin dioxide film is 96.5 nm, the three previously mentioned peaks become more pronounced. With a thickness of 373 nm, the 33.64° with (101) index appears, along with other features that were not visible with thinner films.

In addition to the silicon substrate, XRD spectra were also taken for the tin dioxide films on quartz substrate, with the same thicknesses as on the Si substrate, starting from 41 nm and ending with 908 nm. Figure 15 shows the X-ray diffraction spectra of the SnO_2 films in the range of $2\theta = 0$–80°. The data of bare quartz show a very broad peak at around 21.62°, and the broadening starts from $2\theta = 10$–35° approximately. The lowest ratios of tin dioxide thickness, namely 41 nm, 78 nm, and 96.5 nm, show no evidence of SnO_2 structural peaks. The features of tin dioxide start to appear (i.e., 26.36° (110), 33.92° (101), 51.5° (211)) with the 373 nm thickness film, and become stronger and more pronounced with the 908 nm ratio with values (26.19°, 33.74°, 51.5°). It shows more of the amorphous behaviour due to the smaller size of the particles formed in the film in comparison with SnO_2 on the Si substrate samples. From these data, we were able to calculate the crystallite size (D) using the Deybe-Scherrer formula in Eq. (5), as used by Agrahari et al. [1] and Henry et al. [23]:

$$D = \frac{K\lambda}{\beta \cos\theta} \tag{5}$$

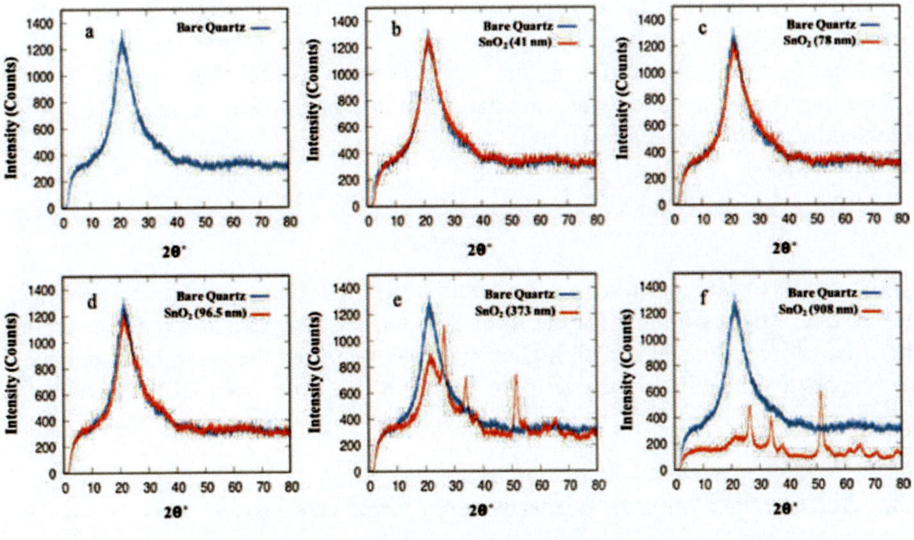

Fig. 15 XRD spectra of SnO_2 thin films on UV-Quartz substrate: **a** bare UV-Quartz substrate, **b** comparison of the Quartz substrate with film of 41 nm thickness of SnO_2, **c** Quartz substrate with 78 nm SnO_2 film thickness, **d** Quartz substrate with 96.5 nm SnO_2 thickness, **e** Quartz substrate with 373 nm SnO_2 film thickness, **f** Quartz substrate with 908 nm SnO_2 film thickness

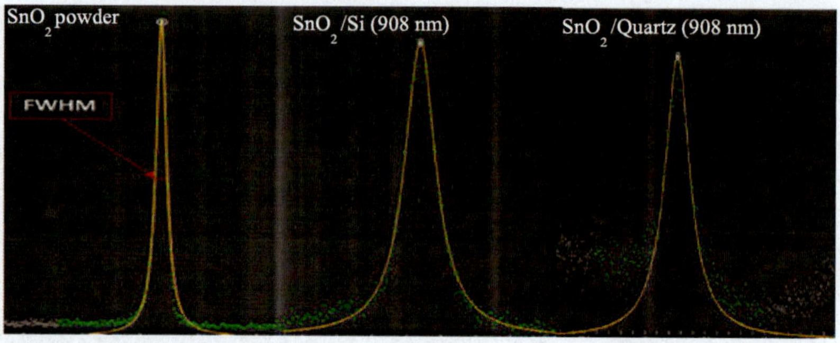

Fig. 16 Lorantzian fit for the first peak (110) orientation for SnO_2 samples

Table 2 Crystallite size (D) and interplanar distance (d) for SnO_2 powder and the most intense peak of SnO_2 with 908 nm thickness

Samples	Crystallite size D (nm)	Interplanar distance d (Å)
SnO_2 powder	19.044	3.380
SnO_2/Si (908 nm)	7.367	3.375
SnO_2/Quartz (908 nm)	6.341	3.382

where K is a shape factor, typically $K = 0.89$ for crystalline spherical solids with a cubic unit cell [18], λ is the wavelength, β is the broadening at half maximum (FWHM) in radians, θ is the Bragg's angle in radians. The most intense peak with a value of $2\theta = 26.78°$ was fitted by using a Cauchy-Lorentz distribution. The FWHM of the first peak was obtained using the Fityk software, as illustrated in Fig. 16.

The interplanar distance was calculated using Bragg's Law given by Eq. (6), obtained by rearranging Eq. (3):

$$d = \frac{\lambda}{2 \sin \theta} \tag{6}$$

The results of both calculations are summarized in Table 3. As shown in Table 2, the crystallite size is smaller for the films than the powder. This is due to the amorphous behavior of the film with higher β values, whereas the interplanar distance has remained almost the same due to the choice of the first peak (110) orientation.

3.3 *Scanning Electron Microscopy (SEM) of SnO_2*

Images of different tin oxide samples were collected at various locations of the samples using the Phenom Pure Scanning Electron Microscope with magnification ranging between 80x to 65,000x. In addition, the Phenom XL Desktop Scanning Electron Microscope was used with electron-optical magnification that ranged between 80x to 100,000x. The SEM images of the powdered sample with the Phenome pure SEM at 2 μm magnification is shown in Fig. 17a, which can be utilized to pick one grain of the SnO_2 particle, and by using the Gwyddion software we can show a 3-dimensional view of the SEM image, as in Fig. 17b.

An average of 4 different spots of the powdered sample was obtained and has been summarized in Table 3. The $Z_{Average}$ can be interpreted as an approximate height of an aggregate of nanoparticles beneath each spot of the SnO_2 sample studied. Also, we obtained the cross-section of the SnO_2 films on Si at 8 μm magnification, which validated the approximate thickness of each sample (as shown in Fig. 18).

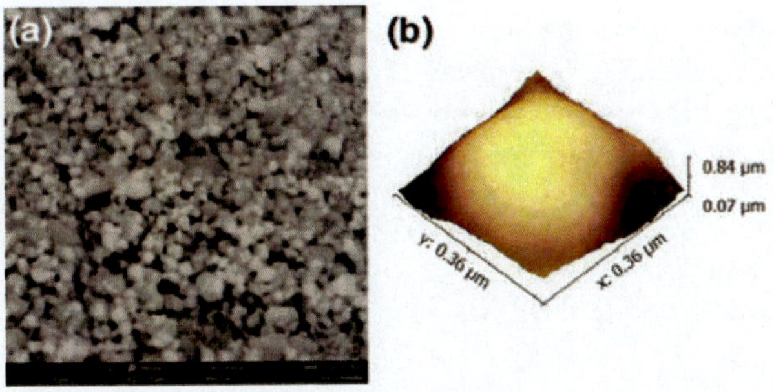

Fig. 17 **a** SEM image of SnO_2 powder at 2 μm magnification, **b** 3D view of SEM image of SnO_2 powder of single grain

Table 3 Averaged dimensions of 4 different spots at 2 μm magnification

Element	$X_{Average}$ (μm)	$Y_{Average}$ (μm)	$Z_{Average}$ (μm)
SnO_2 powder	0.49	0.51	0.75

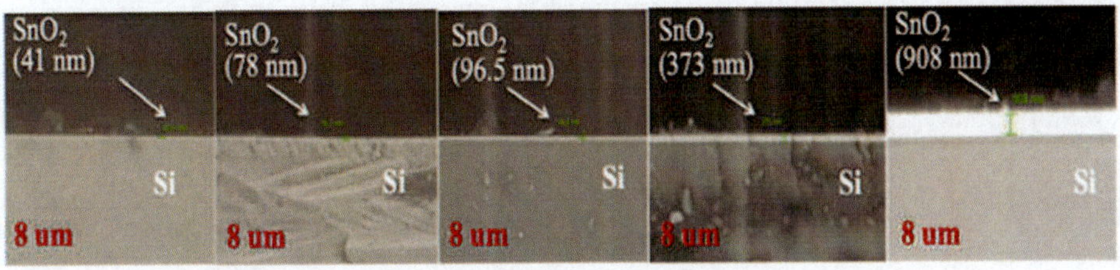

Fig. 18 Cross-sectional images of SnO_2/Si for various thicknesses

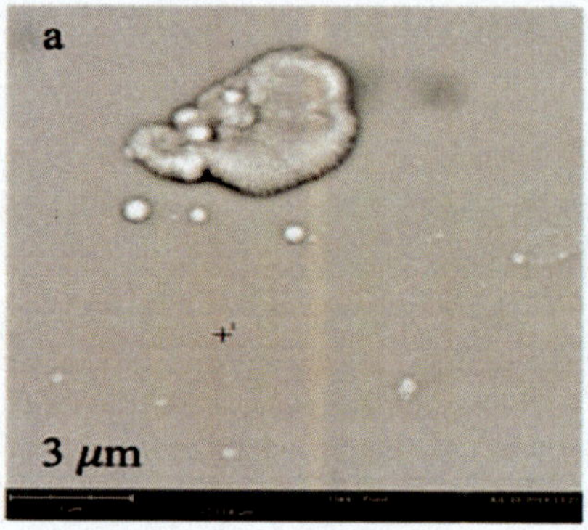

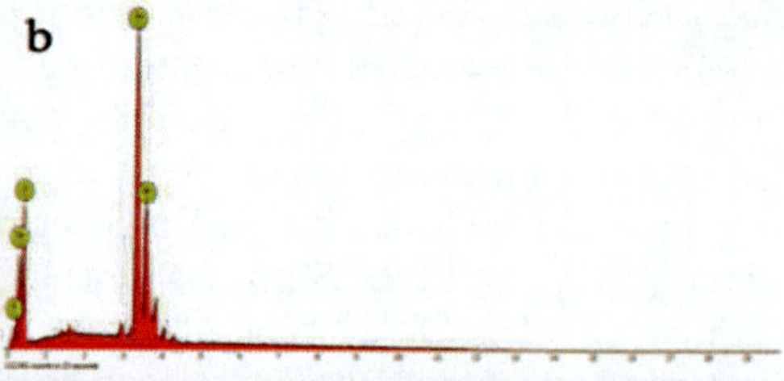

Element Number	Element Symbol	Element Name	Atomic Conc.	Weight Conc.
50	Sn	Tin	16.54	59.56
8	O	Oxygen	82.95	40.26
6	C	Carbon	1.16	0.42

Fig. 19 SEM of SnO_2 on Si with 908 nm thickness at 3 μm magnification (**a**). Corresponding EDS spectrum and composition data with accompanying tabulation of precise concentrations (**b**)

An arbitrary spot on the SnO_2 (908 nm) film on Si was analyzed at 3 μm resolution with the Energy Dispersive X-Ray Spectroscopy (EDS) with the Phenom XL, which allowed analysis of the chemical composition of the film samples, as illustrated in Fig. 19.

The EDS analysis shows the atomic concentration and the weight concentration of SnO_2 to be 16.54 and 59.56, respectively, in comparison to the other elements in the Si sample. The EDS spectrum represents a plot of X-ray versus Energy in keV. The peaks represent a variety of elements in the sample [48].

3.4 *FTIR Spectroscopy of SnO_2*

We used the Thermo Scientific™ Nicolet™ iS50 FTIR Spectrometer, in order to determine the IR -active vibration modes for all tin dioxide samples studied. The ATR technique with diamond crystal was used to collect the Mid-IR spectra in the wavenumber range (400–4500 cm^{-1}). A Smart iTR™ Attenuated Total Reflection (ATR) sampling accessory was used. The spectra were gained after collecting the background spectra and subtracting it from our sample spectra. Figure 20 shows the characterizing bands of pure SnO_2 powder in the range (400–800 cm^{-1}) [26] via peaks 467.36 cm^{-1} (Sn–O stretching vibration) and 569.37 cm^{-1} (Sn–O–Sn asymmetric vibration), in comparison to Sn–O around 530 cm^{-1} [29, 32] and Sn–O–Sn around 600 cm^{-1} [2].

The FT-IR spectra taken for all the silicon samples are shown in Fig. 21. At the lowest thicknesses, there is similar behavior to the silicon substrate results, with peaks at 1107.36 cm^{-1} (the Si–O_2 vibration band), and the 610.72 cm^{-1} feature due to lattice phonon absorption in agreement with Hava et al. [21].

The spectra for the 78 and 96.5 nm tin dioxide films exhibit peaks in the range (400–1150 cm^{-1}). The higher thickness SnO_2 films (373 nm and 908 nm) show

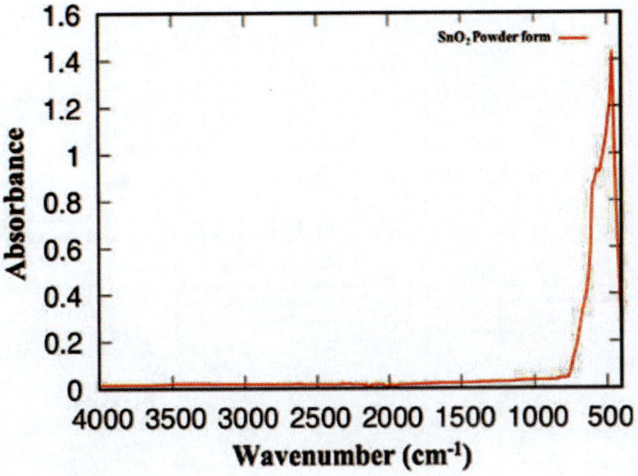

Fig. 20 FTIR spectrum of SnO_2 powder

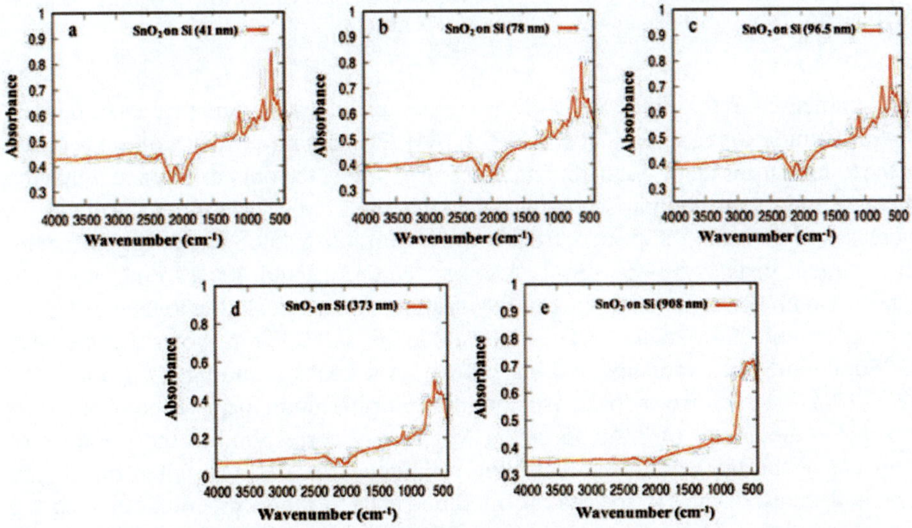

Fig. 21 FT-IR spectra of SnO_2 film on Si substrate: SnO_2 film thickness of 41 nm (**a**), 78 nm (**b**), 96.5 nm (**c**), 373 nm (**d**), 908 nm (**e**)

features of Si that start to disappear in the range (400–700 cm^{-1}). New features begin to appear in the 373 nm film thickness, with peaks around 692.44 cm^{-1} and 565.57 cm^{-1} that relate to the Sn–O stretching and Sn–O–Sn asymmetric vibrations, respectively [18].

The IR spectra for the SnO_2 thin films on quartz were collected to compare the results from different film thicknesses. Figure 22 shows all of the FT-IR data for bare quartz, indicating features that appear at 782.44 and 989.96 cm^{-1}, which represent

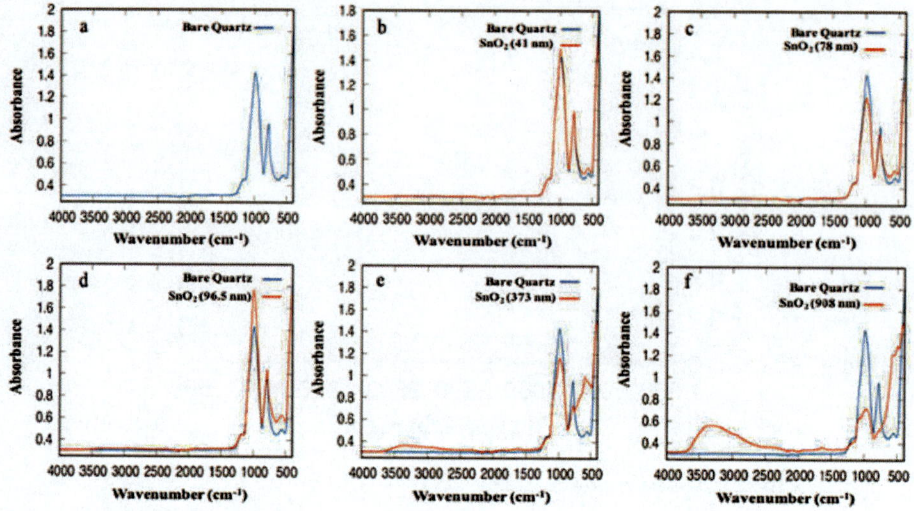

Fig. 22 FT-IR spectra of SnO_2 film on UV-Quartz substrate: Bare Quartz (**a**), SnO_2 film thickness of 41 nm (**b**), 78 nm (**c**), 96.5 nm (**d**), 373 nm (**e**), 908 nm (**f**)

the symmetric stretching of Si–O–Si, and symmetric and asymmetric stretching of Si–OH, similar to the spectra of silica-PLL [44]. There is no evidence of SnO_2 bands in the thinner films up through 96.5 nm thickness, where the only difference is that the baseline is higher in comparison to the two higher thickness films. When the thickness increases to 373 nm, the peak at 782.44 cm^{-1} shifts to 776.68 cm^{-1}, together with an emergent peak at 589.64 cm^{-1}. Another feature around 406.17 cm^{-1} starts to appear, which corresponds to the Sn–O vibration. Also, a peak due to the O–H bond appears around 3348.26 cm^{-1}. The highest thickness of 908 nm shows that the peak at 776.68 cm^{-1} has vanished and instead we have peaks around 494.21 cm^{-1} and 427.79 cm^{-1}, which are characteristic of SnO_2. Furthermore, the O–H bond becomes broader at the center of 3348.55 cm^{-1}, which may correspond to the presence of moisture in the thicker films. According to Dieguez et al., [14], when the crystal size is decreased, the IR spectra are modified due to the interaction between the particles and the electromagnetic radiation depending on the size, shape, and state of aggregation of the crystal.

3.5 Bandgap Energy Calculation for SnO_2

When UV-VIS light impinges on a crystalline solid, a portion (T) is transmitted through the surface of the sample, while some (R) is reflected at the surface, and the rest (A) is absorbed by the solid sample. The absorption pattern of the spectrum demonstrates the band structure of the sample, the energy gap between the lowest conduction band (CB) and the highest valence band (VB), and the nature of the

Table 4 Bandgap calculation for different thicknesses of SnO_2 films on Quartz

Thickness (nm)	Bandgap energy (eV)
78	3.458
96.5	3.383
373	3.562
908	3.519

transition, whether it is direct, indirect, forbidden, or allowed [11]. Using these principles, the UV-VIS transmission data taken with a Shimadzu UV-2600 UV-VIS Spectrophotometer allowed the calculation of the energy bandgap for the specified thicknesses of the SnO_2 films on Quartz as shown in Table 4.

The theoretical formulation to determine the band gap for the semiconductor material is given by Eq. (7):

$$\alpha E_p = k\left(E_p - E_g\right)^{1/2} \tag{7}$$

where α is the absorption coefficient, the k is a constant, and E_g is the band gap energy, and E_p is the discrete photon energy that was calculated using Eq. (8):

$$E_p = hv \tag{8}$$

where h is Planck's constant and v is the frequency of the light. The absorption coefficient α is calculated for the tin dioxide films as the absorbance divided by the thickness of the film. Ultimately, we used a classical Tauc plot to approximate the band gap energy E_g by plotting $\alpha^2 E_p^2$ against E_p. The x-intercept of this line represents the bandgap energy. A python script was written to optimize this process and has been included in the Appendix. The outputs of this script are shown in Fig. 23a–d and Table 5. As the film gets thicker the amount of noise in the absorbance spectra increases as exhibited by the E_p and $\alpha^2 E_p^2$ relationship in the plots, with a noticeable increase in the band gap value.

In an attempt to reduce the noise in the absorption spectra of SnO_2 on UV-Quartz, an annealing process was applied to the two thicker films. The 373 and 908 nm films were annealed up to 360 °C for 15 min. The results of annealing the tin dioxide films in comparison with the unannealed films are illustrated in Fig. 24. The band gap calculation for the annealed sample appeared to be enhanced over that of the as deposited SnO_2 films. The band gap of 373 nm was at 3.562 eV as deposited and after annealing it increased to 3.580 eV. For the 908 nm thickness the as deposited band gap was 3.519 nm and after annealing it increased to 3.552 eV.

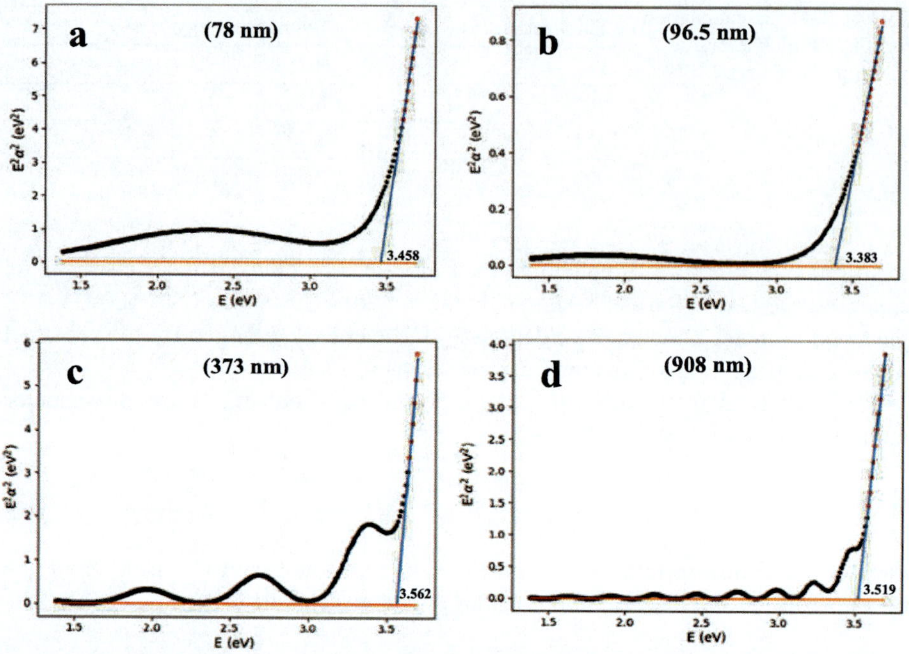

Fig. 23 Bandgap calculation for different thicknesses of SnO_2 films on UV-Quartz: (**a**) 78 nm, (**b**) 96.5 nm, (**c**) 373 nm, (**d**) 908 nm

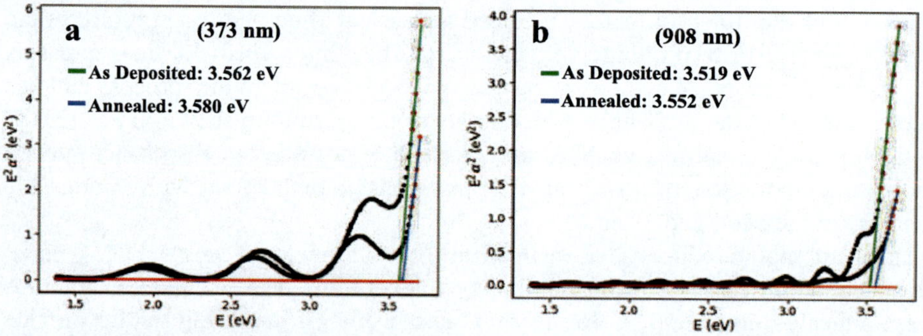

Fig. 24 Bandgap calculation of SnO_2 on UV-Quartz as deposited (green) and annealed (blue) of two thicker films: (**a**) 373 nm, (**b**) 908 nm

4 Composites with Metal Oxides

4.1 *Carbon-Based Composites*

Among all tin dioxide composites, carbon-based materials stand out due to their superior electrochemical properties. A study done by Paek et al. [40] showed improvements in both the reversible capacity and cyclic performance when combining graphene nanosheets (GNS) with the SnO_2 anode material in Lithium-ion batteries. The graphene nanosheet was prepared by the method of chemical reduction of exfoliated oxidized graphite materials. The SnO_2 was prepared by controlled $SnCl_4$ hydrolysis with NaOH. One-dimensional SnO_2 n-type semiconductor coated with a single-wall carbon nanotube (SWCNT) was used to increase the gas sensitivity for gas-sensing applications, in contrast to only p-type semiconducting SWCNT [25]. The hybrid of SnO_2 and SWCNT showed improved sensitivity, especially towards the detection of nitrogen dioxide (NO_2), as reported by Wei et al. [54]. A gas sensor based on hybrid SnO_2/SWNTs has been designed to sense O_3 and NH_3 and has shown enhancement in detecting O_3 and NH_3 in comparison to pure SWCNTs layers at room temperature [20]. Moreover, *n*-type multi-walled nanotube (MWNT) fiber composites with SnO_2 prepared by electrospinning and calcined in air at 500 °C have been able to detect carbon monoxide (CO) at room temperature at 50 ppm concentration, whereas pure SnO_2 nanofibers were found to be not sensitive under these conditions up to 500 ppm [57].

4.2 *Polymer Composites*

Several composites have been produced by using Polyurethanes (PUs), which are an important class of polymer materials, because of their chemical, mechanical, and physical properties. There are many forms of PU: rigid and flexible foams, chemical resistant coatings used as sealants, adhesives, and elastomers. The foam form of PU is favored due to its desirable features, such as enhanced moisture resistance and good sound damping. There are some limitations with the PU foams, such as low mechanical properties and poor thermal stability. The thermal stability of Polyurethanes was improved with the addition of tin dioxide nanoparticles at different densities of PU foam [17]. Also, addition of ceramic powders of tin dioxide to a system of Nafion-based polymer membranes was noted to enhance their chemical and physical properties and led to increased water affinity, higher modulus storage of membrane composites, and furthered polymer ionic channel organizations [46].

5 Engineering Applications of Metal Oxides

5.1 Gas Sensors

Gas sensors based on metal oxides have seen a resurgence in recent years [13]. Tin dioxide in different forms and shapes has been used in a variety of safety, industrial and health applications. It has been used in the production of cost-effective and reliable gas sensors (SMOX) [13] that are relatively simple to fabricate and use [55]. The mechanism of operation of a gas sensor is based on the changes in conductivity of these materials through adsorption and desorption of gases, such as oxygen, on the metal oxide surface. The surface of the metal oxide reacts with the detected gases. The absorbed oxygen gets captured at the grain limit trap sites, and thereby the barrier of the grain boundary potential increases, which leads to an increase in the material resistivity [33]. As a result of this process, the electrical resistivity of a semiconductor metal oxide becomes responsive to any impurities that might be present on its surface or within its volume [22]. The first generation of merchant sensor devices was manufactured in 1968 in Japan by Taguchi Gas Sensor (TGS). They used SnO_2 thick films as their sensing material to detect flammable gases [6]. Certain physical and chemical properties for the gas-sensing materials need to be met in order to have an ideal gas sensor. For example, the sensing material should have high long-term stability over a wide temperature range, with fast and high response to the target gas and low response to ambient gases and humidity that exist in the atmosphere, high selectivity, and cost-effective machinery. However, existing sensors have their pros and cons. Metals, metal oxides, and polymers are frequently employed materials in gas-sensing applications. Metal oxides have shown better sensitivity performance as compared to metals and polymers and exhibit good stability at high temperatures, while metals showed fair stability and polymers had unsatisfactory stability as gas sensors. Metals on the other hand are the most cost-effective sensing materials, in comparison to metal oxides (fair cost) and polymers (intermediate cost). Although metal oxides perform efficiently with almost all the properties mentioned earlier, their selectivity could be enhanced. Consequently, the metal oxides have been tailored to detect diverse gases using functionalized and composites of SnO_2, as summarized in Table 6. Pure SnO_2 has been used by itself in a chemoresistor-based gas sensor in order to detect a variety of reducing gases, such as hydrogen (H_2), carbon monoxide (CO) carbon dioxide (CO_2), and methane (CH_4) [29]. The sensitivity of gas sensors based on metal oxide semiconductors is affected by the film thickness. Reducing the size of SnO_2 particles to a scale of nanometer dimensions and thin films, results in an enhancement in responsiveness, down to the Debye length. The Debye length describes the space-charge size area next to the free carrier concentration surface [16]. Specifically, it has been found that by decreasing the crystallite size (D) of SnO_2 to a range of 5–32 nm the sensitivity towards certain gases, such as (CO), (H_2), and (i-C_4H_{10}), increases [56]. A separate study conducted by Hijazi et al. [24] showed good sensitivity and selectivity of pure SnO_2 and SnO_2 functionalized with 3-aminopropyltriethoxysilane (APTES)

Table 5 Compendium of pure and functionalized tin dioxide materials for detecting gases

Material	Detected gases
[a, b, c]Pure SnO_2	CO, H_2, CH_4, i-C_4H_{10}
[d]Functionalized APTES SnO_2 with alkyl: $C_6H_{11}ClO$ ester: $C_7H_{11}ClO_3$ acid: $C_6H_8Cl_2O_2$	NH_3

[a]Xu et al. [56]
[b]Korotcenkov et al. [29]
[c]Du and George [16]
[d]Hijazi et al. [24]

labeled (SnO_2-APTES) toward ammonia (NH_3) gas in an aim to detect gases from exhaled breath to diagnose diseases. SnO_2-APTES was further functionalized with alkyl ($C_6H_{11}ClO$), acid ($C_6H_8Cl_2O_2$), and ester ($C_7H_{11}ClO_3$) groups. Both (SnO_2-APTES-acid) and (SnO_2-APTES-ester) have shown an increase in conductance (i.e. sensitivity) as compared to (SnO_2-APTES-alkyl) when exposed to 100 ppm of NH_3 gas; whereas pure SnO_2 and SnO_2-APTES have shown no change in conductance in the presence of ammonia gas. In addition, (SnO_2-APTES-acid) exhibited superior selectivity for NH_3 as compared to (SnO_2-APTES-ester), while pure SnO_2 works well for detecting carbon monoxide, hydrogen, methane and isobutane (see Table 5).

5.2 *Lithium-Ion Batteries*

Owing to their superior electrochemical properties, SnO_2-based materials exhibit high specific capacity and cost-effective ingredients. The SnO_2 anode has shown an increase in reversible discharge capacity of 782 mAh/g. However, the regular SnO_2 anode suffers from low cycling ability and cracks due to volume changes with cycling. By reducing the size of the SnO_2 particles, it is possible to increase the ratio of surface-to-volume and lower the change in volume during insertion and extraction of lithium [50]. For instance, nanoparticles of SnO_2 of size 3 nm have shown enhanced electrochemical properties, but with low lithium storage, in comparison to those of 4 nm and 8 nm particle sizes [27]. Song et al. have studied three higher particles sizes with (SnO_2–I = 15 nm), (SnO_2–II = 30 nm), and (SnO_2–III = 70 nm), where SnO_2–I has shown stable electrochemical performance during different cycles. SnO_2–I also showed an improvement in discharge capacity after 30 cycles (i.e. charging and discharging 30 times) with capacity around 460 mAh/g, which means it holds 58.2% of its electrical capacity in comparison to SnO_2–II and SnO_2–III that hold 40.7% and 21.0%, respectively [50]. Also, SnO_2 nanowires were found to be a promising anode material due to their unique structural and electronic properties. However, there were limitations to the use of tin dioxide nanowire anodes in lithium-ion batteries, which was related to the deteriorative effects of catalysts [41]. The electrochemical properties of SnO_2 anode-based battery could be improved by performing appropriate nanostructure enhancements in chemical composition. Theoretically, the tin dioxide anode in lithium-ion batteries was found to be distinctive due to its high capacity [10]. A study conducted by Wang et al. [52] showed that the use of N-doped graphene-SnO_2 anodes in lithium-ion batteries resulted in higher capacity, as well as an increase in rate capability, and improved cycling stability performance. Such enhancements were directly related to the exceptional electronic conductivity and sandwich structure of the doped graphene that facilitated short-length transportation of the electrons and lithium ions.

5.3 *Solar Cells*

There has been an increasing amount of research relating to the enhancement of optical sensors, such as photodiodes, regarding their sensitivity to light in the ultraviolet (UV) region. SnO_2 nanoforms have been applied in dye-sensitized and polymer-based solar cells due to their large band gaps, thermal stabilization, low photocatalytic activity, elevation of electron mobility, and fine anti-reflection properties [37]. In addition, SnO_2 can be used as a coating that improves the sensitivity of traditional GaP and ZnSe based photodiodes in the UV spectral limit. The uncoated photodiodes reach their limit in the range 200–380 nm. Anti-reflection (AR) coating can be used to increase the absorbance of light and reduce the loss of photons. A SnO_2 nanosized layer acts as an optically transparent coating layer that allows the photodiode to detect light at shorter wavelengths. SnO_2 can also be employed as a transparent active electrode that functions in the wavelength range 350–1100 nm. The high transparency of SnO_2 comes from the large magnitude of the band gap energy (Eg ~ 3.6 eV). The structure of the SnO_2 layer leads to refraction that suits the formation of AR coatings. Adding a coating layer of SnO_2 with 54.4 nm thickness leads to a transmission coefficient of 9.6% at 435 nm. In one study, an SnO_2 sheet was used with two different photodiodes, one with zinc selenide (ZnSe) and the other with gallium phosphide (GaP). The AR coating of SnO_2 on ZnSe and GaP based photodiode showed an increase in sensitivity of about 10–15% [15]. The SnO_2 thin films prepared by a sol-gel method were also used in AR coating for solar cells. The sol-gel method created high purity films based on the hydrolysis and the polycondensation of metal-organic precursors. It also offered a nanoscale control of the film structure. The use of tin oxide (SnO_2)/silicon (Si) has been proposed for low-cost photovoltaic devices. The doping in SnO_2 has been used to reduce the sheet resistance and to increase the transparency of the thin film. The doped SnO_2 has proven to provide high optical clarity of the obtained films. Also, due to this layer, the device has excellent technical performance and high reliability, without large fabrication costs [34]. SnO_2 has also been exploited as an electron transfer material (ETM) (i.e. coating material) in perovskite solar cells. Organometallic halide perovskite (ABX_3) has been studied extensively in recent years due to its distinctive properties, such as large absorption coefficient and optimal band gap, which qualify it for absorbing light applications in photovoltaic cells. The SnO_2 nanomaterial is featured with its high electron mobility (0.8 ± 0.14 cm^2/V s), in comparison with zinc oxide (ZnO) and titanium dioxide (TiO_2) with mobilities (0.08 ± 0.01 cm^2/V s) and (0.1 ± 0.03 cm^2/V s), respectively. SnO_2 has been prepared as an ETM in perovskite by a solution processing method in which the SnO_2 films were deposited from $SnCl_2$ solution of 0.1 M concentration by using water and ethanol as solvents. By using ethanol as a solvent, the coating material showed an enhancement in the UV/VIS absorption spectra by 8.38%, with enhanced incident photon-to-electron conversion efficiency (IPCE) and a circuit current density (J_{SC}) of 18.98 mA cm^{-2} [37].

6 Conclusions and Future Research

We have characterized SnO_2 powder, SnO_2/Si and SnO_2/UV-Quartz films by using Raman, FT-IR, XRD, SEM, and UV-VIS spectroscopy techniques. We have demonstrated a red shift in the Raman spectra as the temperature increases from 30 to 170 °C in both A_{1g} and B_{2g} vibrational modes, while E_g exhibited little change. By using a 532 nm laser Raman excitation we were able to record the three Raman active modes for SnO_2 (908 nm thickness) film on quartz, which were difficult to view at Raman excitation wavelengths of 780 and 514 nm. We performed a visual characterization in three dimensions to obtain the morphology of SnO_2 powder (bulk form) and SnO_2 film on silicon substrate utilizing high resolution scanning electron microscopy. Additional trials for temperature-dependence are underway for SnO_2 film samples to study the effects of both decreasing and increasing temperatures. An approach utilizing the local-density approximation by using ab initio software has shown an agreement of phonon frequency relation with IR and Raman spectra [42]. Another approach followed by Lan et al. [31, 30] uses Molecular Dynamics (MD) simulation for the rutile structured TiO_2 to calculate the shifts and broadenings of anharmonic frequency phonon modes at different temperatures ranging between (100–1150 K) by using GULP software. Their MD simulation showed agreement of crystal structure (i.e. lattice parameters), Raman frequencies and thermal expansion with the experimental data. We are studying rutile structured SnO_2 using MD simulation at different temperatures integrated within the Large-scale Atomic/Molecular Massively Parallel Simulator (LAMMPS) software. High-level computations are in progress utilizing MD simulations to compare our experimental data with theoretical results, in order to gain a deeper insight into the effects of temperature on the phonon modes active in rutile tetragonal tin dioxide.

Acknowledgements Financial support from the National Science Foundation (REU Site in Physics at Howard University, NSF Award# PHY-1659224) is gratefully acknowledged. We also would like to acknowledge the assistance of the technical staff and the use of the X-ray diffraction and Raman instrumentation at the Thermo Fisher Scientific facilities in Lanham, MD, for some of the tin dioxide spectral measurements reported here.

Appendix

```
#Python code for calculating the energy bandgap for SnO2 from UV-VIS Spectra

import csv

import numpy as np
import matplotlib.pyplot as plt

import easygui

import sys

x = []

y = []

#Get Data from CSV file

string = easygui.fileopenbox(default = "D:\\D_Documents\\REU_Howard_19\\5_29_19") #this string should be changed based on your directory

with open(string) as csvDataFile:

    csvReader = csv.reader(csvDataFile)

    for row in csvReader:

        x.append(row[0])

        y.append(row[1])

#this reads a CSV file where the first row is the Energy E, and the second is E^2 * Alpha^2

#Alternatively, one can change the numbers on the rows above to read different rows from the file
xx = np.array(x)

yy = np.array(y)
```

```
xx = [float(xx) for xx in xx]

yy = [float(yy) for yy in yy]

#Python Housekeeping

size = np.size(xx)

cut = -565

xx = xx[cut:]

yy = yy[cut:]

#This is where the zooming occurs, you can change this by adjusting
the number "cut"

#It looks at the last "cut" numbers of the array

E = xx

Ealphsq = yy

plt.plot(E, Ealphsq,'k.')

#Changing the variable names to something more intuitive, and here
I plot the data so I'm

 not in the dark

for i in range(0,l-4):

    if abs(abs((Ealphsq[i] - Ealphsq[i+3])/(E[i]-E[i+3])) - sl) < tol*sl:

        line[k] = Ealphsq[i]

        k = k+1

    else:

        sl = np.Infinity

#This loop finds the data we want in the line at the end
```

```
if k < 1:
    print('I died, graph does not terminate in line')
    sys.exit #If you didn't zoom right you'll see this printed, or if you
pick bad data
v = np.zeros(k)
Evec = np.zeros(k)
for i in range(k):
    v[k-i-1] = line[i]
    Evec[i] = E[k-i-1]
plt.plot(Evec,v,'r.')
#plot the points we have chosen to be a part of our line
A = np.column_stack((np.transpose(Evec), np.ones(k)))
AtA = np.matmul(np.transpose(A), A)
v.shape = (k,1)
Atv = np.matmul(np.transpose(A),v)
b = np.linalg.solve(AtA,Atv)
#Normal equations for line of best fit
incpt = -b[1]/b[0]
print('x intercept occurs at Eg = ')
print(incpt)
x = np.linspace(E[0],incpt,150)
y = x*b[0] + b[1]
```

```
plt.plot(x,y)
plt.plot(xx,np.zeros(np.size(xx)))
plt.title('E$_g$ = ' + str(incpt))
plt.xlabel("E (eV)")
plt.ylabel('E$^2$$ \\alpha ^2$ (eV$^2$)')
#print x-intercept
```

References

1. Agrahari, V., Mathpal, M.C., Kumar, S., Kumar, M., Agarwal, A.: Cr modified Raman, optical band gap and magnetic properties of SnO_2 nanoparticles. J. Mater. Sci.: Mater. Electron. **27**, 6020–6029 (2016). https://doi.org/10.1007/s10854-016-4525-2
2. Ansari, S.G., Fouad, H., Shin, H.S., Ansari, Z.A.: Electrochemical enzyme-less urea sensor based on nano-tin oxide synthesized by hydrothermal technique. Chem. Biol. Interact. **242**, 45–49 (2015)
3. Armstrong, P., Knieke, C., Mackovic, M., Frank, G., Hartmaier, A., Goken, M., Peukert, W.: Microstructural evolution during deformation of tin dioxide nanoparticles in a comminution process. Acta Materialia 57(10), 3060–3071 (2009)
4. Batzill, M., Diebold, U.: The surface and materials science of tin oxide. Prog. Surf. Sci. **79**, 47–154 (2005). Reprinted from Progress in Surface Science, Vol 79, Issues 24, Matthias Batzill and Ulrike Diebold, The surface and materials science of tin oxide, Page 70, Copyright (2019), with permission from Elsevier
5. Berthomieu, C., Hienerwadel, R.: Fourier transform infrared (FTIR) spectroscopy. Photosynth. Res. **101**(2–3), 157–170 (2009)
6. Briand, D., Courbat, J.: Micromachined semiconductor gas sensors. Semicond. Gas Sens. 220–260 (2013). https://doi.org/10.1533/9780857098665.2.220
7. Bumbrah, G.S., Sharma, R.M.: Raman spectroscopy—basic principle, instrumentation and selected applications for the characterization of drugs of abuse. Egypt. J. Forensic Sci. **6**(3), 209–215 (2016)
8. Casimir, D., Ahmed, I., Garcia-Sanchez, R., Misra, P., Diaz, F.: Raman spectroscopy of graphitic nanomaterials. In: Raman Spectroscopy. Gustavo M. do Nascimento (ed.) IntechOpen. https://doi.org/10.5772/intechopen.72769 (2017)
9. Cañas-Carrell, J.E., Li, S., Parra, A.M., Shrestha, B.: Metal oxide nanomaterials: health and environmental effects. In: Health and Environmental Safety of Nanomaterials, pp. 200–221. Woodhead Publishing (2014)
10. Chen, J.S., Lou, X.W.: SnO_2-based nanomaterials: synthesis and application in lithium-ion batteries. Small **9**(11), 1877–1893 (2013)
11. Chowdhury, F.R., Choudhury, S., Hasan, F., Begum, T.: Optical properties of undoped and indium-doped tin oxide thin films. J. Bangladesh Acad. Sci. **35**(1), 99–111 (2011)
12. Das, S., Jayaraman, V.: SnO_2: a comprehensive review on structures and gas sensors. Prog. Mater. Sci. **66**, 112–255 (2014)
13. Degler, D.: Spectroscopic insights in the gas detection mechanism of tin dioxide-based gas sensors. Doctoral dissertation, Eberhard Karls Universität Tübingen (2017)

14. Dieguez, A., Romano-Rodriguez, A., Vila, A., Morante, J.R.: The complete Raman spectrum of nanometric SnO2 particles. J. Appl. Phys. **90**(3), 1550–1557 (2001)
15. Dobrovolskiy, Y.G., Perevertailo, V.L., Shabashkevich, B.G.: Anti-reflection coatings based on SnO_2, SiO_2, Si_3N_4 films for photodiodes operating in ultraviolet and visible spectral ranges. Semicond. Phys. Quantum Electron. Optoelectron. **14**(3), 298–301 (2011)
16. Du, X., George, S.M.: Thickness dependence of sensor response for CO gas sensing by tin oxide films grown using atomic layer deposition. Sens. Actuators B: Chem. **135**(1), 152–160 (2008)
17. Esmailzadeh, M., Manesh, H.D., Zebarjad, S.M.: Role of SnO_2 nanoparticles on mechanical and thermal properties of flexible polyurethane foam nanocomposite. J. Porous Mater. **23**(5), 1381–1388 (2016)
18. Ferreira, C.S., Santos, P.L., Bonacin, J.A., Passos, R.R., Pocrifka, L.A.: Rice husk reuse in the preparation of SnO_2/SiO_2 nanocomposite. Mater. Res. **18**(3), 639–643 (2015). https://doi.org/10.1590/1516-1439.009015. Retrieved from SciELO—Scientific Electronic Library Online Publishing with open access source
19. Fierro, J.L.: Metal Oxides Chemistry and Applications, pp. 178–181. Taylor & Francis Group, Boca Raton, FL (2006). ISBN: 978-0-8247-2371-2
20. Ghaddab, B., Berger, F., Sanchez, J.B., Mavon, C.: Detection of O_3 and NH_3 using tin dioxide/carbon nanotubes-based sensors: influence of carbon nanotubes properties onto sensor's sensitivity. Procedia Eng. **5**, 115–118 (2010)
21. Hava, S., Ivri, J., Auslender, M.: Wavenumber-modulated patterns of transmission through one-and two-dimensional gratings on a silicon substrate. J. Opt. A: Pure Appl. Opt. **3**(6), S190 (2001)
22. Henrich, V.E., Cox, P.A.: The Surface Science of Metal Oxides. Cambridge University Press (1996)
23. Henry, J., Mohanraj, K., Sivakumar, G., Umamaheswari, S.: Electrochemical and fluorescence properties of SnO_2 thin films and its antibacterial activity. Spectrochim. Acta Part A: Mol. Biomol. Spectrosc. **143**, 172–178 (2015)
24. Hijazi, M., Stambouli, V., Rieu, M., Tournier, G., Pijolat, C., Viricelle, J.P.: Sensitive and selective ammonia gas sensor based on molecularly modified SnO_2. In: Multidisciplinary Digital Publishing Institute Proceedings, vol. 1, no. 4, p. 399 (2017)
25. Hoa, N.D., Quy, N.V., Wei, L., An, M., Song, H., Kang, Y., Kim, D.: One-dimensional tin-oxide-coated single-wall carbon nanotubes for gas sensor applications. J. Korean Phys. Soc. **54**, 1893 (2009)
26. Juliasih, N., Buchari, Noviandri, I.: Application of SnO_2 nanoparticle as sulfide gas sensor using UV/VIS/NIR spectrophotometer. IOP Conf. Ser.: Mater. Sci. Eng. **188**, 012025 (2017). https://doi.org/10.1088/1757-899x/188/1/012025
27. Kim, C., Noh M., Choi, M., Cho, J., Park, B.: Critical size of a nano SnO_2 electrode for Li-secondary battery. Chem. Mater. **17**, 3297–3301 (2005)
28. Konijnendijk, W.L., Stevels, J.M.: The structure of borosilicate glasses studied by Raman scattering. J. Non-Cryst. Solids **20**(2), 193–224 (1976)
29. Korotcenkov, G., Brinzari, V., Ham, M.H.: Materials acceptable for gas sensor design: advantages and limitations. In: Key Engineering Materials, vol. 780, pp. 80–89. Trans Tech Publications Ltd. (2018)
30. Lan, T., Tang, X., Fultz, B.: Phonon anharmonicity of rutile TiO_2 studied by Raman spectrometry and molecular dynamics simulations. Phys. Rev. B **85**(9), 094305 (2012)
31. Lan, T., Li, C., Fultz, B.: Phonon anharmonicity of rutile SnO_2 studied by Raman spectrometry and first principles calculations of the kinematics of phonon-phonon interactions. Phys. Rev. B **86**(13) (2012). https://doi.org/10.1103/PhysRevB.86.134302
32. Mahadik, D.B., Lee, Y.K., Park, C.S., Chung, H.Y., Hong, M.H., Han, W., Park, H.H.: Effect of water ethanol solvents mixture on textural and gas sensing properties of tin oxide prepared using epoxide-assisted sol–gel process and dried at ambient pressure. Solid State Sci. **50**, 1–8 (2015)

33. Majumder, S.: Synthesis and characterization of SnO_2 films obtained by a wet chemical process. Mater. Sci. **27**(2) (2009). ISSN: 0137-1339
34. Manea, E., Budianu, E., Purica, M., Podaru, C., Popescu, A., Cernica, I., Parvulescu, C.C.: SnO_2 thin films prepared by sol gel method for 'Honeycomb' textured silicon solar cells. Romanian J. Inf. Sci. Technol. **10**(1), 25–33 (2007)
35. Mathew, X., Enriquez, J.P., Mejia-Garcia, C., Contreras-Puente, G., Cortes-Jacome, M.A., Toledo Antonio, J.A., Punnoose, A.: Structural modifications of SnO_2 due to the incorporation of Fe into the lattice. J. Appl. Phys. **100**(7), 073907 (2006)
36. Mukhopadhyay, A.: Measurement of Magnetic Hysteresis Loops in Continuous and Patterned Ferromagnetic Nanostructures by Static Magneto-Optical Kerr Effect Magnetometer. Satyendra Nath Bose National Centre for Basic Sciences, Guwahati (2015)
37. Murugadoss, G., Kanda, H., Tanaka, S., Nishino, H., Ito, S., Imahori, H., Umeyama, T.: An efficient electron transport material of tin oxide for planar structure perovskite solar cells. J. Power Sources **307**, 891–897 (2016)
38. Nasdala, L., Smith, D.C., Kaindl, R., Ziemann, M.A.: Raman spectroscopy: analytical perspectives in mineralogical research. Spectrosc. Methods Mineral. **6**, 281–343 (2004)
39. Oviedo, J., Gillan, M.J.: Energetics and structure of stoichiometric SnO2 surfaces studied by first-principles calculations. Surf. Sci. **463**(2), 93–101 (2000)
40. Paek, S.M., Yoo, E., Honma, I.: Enhanced cyclic performance and lithium storage capacity of SnO_2/graphene nanoporous electrodes with three-dimensionally delaminated flexible structure. Nano Lett. **9**(1), 72–75 (2008)
41. Park, M.S., Wang, G.X., Kang, Y.M., Wexler, D., Dou, S.X., Liu, H.K.: Preparation and electrochemical properties of SnO2 nanowires for application in lithium-ion batteries. Angew. Chem. Int. Ed. **46**(5), 750–753 (2007)
42. Parlinski, K., Kawazoe, Y.: Ab initio study of phonons in the rutile structure of SnO_2 under pressure. Eur. Phys. J. B Condens. Matter Complex Syst. **13**(4), 679–683 (2000)
43. Patil, G.E., Kajale, D.D., Gaikwad, V.B., Jain, G.H.: Preparation and characterization of SnO_2 nanoparticles by hydrothermal route. Int. Nano Lett. **2**(1) (2012). https://doi.org/10.1186/2228-5326-2-17
44. Patwardhan, S.V., Maheshwari, R., Mukherjee, N., Kiick, K.L., Clarson, S.J.: Conformation and assembly of polypeptide scaffolds in templating the synthesis of silica: an example of a polylysine macromolecular "switch". Biomacromolecules **7**(2), 491–497 (2006)
45. Peercy, P., Morosin, B.: Pressure and temperature dependences of the Raman-active phonons in SnO_2. Phys. Rev. B **7**(6), 2779–2786 (1973). https://doi.org/10.1103/PhysRevB.7.2779
46. Scipioni, R., Gazzoli, D., Teocoli, F., Palumbo, O., Paolone, A., Ibris, N., Navarra, M.: Preparation and characterization of nanocomposite polymer membranes containing functionalized SnO_2 additives. Membranes **4**(1), 123–142 (2014)
47. Sergent, N., Epifani, M., Pagnier, T.: In situ Raman spectroscopy study of NO_2 adsorption onto nanocrystalline tin (IV) oxide. J. Raman Spectrosc. **37**, 1272–1277 (2006)
48. Severin, K.P.: Energy Dispersive Spectrometry of Common Rock Forming Minerals, p. 19. Kluwer Academic, Dordrecht, The Netherlands (2004)
49. Sharma, R., Bisen, D.P., Shukla, U., Sharma, B.G.: X-ray diffraction: a powerful method of characterizing nanomaterials. Recent Res. Sci. Technol. **4**(8) (2012). Recreated from Recent Research in Science and Technology Publishing with open access source
50. Song, H., Li, X., Cui, Y., Xiong, D., Wang, Y., Zeng, J., Sun, X.: Controllable lithium storage performance of tin oxide anodes with various particle sizes. Int. J. Hydrog. Energy **40**(41), 14314–14321 (2015)
51. Su, H.-W., Chen, W.-C.: High refractive index polyimide–nanocrystalline-titania hybrid optical materials. J. Mater. Chem. **18**(10), 1139 (2008). https://doi.org/10.1039/b717069f
52. Wang, X., Cao, X., Bourgeois, L., Guan, H., Chen, S., Zhong, Y., Bando, Y.: N-doped graphene-SnO_2 sandwich paper for high-performance lithium-ion batteries. Adv. Funct. Mater. **22**(13), 2682–2690 (2012)
53. Wang, F., Zhou, X., Zhou, J., Sham, T.K., Ding, Z.: Observation of single tin dioxide nanoribbons by confocal Raman microspectroscopy. J. Phys. Chem. C **111**(51), 18839–18843 (2007)

54. Wei, B.Y., Hsu, M.C., Su, P.G., Lin, H.M., Wu, R.J., Lai, H.J.: A novel SnO_2 gas sensor doped with carbon nanotubes operating at room temperature. Sens. Actuators B: Chem. **101**(1–2), 81–89 (2004)
55. Wilson, R., Simion, C., Blackman, C., Carmalt, C., Stanoiu, A., Di Maggio, F., Covington, J.: The effect of film thickness on the gas sensing properties of ultra-thin TiO_2 films deposited by atomic layer deposition. Sensors **18**(3), 735 (2018)
56. Xu, C., Tamaki, J., Miura, N., Yamazoe, N.: Grain size effects on gas sensitivity of porous SnO_2-based elements. Sens. Actuators B: Chem. **3**(2), 147–155 (1991)
57. Yang, A., Tao, X., Wang, R., Lee, S., Surya, C.: Room temperature gas sensing properties of SnO_2/multiwall-carbon-nanotube composite nanofibers. Appl. Phys. Lett. **91**(13), 133110 (2007)
58. Zhu, J.J., Zhu, J.M., Liao, X.H., Fang, J.L., Zhou, M.G., Chen, H.Y.: Rapid synthesis of nanocrystalline SnO2 powders by microwave heating method. Mater. Lett. **53**(1-2), 12–19 (2002)

Perspective Future Development of Nanomaterials

Jamal Akhter Siddique and Arshid Numan

Abstract Engineered nanomaterials (ENMs) and nanotechnology are revolutionising the world. Especially in the twenty-first century, it has gathered considerable interest from the society, academic world, researchers and the industrial sector. Nanotechnology is now an established scientific area that has undergone an exciting growth phase with a wide range of applications in different fields. Although the field of nanomaterials is mature and well established, the scale-up from research and development to commercialisation remains tedious due to the reliability and reproducibility of technology. To overcome the limitation, there is a need to standardise, validate and evaluate each stage of process development. Extensive research has been done at the institutional and investors level for the modernisation and smooth technology transfer. This chapter emphasises the significant progress, applications and challenges in nanotechnologies and ENMs. In the end, the application properties of this specific topic and future directions for research are discussed.

Keywords Engineered nanomaterials (ENMs) · Metrological studies · Synthesis · Future nanomaterials · Challenges and limitations · Environmental hazardous · Commercialisation

1 Introduction

Nanotechnology is a well-informed field of research since the twentieth century [1]. Nanomaterials (NMs) are the mainspring of nanotechnology and nanoscience. According to the International Organization for Standardization (ISO), the prefix

J. A. Siddique
Department of Chemistry, School of Basic and Applied Sciences, Lingaya's University, Faridabad, India

A. Numan (✉)
Graphene & Advanced 2D Materials Research Group (GAMRG), School of Science and Technology, Sunway University, No. 5, Jalan Universiti, Bandar Sunway, 47500 Petaling Jaya, Selangor, Malaysia
e-mail: numan.arshed@yahoo.com

N. M. Mubarak et al. (eds.), *Contemporary Nanomaterials in Material Engineering Applications*, Engineering Materials,
https://doi.org/10.1007/978-3-030-62761-4_12

nano refers to a size ranging approximately from 1 to 100 nm [2]. Nanomaterials are a broad class of materials that include nanoparticles (NPs), which have one dimension less than 100 nm [3]. These materials can be zero to three dimensional, i.e., 0D, 1D, 2D or 3D [4]. The European Commission proposed that a nanomaterial also consists of 50% or more of particles having a size between 1 and 100 nm. On the contrary of the definition given by the European Union agency, the graphene flakes, fullerenes and single-wall carbon nanotubes with one or more peripheral dimensions below 1 nm also be considered as into the category of nanomaterials. The significance of these materials is realised when researchers found that size can control the physiochemical nature of a substance, e.g., electric, mechanical, magnetic, and optical properties, etc. [5].

These activities established an influential noticeable methodology for creating nanoscale dynamism. In the year 2001, the USA started a state-funded National Nanotechnology Initiative (NNI) for the development of nanomaterials in the field of nanotechnology. Later, similar initiatives were followed by the European Union, Japan, and the Republic of Korea (South). Taiwan also implied its role and invested huge funds into such type of National Program on Nanotechnology. The vast number of articles on nanomaterials elucidates the fact that nanoscience and nanotechnology demarcate a wide range of fields, in conjunction with physics, chemistry, biology, materials engineering, medicine, and electronics.

Research on nanomaterials is a rapidly growing topic due to the drastic improvement it brings to the quality of the product by significantly *improving* the characteristic properties of bulk material. Materials developed and designed at such a minute level are considered as engineered nanomaterials (ENMs). Production and demand for ENMs embedded in consumer products are growing significantly in the last few years [6]. These ENMs can instigate exclusive electrical, magnetic, optical, and other material properties. Such promising features have the potential to introduce tremendous impacts in the field of food, cosmetics, medicine, pesticides, electronics, and pharmaceutical research. The worldwide market of nanoscale materials in commercial products has been increased exponentially. Taking conservative estimate by Mordor Intelligence, the global nanomaterials market was valued at about USD 4.1 billion in 2015 and expected to reach USD 11.3 billion by 2020, at a compound annual growth rate (CAGR) of over 22% during the forecast period 2017–2022 [7]. The study done by Woodrow Wilson International Center listing on Nanotechnology Consumer Products has registered 1628 commercial nano-products, including 440 in the European market [8, 9].

The ENMs offer the advantages of easy and flexible modifications technique with small controlled particles. It boosts the commercialisation of the engineered products. Innovations frequently attract attention in industrial and scientific communities. However, the fascinating properties of ENMs come with limitations of delivery of reproducibility in process development, with low safety profile. The safe and reproducible nanomaterials in industries remain a significant concern. To overcome the inadequacies, academics, researchers and industries are working very closely to mitigate the issues in an economical and environmentally friendly way. These objectives address the widespread challenges we face, and most of them are directly affected by the methods involved in manufacturing. Here in this chapter, we specifically focus on ENMs synthesis techniques, limitations, challenges and scale-up issues from research and development to commercialisation.

2 Synthesis Techniques

NMs can be synthesised by using several conventional methods, including biological, physical, chemical, and hybrid techniques [10, 11]. The synthesis of nanomaterials through traditional methods gives harmful byproducts that cause severe environmental distress and challenges to the current researchers. Moreover, such materials could never be utilised for medicinal use due to health hazards, particularly in the clinical field [12]. The researchers associated with ENMs synthesis are working hard to overcome these challenges and trying to find new techniques or methods. The establishment of eco-friendly technologies to develop environmentally favourable nonhazardous products is going to be much more accessible by adopting biotechnological gears and green Nanotechnology [13]. Plenty of adverse restrictions are subsequently eliminated by implementing the different ideas in the manufacturing of ENMs to proceed at physiological temperatures, pH, pressure, and at the same time, a negligible cost [14]. In general, the methods for the **synthesis of nanomaterials can be** categorised **into two main divisions**: "bottom-up method" and the "top-down method". Size, pH of the precursor solution, concentration of the extracts used, concentration of the raw materials, and the protocols that are followed for the synthesis [15] are also treated as challenges in the field of ENMs.

There are numerous ways to synthesise nanomaterials (as shown in Table 1); each synthesis either belongs to bottom-up or top-down has its challenges and limitations. Some of them are showing below in Table 1.

3 Existing Limitations

3.1 Limitations in Method Development

Temperature

Most of the physical methods undergo high temperatures (>350 °C), while the chemical processes can be carried out at low temperatures (<350 °C). At the same time, the synthesis through biological routes would get denatured at a temperature above 100 °C. Such a condition shows the importance of temperature in any synthesis of nanomaterials. The role of temperature in the synthesis determines the nature of nanomaterials formed [36].

Table 1 Different synthesis methods of ENMs, merits, limitations and challenges

Methods	Merits	Limitations and challenges
Top-down methods In this method, a destructive approach is employed. The bulky and larger molecules are used to synthesis of nanomaterials		
Ball milling (Mechanical milling)	Involves the reduction of the particle size with high energy ball milling. John Benjamin developed this method in 1970	It requires high energy; milling time is quite long, steel balls have an issue with contaminating the powder, threat of grinding the sensitive microstructure [16]
Mechanochemical synthesis	Synthesis is based on repeated deformation, welding, and splintering of the reactants mixture	It creates highly sensitive microstructures that can be easily affected by the impurities of atmospheric media. Longer time required in preparing the materials having a size of less than 20 nm [16]
Laser ablation	Uses laser irradiation to minimise the particle size up to nano level. In general, copper vapour lasers are used with Ti: Sapphire (Titanium-doped sapphire) laser Nd: YAG (neodymium-doped yttrium aluminium garnet) laser at 106 μm [17]	It is difficult to realise precise in-bulk processing and the requirement of expensive and large-dimension equipment. Colloidal solution laser path gets blocked, and laser energy gets absorbed by formed nanomaterials despite the target surface. This method also causes a reduction in ablation rate [18]
Ion sputtering method	Preparation of nanomaterials from several metals using magnetron sputtering of metal targets. In this synthesis, collimated beams of the nanomaterials are developed, and nanostructured films are accumulated on the silicon substrates. The entire process needs very low pressure (1 mTorr) [19]	The nature of gasses utilised in sputtering like, He, Ne, Ar, Kr, and Xe, can impose the effect on composition, peripheral morphology, texture, and the optical properties of the films of nanocrystalline metal oxide [20]
Bottom-up methods It is a building up approach. The method employed in reverse, as nanomaterials are formed from simpler substances. In such a technique, Nanomaterials are produced by using tiny particles and adjusting the organisation of atoms		

(continued)

Table 1 (continued)

Methods	Merits	Limitations and challenges
Physical vapour deposition method	Properly controlled vacuum technique like, sputtered deposition and thermal evaporation causes vaporisation of designated material, which later condensed on a substrate [21]. Different kinds of nanomaterials get on the surface of thin films or attached to other nanomaterials	This method is expensive and generates materials in small quantity. To implement this method on the industrial scale, cost minimisation is essential [22]
Chemical vapour deposition method	They are used for the preparation of carbon powder for the colour pigment in the electric lamp and carbon fibre as filaments. The deposition of the thin film is done by chemical reaction of the gaseous molecule containing atoms useful for film formation [23]	Difficult to apply for multicomponent material deposition, [24] and prone to chemical hazards due to corrosive, toxic, and explosive precursor gases
Sol–gel method	(a) Prehydrolysed silica sol leads, direct mixing of nanoparticles or metal and metal oxide. (b) Sol is holding the matrix forming species followed by gel formation, mixing of preformed colloids metal (oxide). (c) Reduction of metal before hydrolysis and complexation of metal with silicone [25]	
Chemical reduction method	The ionic salt is reduced in a suitable medium in the occurrence of surfactant using different reducing agents [26]	All the limitations related to reducing agents such as expensive, toxicity, poor, reducing ability and impurities [27]
Hydrothermal method	Method established on the reaction of aqueous solution vapours with solid material at high pressure and temperature and led to the deposition of small particles	It is difficult to control this process. The main limitations are reproducibility and consistency [16]

(continued)

Table 1 (continued)

Methods	Merits	Limitations and challenges
Biological/Green synthesis method	This synthesis was established to make out the problems like high cost, reaction complications, and safety issue of conventional methods	It can be done only at specific conditions
Nanoparticle synthesis using bacteria	The method involves appropriate utilisation of naturally occurring resources such as microorganism for the synthesis of nanoparticles	Safety risk [28]
Nanoparticle synthesis; fungi	Fungi secrets enzyme and protein worked as a reducing agent that can be used for metal nanomaterial synthesis from metal salt	Safety risk, manufactured materials are of variable size [28]
Nanoparticle synthesis using plant and plant products	yPlant-mediated mediated synthesis of nanoparticle gaining full attention in materials research [29]. Various plant products like extracts are used for the synthesis of different metallic nanomaterials like gold, silver, copper and zinc nanomaterials	Heating conditions are required, which increases the production cost of nanoparticles. Additional temperature is needed, which is possible only in the heating environment [28]
Other methods of nanoparticle synthesis		
Electrochemical deposition	Involves electrodeposition or template synthesis for the production of nanomaterials [30]. Ion etching technique is used for the production of porous alumina membranes [31]	Some limitations in controlling the nanomaterial dimensions and morphologies
Microwave-assisted nanoparticles preparation	The method is more offered over thermal heating, like bacteria along with cellular locations used for the Preparation of NMs	Use of uniform heat in the microwave oven results in shorter crystallisation time and homogeneous nucleation [32]
Supercritical fluid technology	The SCFs have distinguishing physicochemical properties which can be modified between liquid and gaseous states by changing the temperature or pressure and provides a chance to change reaction environment like viscosity, density, surface tension and diffusivity [33]	It operates at high temperature and critical pressure to prepare nanoparticle [32]

(continued)

Table 1 (continued)

Methods	Merits	Limitations and challenges
Ultrasound technique	Ultrasound synthesis techniques consist of two main methods, ultrasonic spray pyrolysis and sonochemistry [34]	The rate of reduction in sonochemical entirely depended on ultrasonic frequency [35]

pH

In green synthesis, pH plays an important factor in determining the rate of reaction. Also, researchers observed that the solution's pH in the medium determines the texture and size of the synthesised nanomaterials [37]. Therefore, the size of nanomaterials can be governed by varying the pH of the solution media. The impact of pH to determine the size and shape of the synthesised silver nanomaterials was established by Soni and Prakash [38].

Time

Time is also an influential factor for the quality and type of nanomaterials synthesis during green technology, where the reaction medium is incubated [39]. The characteristics of the synthesised nanomaterials are also changed with time and significantly dominate the synthesis process. The variation in time may affect the synthesis in many ways like the accumulation of particles due to extended time of storage; particles may grow or shrink during prolong storage; they may have a shelf life, which eventually affects their ultimate potential [16].

Pressure

The pressure applied to the synthesis medium also disturbs the shape and size of the synthesised nanomaterials [40]. The pressure is also a significant challenge for the synthesis of nanomaterials. The reduction rates of metal ions using biological operators have been observed to be much faster at surroundings pressure conditions [41].

Particle size and shape

The dynamic nature and shape of the engineered nanomaterials significantly influence the chemical properties [42]. Particle size shows a notable role in deciding the characteristics of nanomaterials, like the melting point of nanomaterials decreases when the size of the nanomaterials reaches up to nanoscale [43]. Nanomaterials with different structures have similar energy that alters their shape [44]. The type of energy generally used during the study of nanomaterial stimulates the alteration in the form of the nanoparticle.

Pore size

The excellence and solicitation of nanomaterials are considerably subjective by the porosity of the manufactured nanomaterials. Knowledge of biomolecules on nanomaterials has been attained to increase their use in the drug delivery and biomedicals [45].

Preparation cost

Preparation cost is the primary factor in the manufacture engineered nanomaterials. The cost-effectiveness of the synthesis procedure is one of the major challenges that affect the large scale nanomaterials synthesis. Therefore, to simplify the prospective utilisation of nanomaterials in the manufactured goods, the production cost must be controlled and regulated. The physical methods result in high yield within, but it isn't easy to control the morphology, particle size and other structural properties of prepared nanomaterials. In chemical methods, the structural properties of the synthesised nanomaterial can be controlled, but the production of hazardous waste is one of the major concerns. However, biological methods of preparation of nanomaterials are cheaper, environmentally friendly and facile.

Environment challenges

The surrounding conditions always play an essential role in regulating the nature of the manufactured nanomaterials. Environmental influence easily converts the single nanoparticle into core–shell nanomaterials by the association of materials or rejoining with other materials from the environment [46]. In a biological system, the manufactured nanomaterials produce a coating that changes them into massive and more significant sized [17]. Moreover, the environment also influences the physical structure and chemistry of the engineered nanomaterials.

Proximity

Researchers observed when an individual nanomaterial comes into the contact of other materials or comes near the surface of different nanomaterials, the properties of the material get influenced, and sometimes drastic changes occur as well [47]. Such kind of changing behaviour of the nanomaterials can be employed in making more tuned nanomaterials. There are many ramifications of the proximity effect of nanomaterials, such as the particle charging, the substrate infusion, and magnetic properties of the nanomaterials.

3.2 Limitations in Characterisation Technique

There is an agreement that the new functionalities proposed by nanoscale materials will give rise to a flourishing impact of nanoscale products over numerous markets all around [48]. The progression in the characterisation and synthesis of novel nanomaterials includes the study of their morphology, are usually considered vital for the advancement of nanotechnology. Progress in nanotechnology is diligently related to the improvement of nano-metrology. Mostly, the novel nanomaterials, as well as their applications, are regularly growing, which enforce comprehensive analysis of their chemical, morphological, electromagnetic and additional parameters. Thus, Nano-metrology and appropriate characterisation techniques are of utmost importance for regular progress in nanotechnology [49]. In this part of the study, we discuss the challenges, applications, methods, advantages and limitations of the utilisation of the instrumental techniques into the characterisation of nanomaterials.

A summary of similar problems and constraints of the elementary metrology techniques is shown in Table 2 [50] and the applications of these methods to the characterisation of nanomaterials are discussed.

3.3 *Other Limitations*

Numerous living systems, like plants, involved in secondary metabolites that behave as stabilising and reducing entity for the synthesis of nanomaterials. In the same manner, the composition of these metabolites varies with the type of plant, plant part, and the method implied for its extraction [51].

However, different microorganisms generate separate extracellular and intracellular enzymes on changing the amounts that affect the synthesis of nanoparticle [18]. The choice of method to purify the synthesised nanomaterials can impact nanoparticle quantity and quality.

Sometimes, centrifugation is used to isolate the nanomaterials based on gravitational force [13]. In some other cases, chromatography techniques are implemented to separate nanomaterials as per the requirement between the mobile phase and stationary phase coefficients. Well-organised distribution of synthesised nanomaterials is attained by one or more procedures, including the extraction of nanomaterials controlled on their solubility in two different miscible liquid phases (organic solvents and water) followed by electrophoresis or chromatography [52]. Separation of nanomaterials is essential for their application in pharmaceutical and biomedical industries [53].

4 Challenges of Nanomaterials

The safe use of nanomaterials in the consumer products remains a prime issue in the development and implementation of a new roadmap for engineered nanomaterials that goes from the research laboratory, through pilot processing, testing and development, mass production and finally to the public domain [54].

The major challenge in using nanomaterials in the industry is the reliable and reproducible production method of nanomaterials in large quantities. To guarantee the same quality of material every time, it demands precise control of all stages of

Table 2 An overview of basic methods for nano-metrology [50]

Instruments	Method	Nanosystem analysed	Parameter measured	Resolution	Problems
Electron microscopy methods	Transmission electron microscopy (TEM)	Clusters, nanocrystals, nanoparticles, quantum dots, dendrimers, nanowires, nanotubes, and biomolecules	Morphology (particle size and shape); crystallographic information (detection of atomic-scale defects); compositional information (the elements and compounds the sample is composed of); information on the phases present (lattice spacing measurement); topography and elemental mapping (STEM)	The lateral resolution, 0.1 nm	The thickness limit is 200 nm. The method entails time-consuming sample preparation to make it thin; the field of view is relatively small; the sample may be damaged by the electron beam, particularly in the case of biological materials; an ultrahigh vacuum is required for atomic-scale resolution, light atoms are insufficiently contrast
	Scanning electron microscopy (SEM)	Atomic flat surfaces, nanoparticles and nanocrystals, multilayered nanowires, carbon nanotubes, nanofibers, and biomolecules	Topography (topography map, surface characteristics); morphology (particle shape and size); composition (elements and compounds); crystallographic information (the arrangement of atoms)	Depth, 0.5–5 nm; lateral the resolution, 1–20 nm	SEM works only with conductive samples. The method requires high vacuum; the examination of wet materials and biological samples is difficult

(continued)

Table 2 (continued)

Instruments	Method	Nanosystem analysed	Parameter measured	Resolution	Problems
Scanning probe microscopy methods	Atomic force microscopy (AFM) or scanning force microscopy (SFM)	Both conductive and nonconductive nanomaterials. Nanocrystals, nanotubes, and biological samples	Topography, morphology, the elasticity of the surface, frictional characteristics, specific molecular interactions and magnetic characteristics of the surface, the total density of (valence-) electron states up to the Fermi level at the surface	Depth, 0.5–5 nm; lateral resolution, 0.2–10.0 nm	The single scan image size is small. The AFM scanning speed is a limitation. The relatively slow rate of scanning during AFM imaging often leads to thermal drift. AFM images may be affected by hysteresis of the piezoelectric material
	Scanning tunnelling microscopy (STM)	Only conductive including semiconductor particles: nanocrystals, thin films, nanotubes and molecules	Surface topology (size, shape, roughness, defects, electronic structures and local density of states). Surface properties, surface structures and surface reactions and single point defects	Depth, 0.5–5 nm; lateral resolution, 2–10 nm	STM works only with conductive samples. It requires minimisation of all electrical and mechanical noises

(continued)

Table 2 (continued)

Instruments	Method	Nanosystem analysed	Parameter measured	Resolution	Problems
X-ray methods	X-ray photoelectron spectroscopy (XPS)	Nanoparticles, nanowires, thin films, semiconductors, polymers, and biomaterials	Qualitative and quantitative elemental composition in the top 1–10 nm surface layers. The chemical state identification of one or more of the elements in the sample. The binding energy of one or more electronic states. The density of electronic states	Depth, 0.5–10 nm; lateral resolution, 5 nm–50 mm	Considerable errors in analysing chemically heterogeneous surfaces; degradation during analysis; the method requires high vacuum
	X-ray diffraction (XRD)	Nanoparticles, virus particles, polymers, thin films, and biomolecules	Surface topology, crystallographic information (the type of the crystal structure, inters planar differences, ensemble spatial orientation, and charge distribution around atoms). Crystal structure, chemical composition and physical properties of materials and thin films	–	Peak broadening at small crystal sizes makes it difficult to determine the crystalline structure of nanoparticles
Optic methods	Absorption spectroscopy	Colloidal solutions, suspensions, thin films, and powders	The volume concentration of the substance. Indirect determination of particle size	–	Solid particles in a heterogeneous sample scatter light more than they absorb it; hence, the data are doubtful. Restriction to turbid samples

(continued)

Table 2 (continued)

Instruments	Method	Nanosystem analysed	Parameter measured	Resolution	Problems
	Dynamic light scattering (DLS) and fluorescence correlation spectroscopy (FCS)	Colloidal suspensions, micelles, polymer solutions, polymer melts, gels, amino acids, proteins, and liquid crystals	Particle size distributions, the rate of particle diffusion. The Stokes radius or hydrodynamic radius of particles; the hydrodynamic radius of macromolecules or block copolymer micellar systems; changes in the conformation of macromolecules or micelle equilibrium	0.5 nm – micrometre range	Average particle size, which is intensity-biased towards larger or contaminant particles in a sample. In samples with an NP mixture, analysis gravitates towards more refractile particles

the manufacturing process and standardised procedures for quality evaluation. Moreover, the advancement of nanomaterials, coupled with its huge demand in various markets, has raised potential health and environmental risks of ENMs [55]. There is a great challenge to estimate the risks of engineered nanomaterials anticipated from their functional and chemical diversity. The methodology is limited to investigate and determine the effect of these ENMs in complex environments, and ambiguity regarding the extent to which they are or will be used in consumer applications or industries [56].

NMs are being vocalised as the driving force behind the new industrial revolution. To create the base for their further industrial exploitation and to provide significant socioeconomic impact, both private and public-sector spendings are continually increasing to leverage the extensive know-how of NMs. However, some challenges need to be overcome before the full potential of NMs can be realised. The following section presents some of the limitations and challenges in the large scale deployment of NMs.

4.1 Commercialisation Challenges

A revolutionary development has been registered in the field of nanotechnology innovations. Plenty of nanotechnology applications have been developed from the side by side, in the convergence of novel materials with new device buildings [57]. Numerous commercial products like Nano-devices tools for surgical and drug delivery [58], titanium dioxide nanoparticles based sunscreen [59], early diagnosis kit [60], nanocomposites utilising in cars [61], medical imaging tools, quantum dots [62], sunglasses having nanocoating [63], and carbon nanotubes help in field emissive displays [64] are available in the market. This R&D movement has established commercial opportunities and encouraged governments, along with commercial organisations, to implement rapid commercialisation strategies [65].

Challenges to technology commercialisation

The commercialisation of nanotechnologies is entirely based on what technology has to offer today [66]. As per the business analysts, the essential elements effectively commercialising nanotechnology include market size, market potential, and the current economic scenario [67].

Lack of infrastructure

Nanotechnology research is helpless without expensive instruments. Therefore, the absence of infrastructure restricts and delays the growth of nanotechnology. Additionally, equipment becomes outdated in a short duration, making it more robust for researchers to keep well-informed of the competition [68].

Valley of Death

This is the significant pause between a scientist obtaining promising results and securing funds for commercialisation and proto-typing [68]. Studies indicate that for every single penny that is devoted to the necessary research, more than ten times must also be invested in commercialising of the competitive product [68]. Frequently, scientists are not getting involved in commercialisation, which harms the technologies and ultimately turns out to be a loss.

Time Lag

The regular time invested between completion of research and commercialisation of a product is approximately three to ten years [68]. The capitalist's venture and other commercial foundations of funding find this gap to be a significant disadvantage since their capital is getting blocked [68].

Lack of standard for evaluation

A significant hurdle for industrialisation of nano-products is the lacking of a bridge between researches and organisation that develops standards and issues certificates to speedup. Ultimately, the controlling standards by which nanotechnologies can be evaluated are missing.

Bureaucratic delays

Patent regularities take around three years to respond to a single requisition, a severe issue when even an insignificant delay can be disadvantageous. There is a real absence of a rational policy on technology transfer from universities to start-up businesses because of that a considerable bureaucracy must be dealt with for any such transfer [69].

The dearth of funding

As it is a capital prone issue (involves attaining the latest instruments and updated facilities), research in nanomaterials faces challenges in gaining funds. Commercialisation needs a large amount of investment mostly, and the mediocre firms are often unable to secure.

Lack of trained professionals

Trained professionals are the backbone of any industry; the field also suffers from such glitches. The scarcity of capable, trained scientists, technicians, engineers, and researchers are also barriers [68] in the development of these industries. A typical prospect accepts that this is because of a lacking curriculum revision and inclusion of nanoscience and technology in the engineering and science syllabus.

Public support

Companies with recognised brands considered as 'brand image' to influence public sentiments in their favour. Still, mediocre firms and start-ups that presently constitute a foremost chunk of nanotechnology, nanomaterials firms do not have that excellence and are more exposed to misdirected public clamour.

Both the methodological community and the popular press carry focus upon nanotechnologies. Many predictions exist for the role of nanomaterials in future products and frequent activities throughout the globe need to cheer-up in the development of nanomaterials and nano-enabled materials.

4.2 Environmental Challenges

ENMs are the most promising and grooming field in the current scenario. Regardless of such advancement in NMs technology, the information related to the possible impact of NMs on humanity is still not enough. The ENMs causes multiple types of environmental hazards, which remain unnoticeable after release into the environment. Hence, the study needs to be done in such aspects, especially to explain the structure–function of ENMs with respect to basic chemistry, like functionality and toxicity. Therefore, hazardous detail assessments should be implemented on NMs that shows correct exposure of danger during manufacture or use. Hereafter, nanoscience has been suggested to minimise conceivable environmental threats to humanity by the deployment of NMs. Advance deposing of current items with new nano-products that are additional ecologically concerned [70]. These efforts may be useful for appropriate development of tenders and research inclination toward further development of nanotechnology.

4.3 Environmental Challenges

Cumulative application of ENMs in commercial utilisation, as well as several consumer products, inclines to escalate the probability of their exposure t humans in both direct and indirect means [71]. The alarm has also been made whether the production of ENMs may generate additional hazards. With the help of such significance, the positive and negative impact of nanotechnology on the environment has been discussed in the following sections.

Positive impact on the environment

Nanotechnology promises remarkable environmental, social, and financial benefits. As an example, ENMs are used in aircraft as a replacement of outdated composites to reduce the overall aircraft weight significantly, this would save thousands of tons of fuel [72]. Additionally, ENMs are useful in wind turbine blades to create stronger and lighter ones than earlier and keep them more effective in energy conversion efficiency [73]. Nanomaterials are gradually applied in petroleum refining and automotive exhaust systems for the enhancement of chemical reactions by reducing pollution and expenses [74]. Researchers have put massive efforts into producing carbon nanotube "scrubbers" to separate CO_2 from power plant emissions [75].

Moreover, researchers are working hard to introduce low resistance wires with the help of carbon nanotubes, helpful to lessen transmission power loss [76]. EMNs are playing a dynamic role in minimising the engineering cost and enhance the effectiveness of solar panels, which are considered as the freshest energy sources. Additionally, with ENMs, solar panels can be made in malleable rolls against the conventional panels and are even printable [77]. ENMs have been successfully applied for the sanitisation of water and air by the help of filtration, adsorption, and oxidation with better competence than conventional ongoing techniques. One of the exclusive qualities of ENMs is that they can be employed to retort to pollutant agents, eventually relevant to alteration into nontoxic units. The coating on ENMs can also fight with pollutants and have an auto-cleaning feature [78]. A self-cleaning ENMs surface coating technology might spare energy, water, and cleaning agents. Likewise, nanotechnology has been implemented to make more efficient and environmentally friendly batteries [79]. Furthermore, by cracking of oil into biodegradable complexes, ENMs may deliver a significant role in cleaning up oil spills [80].

Adverse effects on the environment

According to the US agency concerned with Environmental Protection, "the toxicity of NMs is difficult to identify because they have unique chemical properties, high reactivity, and does not get dissolved in liquid" [81]. As ENMs are incredibly reactive, even the properties of ENMs in environmental samples could alter during collecting and analysing the samples [81]. The transfer of some specific nanoparticles like TiO_2 was also observed to cause a genotoxic effect (a small dose of 0.25 mM) to mutilation DNA (at higher quantity) on the plants, for example, Nicotiana tabacum and Allium cepa [82]. Instance, nano Zn oxide and nano-Zn particles applied adverse effects on the seed germination and growth of ryegrass/corn, respectively [83]. After the discharge of ENMs to the environment, it may intermingle with other pollutants and create fusion materials. According to the toxicological examinations, ENMs may harm unicellular aquatic beings and individuals like Daphnia and fishes [83]. The carbon nanotubes and their derivatives were also found to release toxic effects on marine entities, such as Thalassiosira pseudonana, and Oryzias melastigma [84]. ENMs are also found to exert an adverse impact on the air system. The ENMs played a significant role in the construction of dust clouds after being freed into the environment [85]. The ENMs can affect all the environmental media, i.e., air, water, and land. Human and other mammals' health also gets affected due to exposure to ENMs. The toxicity and exposure of NMs move through and get absorbed in the cell membranes of mammalians. The immersion rate of the ENMs in cells is directly related to their size, accumulation, and sedimentation characteristics [86]. Disclosure routes of NMs are diverse and include oral, inhalation, dermal, and gastrointestinal tract while consuming goods such as skincare, sunscreen products, paint and coating products, food and health supplements, food additives, and food colourings [87, 88].

Furthermore, in some cases, ENMs are deliberately injected into the human body for medical applications. Scientists have found relations between ENMs exposure during neonatal (the first month after birth) with increased decreased lung function, asthma exacerbation, wheezing and coughing without infection [89]. Inhaled NMs get dispensed throughout the body via the bloodstream [90]. Epidemiological studies have reported that dangerous cardiovascular disadvantages like changing blood coagulation may create oscillation in cardiac frequency and function could be due to ENMs acquaintance [91–93]. Becker et al. [94] established that carbon nanotubes and TiO_2 at a nanoscale level might cause tumours.

There is no hesitation in accepting that nanotechnology has been effectively utilised in various fields for the welfare of humanity. Though, any new undocumented technology comes with a few disadvantages. The future studies with sincere attention on the fast-tracked or durable consequence of ENMs help in approximating the exact toxicological sketch of these materials on the environment. As the ENMs are susceptible structures, these structures could intermingle with other toxins to generate hazardous toxic anomalies. These studies should also be covered in future research to establish the disposal guidelines of NMs. Environmental engineers, scientists, authorities, nongovernment and government societies can only assume about the impact of NMs. Therefore, it is essential to conduct proper life cycle evaluation and risk assessment examination for NMs before extensive application. Much exploration is needed in this arena as harmless bulk materials could become poisonous and sensitive ingredients at nanoscale. Through reducing the massive breaches in knowledge about the nature of interactions of NMs, we will have appropriate strategies concerning the processing, applications, and regulation of NMs in the future.

5 Applications of Nanomaterials

Foretelling the fate of any up-growing technology is challenging. However, there is a prevalent natural tendency to undervalue the power of technology and the speed of its expansion. The development of engineered nanomaterials is already surpassing the predictions made when the National Nanotechnology Initiative (NNI) was established in 2000. The technologies based on nanomaterials are an emerging science that is predicted to have swift and durable future growth. ENMs are assumed to contribute prominently to economic development as well as in job establishment around the world in the coming decades. According to the researchers, nanotechnology is proposed to have four distinct generations of progression. The focus of the technology in the expansion of the 'Safe-by-design' model for engineered nanomaterials is presently under examination by the researchers. The simple idea involves first to assure the safety testing of the material then start commercialisation of the product; the safety measures should be taken and pooled with the design and invention stage of an engineered nanomaterial development.

The studies highlighted the enormous potential impact of nanomaterials on new technologies. The opportunities for nanomaterials research and development have been defined for a broad spectrum of technology applications. To meet our goal of eco-friendly engineered nanomaterial, we need to work on basic needs in such field of research. There are numerous desirable needs of material in different sectors of research, which are discussed as under.

Electronics and Information Technology

ENMs are significantly involved in the development of computing and electronics, contributing to smaller, faster, and more compact systems that can achieve and store more significant and larger volumes of information. Moreover, the future fortune of IT sector engineering is intensely based on further positive downscaling procedures to construct new devices with advanced functionalities, better flexibility and dependability, and enhanced performance. The new nano-electronics will give a step modification in functionality, higher speed, minimised power consumption, compact complexity that will deliver novel photonics, computer, and informatics tenders over the next 20 years.

Manufacturing engineering

Nanomaterials must achieve the need of looking for new structural materials, in the reduction of consumption and bargain the life-cycle costs. Improvements minimise the fuel consumption and operating expenditures, decrease or zero emissions, developed capacity and luxury and noise reduction also.

Health: biological and medical applications

The latest drug systems with controlled release can be manufactured by using nanoengineering. The release of drugs into the human body is simplified by developing nanoscale features, i.e., supplying just the number of medications needed to patients helps in minimising the amount of drugs. New ointments are required to be prepared with nanoscale binders, which will have offshoots into the cosmetics industry. Such changes get able to achieve the research and development needed for the full exploitation of nanomaterials in healthcare.

Microelectronics and photonics

Expansion of quantum communication and computing, data storage implemented new materials, and ideas like magnetic spintronics and semiconductors will transform the capabilities. Mostly, such concepts are depending on nanostructures, and in general cases, structural concepts using synchrotron and neutron sources are vital.

Nanomaterials in energy technologies

Climate changes and the shortage in energy supply are the utmost pressing concerns faced by the developed and developing countries. Energy consumption and associated problems can be solved only by the use of renewable sources and developing nuclear energies. Saving energy and resourceful use are the basic necessities in this development.

Environment

In this human-made and natural environment, ENMs can help to resolve issues like soil and groundwater cleaning, air cleansing, pollution level and sense. The need of the hour is human-made waste reduction like nuclear waste, along with the urgent requirement of developing simple geological dispose-off techniques acceptable for society. A healthier expectation of climate change is directly linked to the understanding of aerosols effect (nanoparticles) in the atmosphere.

Security and safety

ENMs will come with the solutions to the anticipation and protection against extremism threats, including natural and industrial accidental threats. Technology will also provide a satisfactory response to the security and safety of critical installations and the environment. Apart from all compulsorily required benefits, the risks involved with nanoscale have also to be addressed. History shows that upcoming science and technology mostly cause anxiety and misunderstanding in society. Considerable research efforts should be made to investigate health impacts and methods to prevent them.

Medical and healthcare applications

Nanotechnology is already broadening the medical tools, knowledge, and therapies currently available to clinicians. Such as Nanomedicine is an application of nanotechnology in the field of medicine. It helps in development on the natural scale of biological phenomena to produce exact solutions for disease prevention, diagnosis, and treatment.

Future transportation benefits

Nanotechnology promises in developing multifunctional materials that will take part in the development and maintaining safer, lighter, smarter, and more capable aircraft, vehicles, spacecraft, and ships. Furthermore, nanotechnology proposes various means to expand the transportation infrastructure.

Energy applications

The environmental pollution due to extensive use of fossil fuels and their rapid depletion catalysed the research for the development of alternative renewable energy resources. Scientists are working on the ways to establish clean, sustainable, and renewable energy sources, along with the efforts to decrease energy consumption and diminish the toxicity threat to the environment. ENMs are playing a significant role in the development of alternative energy conversion and storage devices.

Environmental remediation

In addition to the ways that technology can help to expand energy efficiency, there are still many techniques that can help to identify and clean up environmental pollutants. ENMs can be applied for the efficient removal of chemical and biological pollutants. Various kinds of nanomaterials are used as adsorbent and catalysts for the detection and removal gases (SO_2, CO, NOx, etc.), hazardous chemicals (arsenic, iron, manganese, nitrate, heavy metals, etc.), organic chemicals (organic pollutants (aliphatic and aromatic hydrocarbons) and biological substances, such as viruses, bacteria, parasites and antibiotics. Recently, significant advances are made in the fabrication of novel nanoscale materials and processes for the treatment of drinking water and industrial wastewater contaminated by toxic metal ions, radionuclides, organic and inorganic solutes, bacteria and viruses and the treatment of air.

6 Conclusion

In this chapter, we discussed the ENMs their processing technologies and potential applications. Although ENMs are continuously conquering every field of life, there are still many challenges which need to be overcome. ENMs are valuable for various applications, yet, the exposure of the nanotech industry is restricted due to their uncontrollable use and discharge to the natural environment, which creates health hazard alarms. Most of the nanotechnology R&D remains at a theoretical level, and essential research has not been robustly linked to viable commercial production and large-scale manufacturing. Risk and safety studies are also compulsory to ensure public acceptance. There should be a policy which emphasis on infrastructure, education, standards and workforce preparation. Nanotechnology commercialisation can be improved by the implementation of new regulations with a proper roadmap, which should appeal to investors and facilitate marketing. Government agencies should coordinate their efforts.

ENMs hold potential for solving several current and future challenges from climate change, energy scarcity, medical and industrial requests, that will improve our standard of living while maintaining and stimulating the competitiveness of the industry. We need to develop more natural ways and instruments to perform metrological studies swiftly and in an accessible manner. Also, more work is needed on identifying ethical issues, studying public perception and regulatory issues. Integrating nanoscience and nanotechnology into the engineering and science curriculum will enable the preparation of the future generation of scientists and engineers.

Results of studies and improvements in ENMs fields are inflowing into all-around of our lives, such as energy, agriculture, defence, materials science, aerospace, environmental science, and medicine. Nanotechnology is emerging quickly, with its applications being examined via a growing number of scientists, engineers and clinicians, government agencies, venture and capitalists. Some active research areas include nanoporous materials, nanorobotics, nanodevices, nanolithography, nanopowders, nanocomputers, and nanostructured catalysts, molecular nanotechnology, molecular manufacturing, nano-layers, medicines and nano-biology (e.g., diseases prediction, prevention, and treatment). Worldwide, biomedical engineers and clinical researchers are working on disease preclusion and early delivery of care. The current technologies depend on processes that take place at the nanoscale. Some instances of these technologies are lithography, ion exchange, adsorption, drug design, catalysis, composites, and plastics.

References

1. Dagar, G., Bagchi, G.: Nanoparticles as potential endocrine disruptive chemicals. In: Saxena, S., Khurana, S. (eds.) NanoBioMedicine. Springer, Singapore (2020)
2. ISO/TS 27687: Nanotechnologies terminology and definitions for nano-objects and nanoparticle, nanofibre and nanoplate. International Organization for Standardization (2008)
3. Darya, et al.: One-step decoration of TiO_2 nanotubes with Fe_3O_4 nanoparticles: synthesis and photocatalytic and magnetic properties. ACS Appl. Nano Mater. **3**(2), 1553–1563 (2020)
4. Tiwari, et al.: Zero-dimensional, one-dimensional, two-dimensional and three-dimensional nanostructured materials for advanced electrochemical energy devices. Prog. Mater. Sci. **57**, 724–803 (2012)
5. Potocnik, J.: Commission recommendation of 18 October 2011 on the definition of nanomaterial (Text with EEA relevance) (2011/696/EU). In: Commission, TE (ed.) Official Journal of the European Union (2011)
6. Maria, et al.: Nanoparticle exposure and hazard in the ceramic industry: an overview of potential sources, toxicity and health effects. Environ. Res. **184**, 109297 (2020)
7. Nanomaterials Market—Global Trends, Investment Analysis and Future Scope to 2022. Mordor Intelligence, Hyderabad, Telangana, Mar 2017
8. Project on Emerging Nanotechnologies: Analysis—PEN Consumer Products Inventory. Woodrow Wilson International Center for Scholars. www.nanotechprojet.org (2014). Accessed 5 Nov 2014
9. ICON: Towards predicting nano-biointeractions: an international assessment of nanotechnology environment, health and safety research needs. International Council on Nanotechnology, Rice University, Houston, TX (2008)
10. Zhang, et al.: Time temperature indicator for perishable products based on kinetically programmable Ag overgrowth on Au nanorods. ACS Nano, **7**, 4561–4568 (2013)
11. Pervikova, et al.: Nano alloying of clusters of immiscible metals and the formation of bimetallic nanoparticles in the conditions of non-synchronous explosion of two wires. Powder Technol. **360**(15), 855–862 (2020)
12. Zheng, et al.: Biosynthesis, characterisation and antibacterial activity of silver nanoparticles by the Arctic anti-oxidative bacterium Paracoccus sp. Arc7-R13. Artif. Cells Nanomed. Biotechnol. **47**(1), 1488–1495 (2019). https://doi.org/10.1080/21691401.2019.1601631
13. Chauhan, et al.: Methodological advancements in green nanotechnology and their applications in biological synthesis of herbal nanomaterials. Int. J. Bioassays **1**(7), 6–10 (2012)
14. Ingale, A.G., Chaudhari, A.N.: Biogenic synthesis of nanomaterials and potential applications: an eco-friendly approach. J. Nanomed. Nanotechnol. **4**, 2 (2013)
15. Baker, et al.: Plants: emerging as nanofactories towards facile route in synthesis of nanomaterials. BioImpacts **3**(3), 111–117 (2013)
16. Baer, D.R.: Surface characterisation of nanomaterials: critical needs and significant challenges. J. Surf. Anal. **17**(3), 163–169 (2011)
17. Alexander, et al.: Protein-loaded soluble and nanoparticulate formulations of ionic polyphosphazenes and their interactions on molecular and cellular levels. Mater. Sci. Eng.: C **106**, 110179 (2020)
18. Novo, et al.: From phyto to agromining: past, present, and future scope. In: Kallel, A., Ksibi, M., Ben Dhia, H., Khélifi, N. (eds.) Recent Advances in Environmental Science from the Euro-Mediterranean and Surrounding Regions. EMCEI 2017. Advances in Science, Technology & Innovation (IEREK Interdisciplinary Series for Sustainable Development). Springer, Cham (2018)
19. Kowalczyk, Lagzi, I., Grzybowski, B.A.: Nanoseparations: strategies for size and/or shape-selective purification of nanomaterials. Curr. Opin. Colloid Interface Sci. **16**(2), 135–148 (2011)
20. Fedotkina, et al.: Calcium phosphate coating deposition by radio frequency magnetron sputtering in the various inert gases: the pilot study. Mater. Chem. Phys. **235**, 121735 (2019)
21. Pandey, et al.: Physical vapor deposition of metal nanoparticles on chemically modified graphene: observations on metal–graphene interactions. Small, **7**(22), 3202–3210 (2011)

22. Willems: Roadmap Report on Nanoparticles. W&W Espana sl, Barcelona, Spain, p. 157 (2005)
23. Pedersen, H., Elliott, S.D.: Studying chemical vapor deposition processes with theoretical chemistry. Theor. Chem. Acc. **133**(5), 1476 (2014)
24. Reina, et al.: Large area, few-layer graphene films on arbitrary substrates by chemical vapor deposition. Nano Lett. **9**(1), 30–35 (2008)
25. Cushing, et al.: Recent advances in the liquid-phase syntheses of inorganic nanoparticles. Chem. Rev. **104**(9), 3893–3946 (2004)
26. Guzman, et al.: Synthesis of silver nanoparticles by chemical reduction method and their antibacterial activity. Int. J. Chem. Biomol. Eng. **2**(3), 104–111 (2009)
27. Zhang, et al.: preparation of copper nanoparticles by chemical reduction method using potassium borohydride. Trans. Nonferrous Met. Soc. China **20**, s240–s244 (2010)
28. Pantidos, N., Horsfall, L.E.: Biological synthesis of metallic nanoparticles by bacteria, fungi and plants. J. Nanomed. Nanotechnol. **5**(5), 1 (2014)
29. Prathna, et al.: Biomimetic synthesis of silver nanoparticles by Citrus limon (lemon) aqueous extract and theoretical prediction of particle size. Colloids Surf. B Biointerfaces **82**(1), 152–159 (2011)
30. Dikusar, et al.: On limiting rate of dimensional electrodeposition at meso-and nanomaterial manufacturing by template synthesis. Surf. Eng. Appl. Electrochem. **45**(3), 171–179 (2009)
31. Gayathri, et al.: Advances in nanoporous anodic alumina-based biosensors to detect biomarkers of clinical significance: a review. Adv. Healthc. Mater. **7**, 1700904 (2018)
32. Ali, et al.: Characterisation, applications, and challenges of iron oxide nanoparticles. Nanotechnol. Sci. Appl. **9**, 49 (2016)
33. Wang, S., et al.: Supercritical hydrothermal synthesis of inorganic nanomaterials. In: Supercritical Water Processing Technologies for Environment, Energy and Nanomaterial Applications. Springer, Singapore (2020)
34. Natarajan, et al.: Sonochemical synthesis of sulfur doped reduced graphene oxide supported CuS nanoparticles for the non-enzymatic glucose sensor applications. Sci. Rep. **7**(2494), 1–10 (2017)
35. Bang, J.H., Suslick, K.S.: Applications of ultrasound to the synthesis of nanostructured materials. Adv. Mater. **22**(10), 1039–1059 (2010)
36. Supriya, et al.: The role of temperature on physical-chemical properties of green synthesized porous carbon nanoparticles. Waste Biomass Valor (2019). https://doi.org/10.1007/s12649-019-00675-0
37. Singh, R., Dutta, S.: The role of pH and nitrate concentration in the wet chemical growth of nano-rods shaped ZnO photocatalyst. Nano-Struct. Nano-Objects **18**, 100250 (2019)
38. Soni, N., Prakash, S.: Factors affecting the geometry of silver nanomaterials synthesis in Chrysosporium tropicum and Fusarium oxusporum. Am. J. Nanotechnol. **2**(1), 112–121 (2011)
39. Darroudi, et al.: Time-dependent effect in green synthesis of silver nanomaterials. Int. J. Nanomed. **6**(1), 677–681 (2011)
40. Abhilash, B.D.P.: Synthesis of zinc-based nanomaterials: a biological perspective. IET Nanobiotechnol. **6**(4), 144–148 (2012)
41. Tran, et al.: Silver nanomaterials: synthesis, properties, toxicology, applications and perspectives. Adv. Nat. Sci.: Nanosci. Nanotechnol. **4**, Article ID 033001 (2013)
42. Baer, et al.: Surface characterisation of nanomaterials and nanomaterials: important needs and challenging opportunities. J. Vac. Sci. Technol. A, **31**(5), Article ID 050820 (2013)
43. Akbari, et al.: Particle size characterisation of nanomaterials—a practical approach. Iran. J. Mater. Sci. Eng. **8**(2), 48–56 (2011)
44. Marks, L.D., Peng, L.: Nanoparticle shape, thermodynamics and kinetics. J. Phys.: Condens. Matter **28**, 5 (2016)
45. Talrejaa, et al.: Phenol-formaldehyde-resin-based activated carbons with controlled pore size distribution for high-performance supercapacitors. Chem. Eng. J. **379**, 122332 (2020)
46. Cheng, et al.: Environmental transformations and ecological effects of iron-based nanoparticles. Environ. Pollut. **232**, 10–30 (2018)

47. Baer, et al.: Characterisation challenges for nanomaterials. Surf. Interface Anal. **40**(3–4), 529–537 (2008)
48. https://www.innovateuk.org/_assets/pdf/corporatepublications/nanoscaletechnologiesstrategy.pdf
49. Postek, M.T.: Report of the National Nanotechnology Initiative, Gaithersburg (2004)
50. Linkov, et al.: Comparative advantages and limitations of the basic metrology methods applied to the characterisation of nanomaterials. Nanoscale **5**, 8781 (2013)
51. Park, et al.: Polysaccharides and phytochemicals: a natural reservoir for the green synthesis of gold and silver nanomaterials. IET Nanobiotechnol. **5**(3), 69–78 (2011)
52. Hanauer, et al.: Separation of nanomaterials by gel electrophoresis according to size and shape. Nano Lett. **7**(9), 2881–2885 (2007)
53. Jafarizadeh, et al.: Commercialisation consideration. In: Nanobiotechnology in Food: Concepts, Applications and Perspectives. Springer, Cham (2019)
54. Nel, et al.: Where are we heading in nanotechnology environmental health and safety and materials characterization? ACS Nano **9**, 5627–5630 (2015)
55. Bouwmeester, et al.: Minimal analytical characterisation of engineered nanomaterials needed for hazard assessment in biological matrices. Nanotoxicology **5**, 1–11 (2011)
56. Pelaz, et al.: The state of nanoparticle-based nanoscience and biotechnology: progress, promises, and challenges. ACS Nano **6**, 8468–8483 (2012)
57. F. Fankhauser, Klinische Monatsblatter: Application of static and dynamic light scattering—a review. Clin. Mon. Sheets Ophthalmol. **227**, 194–198 (2010)
58. Sargen, J.F.: USET: the national nanotechnology initiative (operating report) (2013)
59. Paradise, et al.: Exploring emerging nano-biotechnology drugs and medical devices. Food Drug Law J. **63**, 407–420 (2008)
60. Smijin, T.G., Pavel, S.: Titanium dioxide and zinc oxide nanoparticles in sunscreens: focus on their safety and effectiveness. Nanotechnol. Sci. Appl. **4**, 95–112 (2011)
61. Choi, Y., Kwak, J., Park, J.: Nanotechnology for early cancer detection. Sensors **10**(428), 55 (2010)
62. Kiziltas, et al.: Rheological and thermal properties of exfoliated graphite nanoplatelets-filled impact modified polypropylene nanocomposites. Polym. Compos. **39**, E1512–E1519 (2018). https://doi.org/10.1002/pc.24400
63. Mokter, et al.: Improvement of mechanical strength of hydrophobic coating on glass surfaces by an atmospheric pressure plasma jet. Surf. Coat. Technol. **357**, 12–22 (2019)
64. Tekaya, et al.: Effect of Ti/TiN multilayer protective nanocoatings on Zr-based metallic glasses mechanical performance. Thin Solid Films **539**, 215–221 (2013)
65. Berger, M.: Well over 2000 companies worldwide already involved in nanotechnology (2010)
66. Aithal, P.S., Aithal, S.: Nanotechnology innovations and commercialisation—opportunities, challenges & reasons for delay. Int. J. Eng. Manuf. (IJEM) **6**(6), 15–25 (2016)
67. Hobson, D.W.: Commercialisation of nanotechnology. Wiley Interdiscip. Rev. Nanomed. Nanobiotechnol. **1**, 189–202 (2009)
68. McNeil, R.D., Lowe, J., Mastroianni, T., Cronin, J., Ferk, D.: Barriers to nanotechnology commercialisation. US Department Of Commerce Technology Administration (2007)
69. (a) Beer, et al.: Toxicity of silver nanoparticles - nanoparticle or silver ion? Toxicol. Lett. **208**, 286–292 (2012). (b) Patra, J.K., Gouda, S.: Application of nanotechnology in textile engineering: an overview. J. Eng. Technol. Res. **5**, 104–111 (2013)
70. Iavicoli, et al.: Opportunities and challenges of nanotechnology in the green economy. Environ. Health **13**, 78–84 (2014)
71. Pattan, G., Kaul, P.: Health hazards associated with nanomaterials. Toxicol. Ind. Health **30**, 499–519 (2014)
72. Kausar, A., Rafique, I., Muhammad, B.: Aerospace application of polymer nanocomposite with carbon nanotube, graphite, graphene oxide, and nanoclay. Polym. Plast. Technol. Eng. **56**(13), 1–19 (2017)
73. Patel, V., Mahajan, Y.R.: Techno-commercial Opportunities of Nanotechnology in Wind Energy. Wiley VCH Verlag GmbH & Co. KGaA (2017). https://doi.org/10.1002/9783527696109

74. Etim, U., Peng, B., Yan, Z.: Nanotechnology Applications in Petroleum Refining, pp. 37–65 (2018)
75. Xiao, et al.: Crystallographic visualization of postsynthetic nickel clusters into metal-organic framework. J. Am. Chem. Soc. **141**(34), 13654–13663 (2019)
76. Bloch, et al.: Post-synthetic structural processing in a metal–organic framework material as a mechanism for exceptional CO_2/N_2 selectivity. J. Am. Chem. Soc. 130702093658003 (2013)
77. Azmi, W.H., Sharif, M.Z., Yusof, T.M.: Potential of nanorefrigerant and nano-lubricant on energy saving in refrigeration system—a review. J. Renew. Sustain. Energy Rev. **69**, 415–428 (2017)
78. Sumaiya, S., Kardel, K., El-Shahat, A.: Organic solar cell by inkjet printing—an overview. Technologies **53**, 1–18 (2017)
79. Nica, et al.: Innovative self-cleaning and biocompatible polyester textiles nano-decorated with Fe–N-Doped titanium dioxide. Nanomaterials **6**, 214–223 (2016)
80. Sun, et al.: Graphene oxidewrapped $ZnMn_2O_4$ as a high performance lithium-ion battery anode. Nanotechnology **28**, 455401 (2017)
81. Daza, et al.: Multi-shell nano-carboscavengers for petroleum spill remediation. Sci. Rep. **7**, 41880 (2017)
82. US Environmental Protection Agency (USEPA): Nanotechnology White Paper. https://www.epa.gov/osa/nanotechnology-white-paper (2016)
83. Ghosh, M., Bandyopadhyay, M., Mukherjee, A.: Genotoxicity of titanium dioxide (TiO_2) nanoparticles at two trophic levels: plant and human lymphocytes. Chemosphere **81**, 1253–1262 (2010)
84. Lin, D., Xing, B.: Phytotoxicity of nanoparticles: inhibition of seed germination and root growth. Environ. Pollut. **150**, 243–250 (2007)
85. Gao, M.L., Zhang, Z., Lv, M.T., Song, W.H., Lv, Y.H.: Toxic effects of nanomaterial adsorbed cadmium on Daphnia magna. Ecotoxicol. Environ. Saf. **148**, 261–268 (2018)
86. Kwok, et al.: Chronic toxicity of double-walled carbon nanotubes to three marine organisms: influence of different dispersion methods. Nanomedicine **5**, 951–961 (2010)
87. Turkevich, et al.: Potential explosion hazard of carbonaceous nanoparticles: explosion parameters of selected materials. J. Hazard. Mater. **15**(295), 97–103 (2015)
88. Binderup, M.-L., Bredsdorff, L., Beltoft, V.M., Mortensen, A., Loeschner, K., Larsen, E.H., Eriksen, F.D.: Systemic Absorption of Nanomaterials by Oral Exposure (Report to the Danish Environmental Protection Agency, Environmental Project No. 1505). The Danish Environmental Protection Agency, Copenhagen (2013).
89. Ma, et al.: Effects of amorphous silica coating on cerium oxide nanoparticles induced pulmonary responses. Toxicol. Appl. Pharmacol. **288**, 63–73 (2015)
90. Mackevica, A., Foss, H.S.: Release of nanomaterials from solid nanocomposites and consumer exposure assessment—a forward-looking review. Nanotoxicology **14**, 1–50 (2015)
91. Meldrum, et al.: Mechanistic insight into the impact of nanomaterials on asthma and allergic airway disease. Part. Fibre Toxicol. **14**, 45–53 (2017)
92. Rajagopalan, et al.: Air pollution and cardiovascular disease JACC state-of-the-art review. J. Am. Coll. Cardiol. **72**(17) (2018)
93. Yang, L., Ji, W., Huang, J., Xu, G.: An updated review on the influential parameters on thermal conductivity of nano-fluids. J. Mol. Liq. (2019) (Elsevier)
94. Becker, H., Herzberg, F., Schulte, A., Kolossa-Gehring, M.: The carcinogenic potential of nanomaterials, their release from products and options for regulating them. Int. J. Hyg Environ. Health **214**, 231–238 (2011)
95. Gorga, R.E.: Nanotechnology in textiles. Text World 3, Nov/Dec 2010. Nanomaterials Definitions, Classifications, and Applications 33